Studies in Computational Intelligence

Volume 480

Series Editor

J. Kacprzyk, Warsaw, Poland

For further volumes:
http://www.springer.com/series/7092

Gourab Sen Gupta · Donald Bailey
Serge Demidenko · Dale Carnegie
Editors

Recent Advances in Robotics and Automation

Springer

Editors
Gourab Sen Gupta
Donald Bailey
School of Engineering and Advanced Technology
Massey University
Palmerston North
New Zealand

Dale Carnegie
School of Engineering and Computer Science
Victoria University of Wellington
Wellington
New Zealand

Serge Demidenko
RMIT International University Vietnam
Ho Chi Minh City
Vietnam

ISSN 1860-949X ISSN 1860-9503 (electronic)
ISBN 978-3-642-42688-9 ISBN 978-3-642-37387-9 (eBook)
DOI 10.1007/978-3-642-37387-9
Springer Heidelberg New York Dordrecht London

Printed on acid-free paper

Springer is part of Springer Science+Business Media (www.springer.com)

Guest Editorial

This special issue titled "Recent Advances in Robotics and Automation" in the book series of "Studies in Computational Intelligence" contains the extended version of the papers selected from those presented at the 5th International Conference on Automation, Robotics and Applications (ICARA 2011) which was held in Wellington, New Zealand from 6–8 December, 2011. A total of 89 papers were presented at ICARA 2011, out of which this special issue contains only 26 papers.

The collection of papers presented in this book deals with the recent advancements in design methodologies, algorithms and implementation techniques to incorporate intelligence in *robots and automation systems*. Several articles deal with navigation, localisation and mapping of mobile robots, a problem that engineers and researchers are grappling with all the time. Fuzzy logic, neural networks and neuro-fuzzy-based CI techniques for real-world applications have been detailed in a few articles.

The first group of articles deal with fuzzy systems and neural networks. In the first article, **Benjamin Johnen et al.**, have presented the design and evaluation of artificial neural networks for learning static and dynamic position behaviour of an industrial robot. Different layouts and configurations of feed-forward networks with back propagation learning algorithms were examined resulting in a multi-layer network based on the calculation of the forward transformation. **Khalid Isa and M. R. Arshad** have presented a mathematical model and motion control analysis of a buoyancy-driven underwater glider. The glider mathematical model uses the Newton–Euler method. In order to predict and control the glider motion, a neural network control has been used as a Model Predictive Control (MPC) as well as a gain tuning algorithm. **Miguel A. Olivares-Mendez et al.**, have described the development of a visual line guided system for use on board an Autonomous Guided Vehicle (AGV) commercial car, controlling the steering and using just the visual information of a line painted below the car. In order to implement the control of the vehicle, a Fuzzy Logic controller has been implemented that is robust against curvature changes and velocity changes. **O. Hassanein et al.**, have researched on the system identification of Autonomous Underwater Vehicles (AUV) dynamics to obtain its coupled nonlinear dynamic model. Adaptive fuzzy technique has been used to overcome the uncertain external disturbance and the difficulties of modelling the hydrodynamic forces of the AUV.

The next group of five articles deal with localisation, mapping and path planning. **Munir Merdan et al.**, have presented their research findings into a flexible multi-agent system architecture for task allocation in a complex and large industrial batch process system. **Sobers Francis et al.**, have reported the importance of kinematic and dynamic constraints in a path planning module for AGV. The model has been simulated and the behaviour has been compared with a real robot. A Simultaneous Localization and Mapping (SLAM) algorithm, based on Incremental Maximum Likelihood (IML) has been implemented by **Oussama El Hamzaoui et al.**, to give a robot the ability to navigate autonomously in an unstructured environment. **Christoph Rasche et al.**, have presented an approach to coordinate Unmanned Aerial Vehicles (UAV) by setting up formation patterns using bifurcation theory. It has been combined with potential field approach for exploration, based on harmonic functions. **Felix Hackbarth** has presented the experimental results on localisation of indoor mobile robots with limited sensor capabilities. The robot is able to locate itself near obstacles on the basis of a Bayesian approach with probabilistic global occupancy knowledge of the environment.

Vision systems are being incorporated in many intelligent systems for machine inspection, surveillance, ranging and control. **Luis Mejias et al.**, have presented an Omni-directional camera-based control system for micro aerial vehicles. The system has been tested using small quad rotors and shown to be effective in avoiding collision. **Ben Drayton et al.**, have designed a compact, configurable, indirect time of flight imager for mobile robotic applications. Changing lighting conditions pose significant challenges to vision-based localisation algorithms. **Jared Le Cras et al.**, have investigated the application of a colour model which separates brightness from chromaticity to eliminate features that may be caused by dynamic illumination. **Gontje C. Classen et al.**, have presented a data fusion algorithm to integrate data from optical and inertial sensors to implement a tracking system for a servo-controlled handheld tool in a computer-assisted surgery system. A mobile robotic system used to image vehicle undercarriages has been described by **Robert Ross et al.** The system uses a wide-angle catadioptric camera and generates a mosaic view of the vehicle undercarriage. Two cameras and a controlling PC have been employed by **Holger H. Rapp** to predict the trajectory of a ping-pong ball using a novel fluid-dynamical model and control a standard industrial robotic arm to catch it. The proposed algorithm has been shown to drastically increase the intelligence and performance of the system.

An interesting concept of 'smart floors' has been introduced by **Soo Hyeok Kang et al**. A mobile robot is able to reach its target position using the information stored in the smart floor. **Michael Schnurmacher et al**., have combined data from a radar and lidar to obtain accurate velocity and position information for car-following on highways.

In the area of intelligent and precise control, **Christian Ziegler and Jorg Franke** have used a low-cost stereoscopic measuring system in a closed-loop controller to position a patient couch with high accuracy. **Kelvin Gong and Allan I. McInnes** have implemented a modular, hierarchical control scheme for a mobile manipulator to coordinate the motion of the manipulator and the base. The controller can be extended to avoid obstacles without requiring redesign of the rest of the controller. There is a trend to shift functional behaviour of industrial automation systems from hardware to software to increase flexibility, but it increases the complexity of test-case generation for automated testing. **Reinhard Hametner et al**., have introduced an automated test case generation where the test cases are specified by UML state chart diagrams.

Yao Fu et al., have presented a realistic simulator for humanoid robot soccer which bridges the gap between the simulated environment and the real world. Another simulation work has been described by **Mouna Souissi et al**., for a humanoid robot movement. They have reported the outcome of their study of having a humanoid robot equipped with backbone pitch joints. The study concludes that a number of two pitch joints is a good trade-off in matters of work at the hip and the thorax inclination.

Andreas Kamagaew et al., have shown how a swarm of autonomous vehicles, with decentralized control architecture, can increase flexibility in a facility logistics system. **Gottfried Koppensteiner et al**., have researched into the application of knowledge-driven mobile robots for disassembly tasks. The proposed system has mobile robots with particular skills which are supervised by an agent with related objectives and knowledge. **Munie Merdan et al**., have presented their research which deals with automation of power distribution network. A multi-agent approach has been taken to realize and implement a reliable and efficient electric energy system.

In the last chapter of the book, **Rini Akmeliawati et al**., have presented an intelligent algorithm to estimate 2D human pose for action recognition. This work is of great significance in human–robot interaction.

We do hope that the readers would find this special issue informative and useful in augmenting their research or practical engineering work in the area of robotics and automation. We are very happy to be able to offer the readers such a diverse special issue both in terms of its topical coverage and geographic representation.

Finally, we would like to thank all the authors wholeheartedly who have contributed their work to this special issue. We are grateful to all reviewers from all around the world who have generously devoted their precious time to review the manuscripts.

A/Prof. Gourab Sen Gupta
g.sengupta@massey.ac.nz

A/Prof. Donald Bailey
d.g.bailey@massey.ac.nz

Prof. Serge Demidenko
serge.demidenko@rmit.edu.vn

Prof. Dale Carnegie
Dale.Carnegie@vuw.ac.nz

Contents

Neural Network Development and Training for the Simulation of Dynamic Robot Movement Behavior 1
Benjamin Johnen, Carsten Scheele and Bernd Kuhlenkötter

Neural Network Control of Buoyancy-Driven Autonomous Underwater Glider 15
Khalid Isa and M. R. Arshad

Autonomous Guided Car Using a Fuzzy Controller 37
Miguel A. Olivares-Mendez, Pascual Campoy, Ignacio Mellado-Bataller, Ivan Mondragon, Carol Martinez and Jose Luis Sanchez-Lopez

On-Line Adaptive Fuzzy Modeling and Control for Autonomous Underwater Vehicle. 57
O. Hassanein, Sreenatha G. Anavatti and Tapabrata Ray

Process Rescheduling and Path Planning Using Automation Agents 71
Munir Merdan, Wilfried Lepuschitz, Benjamin Groessing and Markus Helbok

Model Based Path Planning Module. 81
Sobers L.X. Francis, Sreenatha G. Anavatti and Matthew Garratt

Autonomous Navigation and Mapping with CoreSLAM 91
Oussama El Hamzaoui, Jorge Corsino Espino and Bruno Steux

A 3D Path Planning Approach Extended by Bifurcation Theory for Formation Flights 103
Christoph Rasche, Claudius Stern, Lisa Kleinjohann and Bernd Kleinjohann

Self-enhancing Featureless Robot Localization with Fixed Memory Tags 115
Felix Hackbarth

Vision Based Control for Micro Aerial Vehicles: Application to Sense and Avoid 127
Luis Mejias, Iván F. Mondragón Bernal and Pascual Campoy

Design and Characterisation of a Full-Field Range Imaging Camera for Use in Mobile Applications 143
Ben Drayton, Dale A. Carnegie and Adrian A. Dorrington

Improving Robustness of Vision Based Localization Under Dynamic Illumination 155
Jared Le Cras, Jonathan Paxman and Brad Saracik

Optical-Inertial Tracking System with High Bandwidth and Low Latency 171
Göntje C. Claasen, Philippe Martin and Frédéric Picard

VUIR: A Vehicle Undercarriage Inspection Robot 183
Robert Ross, John Devlin and Anthony de Souza-Daw

Improving System Performance Via Reevaluation of Models 193
Holger H. Rapp

Smart Floor with Learning Capability for Mobile Robot System 205
Soo Hyeok Kang, Yong Ho Kim, Byung-Cheol Min, Soon-Geul Lee, Jinung An and Dong Han Kim

High Level Sensor Data Fusion of Radar and Lidar for Car-Following on Highways 217
Michael Schnürmacher, Daniel Göhring, Miao Wang and Tinosch Ganjineh

Closed-Loop Control of Patient Handling Robots 231
Christian Ziegler and Jörg Franke

A Modular Hierarchical Control Scheme for Mobile Manipulation 243
Kelvin Gong and Allan I. McInnes

Automated Test Case Generation for Industrial Control Applications 263
Reinhard Hametner, Benjamin Kormann, Birgit Vogel-Heuser, Dietmar Winkler and Alois Zoitl

A Realistic Simulator for Humanoid Soccer Robot Using Particle Filter 275
Yao Fu, Hamid Moballegh, Raúl Rojas, Longxu Jin and Miao Wang

Simulation of Humanoid Robot Vertebra 289
Mouna Souissi, Vincent Hugel and Pierre Blazevic

Flexible Automation in Facility Logistics 301
Andreas Kamagaew, Jonas Stenzel and Michael ten Hompel

Application of Knowledge Driven Mobile Robots for Disassembly Tasks 311
Gottfried Koppensteiner, Christoph Krofitsch, Reinhard Hametner, David P. Miller and Munir Merdan

Power Distribution Control Using Multi-Agent Systems 323
Munir Merdan, Alexander Prostejovsky, Ingo Hegny, Wilfried Lepuschitz, Filip Andrén and Thomas Strasser

Human Upper Body Pose Region Estimation 335
Sara Bilal, Rini Akmeliawati, Amir A. Shafie and Momoh J. E. Salami

Editors Biography

Gourab Sen Gupta received his B.E. (Electronics) degree from the University of Indore, India, in 1982 and Masters of Electronics Engineering (MEE) degree from the Philips International Institute, Technical University of Eindhoven, Holland, in 1984. After working for 5 years as a Software Engineer in Philips India in the Consumer Electronics division he joined Singapore Polytechnic in 1989 where he worked as a Senior Lecturer in the School of Electrical and Electronic Engineering. In 2008 he was awarded a Ph.D. in Computer Systems Engineering for his research on Intelligent Control of Multi-Agent Collaborative Systems. Since September 2002 he has been with the School of Engineering and Advanced Technology (SEAT), Massey University, New Zealand where he is an Associate Professor.

His current research interests are in the area of embedded systems, robotics and automation, real-time vision processing and sensor applications. He has published and presented over 110 research papers in various journals and conferences. He has been a guest editor for leading journals such as IEEE Sensors Journal, International Journal of Intelligent Systems Technologies and Applications (IJISTA), and Studies in Computational Intelligence (Special Issue on Autonomous Robots and Agents) by Springer-Verlag. He is a Senior Member of IEEE and a Distinguished Lecturer of the IEEE Instrumentation and Measurement Society (2013–2015).

Donald G Bailey received the B.E. (Hons) degree in Electrical Engineering in 1982, and the Ph.D. degree in Electrical and Electronic Engineering from the University of Canterbury, New Zealand in 1985. From 1985 to 1987, he applied image analysis to the wool and paper industries within New Zealand. From 1987 to 1989, he was a Visiting Research Engineer at University of California at Santa Barbara. He joined Massey University in Palmerston North, New Zealand as Director of the Image Analysis Unit at the end of 1989. In 1996, he was a Visiting Research Fellow at the University of Wales, in Cardiff, and in 2002 as a Visiting Associate Professor at University of California at Santa Barbara. In 2008, he was a Visiting Research Fellow at Imperial College, London. He is currently an Associate Professor at Massey University in the School of Engineering and Advanced Technology. Professor Bailey is Leader for the Electronics, Information and Communications Systems Cluster within the School, and is leader of the Image and Signal Processing Research Group.

His primary research interests include applications of image analysis, machine vision and robot vision. One area of particular interest is the application of FPGAs to implementing real-time image processing algorithms. He has published over 200 papers, and is the author of the book "Design for Embedded Image Processing Using FPGAs". He is a senior member of IEEE.

Serge Demidenko is professor and Head of Centre of Technology as well as Centre of Communication and Design at RMIT International University Vietnam. He is also an adjunct academic staff with Monash University and RMIT University (both—Australia). Areas of his professional specialisations include electronic design and testing, digital signal processing, instrumentation and measurement. He has been an academic and research staff of institutions of higher learning in a number of countries in Europe, Asia and Australasia. Professor Demidenko has enjoyed working in close collaboration with major industrial companies such as Texas Instruments, Free scale Semiconductors, Flextronics and Intel. His publication list includes 3 co-authored research monographs, 14 book chapters, more than 30 refereed journal and over 100 refereed conference papers, 14 edited books and journal special issues and 25 engineering patents.

He is a Fellow of the Institute of Electrical and Electronics Engineers (IEEE) as well as a Fellow of the Institution of Technology and Engineering and a UK Chartered Engineer.

He is an associate editor for six international peer-reviewed journals. Professor Demidenko also devotes time and energy to professional societies and professional registration and accreditation bodies, having held leadership roles in IEEE Instrumentation and Measurement Society, Test Technology Technical Council of IEEE Computer Society, Institution of Engineering and Technology, etc. He has also contributed to organisation of over 100 international conferences as a general and technical program chair and committee member.

Dale Carnegie has a multi-disciplinary approach to research. In 1987 he graduated with a B.Sc. in Theoretical Physics and Applied Mathematics, in 1989 he attained an M.Sc. with first-class honors in Applied Physics and Electronics, and in 2000 he was awarded a Ph.D. in Computer Science. After working at the lecturer and then senior lecturer level at the University of Waikato's Department of Physics and Electronic Engineering, he took up a position at Victoria University of Wellington, New Zealand, where he helped establish a new Engineering Programme. He also established Victoria University's first Mechatronics Research Group. A full Professor since 2008, he is now the Head of School of Engineering and Computer Science.

Professor Carnegie's current research interests are in the area of Mechatronics, autonomous mobile robots, sensors, embedded systems, adaptive control and engineering education. Specific areas of on-going research include autonomous search and rescue robots and full field image ranging systems. He has published and presented over 150 research papers in various journals, book chapters, conferences and patents. In 1994, Professor Carnegie founded the Electronics New Zealand Conference series which is still held annually. He was a foundation member of the New Zealand Robotics Alliance and the NZ Mechatronics Forum. He has been a senior member of the IEEE since 2002, and is currently a member of the Robotics and Automation Society, the Circuits and Systems Society, the Systems, Man and Cybernetics Society and the Instrumentation and Measurement Society.

Neural Network Development and Training for the Simulation of Dynamic Robot Movement Behavior

Benjamin Johnen, Carsten Scheele and Bernd Kuhlenkötter

Abstract In this chapter the design and evaluation of artificial neural networks for learning static and dynamic positioning behavior of an industrial robot are presented. For the collection of training data, an approach based on the Levenberg–Marquardt algorithm was used to calibrate the robot and the coordinate measuring machine to a common reference system. A sequential approach for the network design development is presented. The network was verified by measuring different robot path segments with varying motion parameters, e.g. speed, payload and path geometry. Different layouts and configurations of feed-forward networks with backpropagation learning algorithms were examined resulting in a multi-layer network based on the calculation of the forward transformation.

1 Introduction

Industrial robots come with their own set of coordinate systems to express Cartesian positions in their domain. A typical set includes definitions for world, tool and work object frames. These frames describe the position and orientation (pose) of the robot

Based on *Learning Robot Behavior with Artificial Neural Networks and a Coordinate Measuring Machine*, by Benjamin Johnen, Carsten Scheele and Bernd Kuhlenkötter which appeared in the Proceedings of the 5th International Conference on Automation, Robotics and Applications (ICARA 2011). © 2011 IEEE.

B. Johnen (✉) · C. Scheele · B. Kuhlenkötter
Industrial Robotics and Production Automation (IRPA), TU Dortmund University, Dortmund, Germany
e-mail: benjamin.johnen@tu-dortmund.de

C. Scheele
e-mail: carsten.scheele@tu-dortmund.de

B. Kuhlenkötter
e-mail: bernd.kuhlenkoetter@tu-dortmund.de

G. Sen Gupta et al. (eds.), *Recent Advances in Robotics and Automation*,
Studies in Computational Intelligence 480, DOI: 10.1007/978-3-642-37387-9_1,

in the overall setup, the pose of the tool and the pose of the work object. The robot establishes its own TCP pose by solving the forward kinematics problem: This calculation converts the representation of the TCP pose in joint space (which is e.g. described by a vector of the current values of all joints) into the Cartesian space (cf. [1]). This approach is prone to inaccuracies from various factors. In the static case, these factors are e.g. transmission tolerances and fabrication tolerances such as arm length. When the robot is moving (dynamic case), positioning errors may arise from joint friction or robot payload (cf. [2]). For an impartial acquisition of these inaccuracies, an external measurement system is required.

The precise measurement of the robot deviation opens up the possibility to apply learning methods for creating a robot simulation that will be able to predict these deviations. In this chapter, we present an approach where we design and train artificial neural networks to build static and dynamic models of the robot behavior. The robot used in the experiments is an ABB IRB 2400/16 (handling weight 16 kg, see Sect. 2); the coordinate measuring machine is a Nikon Metrology K610 (see Sect. 3). Inherent to its functional principle, a coordinate measuring machine initially expresses the measured positions in its own coordinate system. This gives rise to the fundamental problem to correlate measuring and robot system. Because of measuring inaccuracies, this problem turns out to be an optimization problem minimizing the residual error. One approach is to express an existing coordinate system of one device in the reference system of the other device (e.g. expressing the robot base frame in the measuring system coordinates). The other approach is to define a matching pair of coordinate systems in both devices. Expressing the robot base frame in the measuring system coordinates can be done by directly measuring mechanical features of the robot base. The solution quality of this strategy is mainly influenced by fabrication tolerances of the robot and measuring inaccuracies of the measurement system. Additionally, not all robots exhibit the necessary geometrical properties (e.g. planes or boreholes with well-defined positioning relative to the robot base frame).

The other approach of defining a matching pair of coordinate systems in both worlds has the advantage of allowing choosing a reference frame, which is convenient to measure. In this chapter, we present an approach which bases on the flange coordinate system of the robot. The robot flange is precisely machined and has narrowly toleranced mating dimensions. Obviously, choosing the flange system also has some disadvantages: On the one hand, it is subject to all the inaccuracies of the kinematic chain. On the other hand, we still encounter the uncertainty of the measurement system. In the following, we will show that this is nevertheless a feasible approach for solving the matching problem.

2 ABB IRB 2400/16

The IRB 2400 is a typical industrial six-axis robot. The specified handling weight of the model used in these experiments is 16 kg. The position repeatability specified by the manufacturer is 0.06 mm. During the experiments, a payload of

20 kg has been mounted to the robot. The excessive load has been chosen to heighten the dynamic effects and because later versions of the IRB 2400 are actually specified with a handling weight of 20 kg.

3 Nikon Metrology K610

The Nikon Metrology K610 coordinate measurement machine is an optical measurement system for static and dynamic measurements in Cartesian space. The system estimates the position of infrared light emitting diodes (LEDs) by triangulation using three infrared line scan cameras. Each camera calculates the position of a LED in a plane. Cutting the three planes results in the 3D position of the LED relative to the camera. For measuring a full frame with position and orientation, a setup with at least three LEDs is required. The diodes are to be attached onto the object of interest (cf. Fig. 1). The measurement software allows defining virtual measurement frames with respect to a given set of LEDs. As long the relative positions of the diodes are fixed, the positions of the LEDs can be converted to position and orientation of the previously defined measurement frame. If more than three LEDs are employed to define a frame, the surplus LEDs are used for redundancy (not all LEDs have to be visible to the camera at the same time) or error minimization (with help of the additional data). For calibrating fixed objects, a special measuring probe “SpaceProbe” can be used. It is equipped with a probe head and three sets with each three LEDs for measuring position and orientation of the probe head. The surplus 6 LEDs are exclusively used for error minimization. The measurement system can touch up multiple points of a geometric shape (e.g. a plane or a cylinder to determine its defining parameters). The product specification of the Nikon Metrology K610 state a volumetric precision of down to 60 μm, depending on the measurement situation (distance and inclination towards the camera, ambient light conditions). The sampling rate depends on the number of the LEDs in the setup, because the diodes are lit sequentially. For three

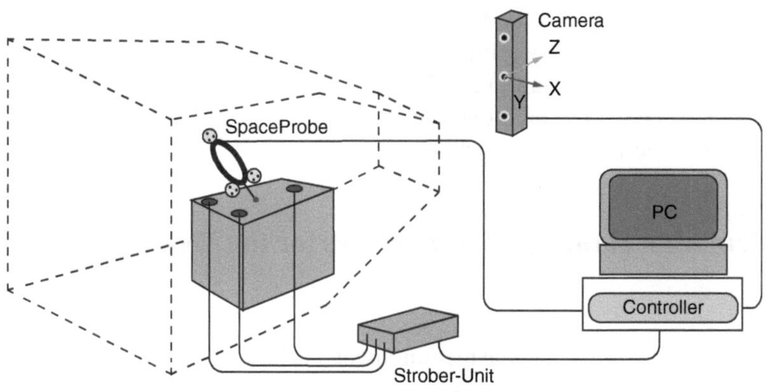

Fig. 1 Layout of the K-series measurement system

LEDs that already allow for a position and orientation measurement of an object, frequencies of up to 1,000 Hz are possible. With six sets of three LEDs (e.g. one set for each arm for the most common industrial robots), a frequency of up to 194 Hz can be realized.

4 Coordinate System Definition

A matching pair of coordinate systems is defined based on the flange coordinate system at a single robot position. By measuring the cylindrical surface, the mounting surface and the fit bore of the flange with the SpaceProbe, the frame center and orientation of the flange coordinate system is determined with the measuring system. For definition of a matching coordinate system for the robot, a new work object coordinate system at the current position with the current orientation of the flange is defined in the robot controller.

For reduction of the deviation of the defined reference coordinate systems, the initially defined coordinate system of the measurement system is subsequently corrected based on the measured positions of a regular grid. Therefor an optimization problem is defined. The objective function is the sum of all Euclidean distances between the position vectors $\mathbf{r}_i$ of the robot and the measured position $\mathbf{m}_i$ for all points i of the regular grid. The intention is to find a translation vector $\mathbf{t}$ and a rotation matrix $\mathbf{R}$ that transforms the measured points so that the error is minimized. This leads to the formulation of the following optimization function:

$$\begin{aligned} F\left(t_x, t_y, t_z, \alpha, \beta, \gamma\right) &= \sum\nolimits_i \left|\mathbf{r}_i - \mathbf{t} - \mathbf{R} \cdot \mathbf{m}_i\right|, \\ \mathbf{R} &= \mathbf{R}_z \cdot \mathbf{R}_y \cdot \mathbf{R}_x. \end{aligned} \tag{1}$$

The matrixes $\mathbf{R}_x$, $\mathbf{R}_y$, and $\mathbf{R}_z$ are the standard rotation matrixes around the axes of the respective indices. To solve the optimization problem the Levenberg–Marquardt algorithm [3] is used, a numerical solution for nonlinear least squares problems. The calculation result depends on a sufficient starting vector. Starting point of the optimization is the frame calibrated with the measuring probe. Due to the high accuracy of the measurement system, the optimization results are to be expected smaller than 1 mm and 0.2° respectively. Therefore, a null vector is a sufficient start value. The employed implementation of the Levenberg–Marquardt algorithm was the library Levmar [4].

5 Calculation of the Spatial Distances for the Measurements

Both the robot and the coordinate measurement machine deliver their respective data as a series of discrete position and orientation data. Both series can be considered as a discrete approximation of a continuous movement in Cartesian space. Figure 2

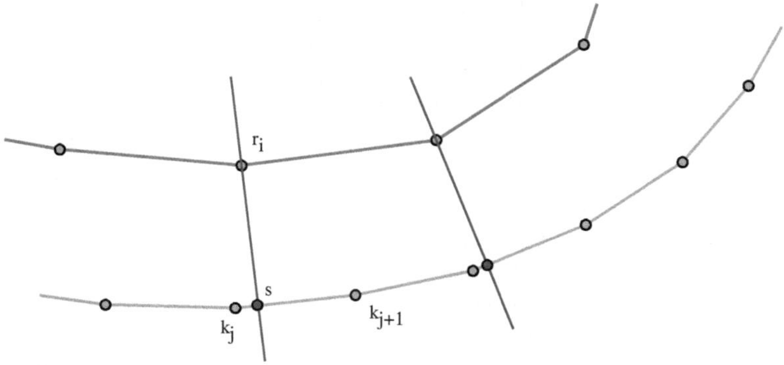

Fig. 2 Two-dimensional visualization of the minimal distance calculation between robot and measurement path

shows the approximation of the robot movement and a belonging measurement with a linear interpolation between the sampling points. To calculate the distance between the two paths, the linear interpolation is used to calculate the local distance for each of the sampling points of the upper path. For each point $\mathbf{r}_i$ the closest point on the connection between the points $\mathbf{k}_j$ and $\mathbf{k}_{j+1}$ can be determined as follows: Let $\mathbf{n}$ be the vector from $\mathbf{k}_j$ to $\mathbf{k}_{j+1}$. Then $\mathbf{s}$ is the intersection of the plane, which is spanned by the normal vector $\mathbf{n}$ and the point $\mathbf{r}_i$ and the line from $\mathbf{k}_j$ to $\mathbf{k}_{j+1}$ and represents the minimal distance to the sampling point $\mathbf{r}_i$. The following equation holds for the minimal distance $|\mathbf{s}-\mathbf{r}_i|$ and the intersection points $S(\cdot,\cdot)$:

$$|\mathbf{s} - \mathbf{r}_i| = \min_{j=0,\ldots,m-1} \left|S(\mathbf{k}_j, \mathbf{k}_{j+1}) - \mathbf{r}_i\right|. \tag{2}$$

6 Artificial Neural Networks

Artificial Neural Networks (ANN) are in many cases successfully applied to calculate forward and backward kinematic. The most commonly used network architectures are feed-forward networks. In the case of the forward kinematic, mostly parallel kinematic robots are inspected (e.g. [5–7]). For these types of robots, the computation complexity for the forward kinematic is higher than that for the backward kinematic. Plain kinematics like a robot with two rotational joints and a planar workspace in [8] are investigated as well as six-joint robots [9]. In case of the backward transformation, there is a broader range of research examining six-joint robots (e.g. [10–12]). In addition, other kinematics are considered e.g. three rotational joints in [13]. Although developed for similar purposes, the ANN presented in these works are very different with respect to the number of neurons and layers and the functional design of the single neurons. Nevertheless, the majority of all approaches employ a common technique for function approximation with ANNs:

There are one or more hidden layers with sigmoid activation functions and a subsequent layer with linear activation functions. This structure can transform the output values into the desired range. For the input encoding, different approaches were pursued: Besides using the unaltered joint values and Cartesian coordinates, the input can be normalized to the interval [0,1] (cf. [6]). The applied learning rule is in most cases the standard backpropagation algorithm; sometimes in altered versions like the momentum term variant (cf. [10]).

The developed Artificial Neural Network calculates the forward transformation and additionally incorporates the inaccuracies of the robot kinematic. Based on the joint values of a programmed position, the ANN calculates an estimation of the corresponding measured Cartesian position. Consequently, the ANN works as a black-box-model for all influencing factors of the deviation between the measured Cartesian positions and the position calculated out of the joint values by the robot controller. The architecture of the ANN was initially developed based on static position measurements on a regular grid. The basic structure was limited to feed-forward networks with standard backpropagation learning algorithm. Afterwards, network modifications and different learning rules were examined to optimize the network result for data of local path segments. Because of the basic principal of the forward transformation, six input neurons for the joint values and three output neurons representing the Cartesian coordinates were used. The network design and its parameters were developed towards optimizing the quality of results. Execution time was only a minor criterion.

The examined network parameters were: input and output format, the network topology, i.e. number of hidden layers and number of neurons per layer, activation functions and the learning parameter of backpropagation. Because the parameters interact with each other, finding an optimal solution for all parameters is a complex problem. Therefore, a sequential approach was used by optimizing one parameter at a time to find a near optimal solution which calculates robot positions with deviations lower than 1 mm in the static case. Criterion for the optimization steps was the minimization of the mean square error (MSE) of the network output.

Figure 3 gives an example of the learning process for different input formats and activation functions. All graphs illustrate the learning curves for the same six different training sets. The left column shows in each case the training error. The right column displays the corresponding validation error of the neural network. The comparison of the intervals [0,1] and [−1,1] illustrates, that for the network with hyperbolic tangent activation function and even more for the one with logistic activation function the validation error is scattering. Additionally, one can see that differences between logistic and hyperbolic tangent activation function gets less distinct after increased number of training cycles. Nevertheless, the later one achieves in general slightly better results.

The resulting ANN is a multilayer feed-forward network with output splitting. Like the in [8] presented technique, the output splitting connects the neurons in a layout that the output neurons presenting the Cartesian coordinates are completely separated from each other. Thus, the network consists of three subnets for each output value all connected to the input neurons. Figure 4 represents this layout

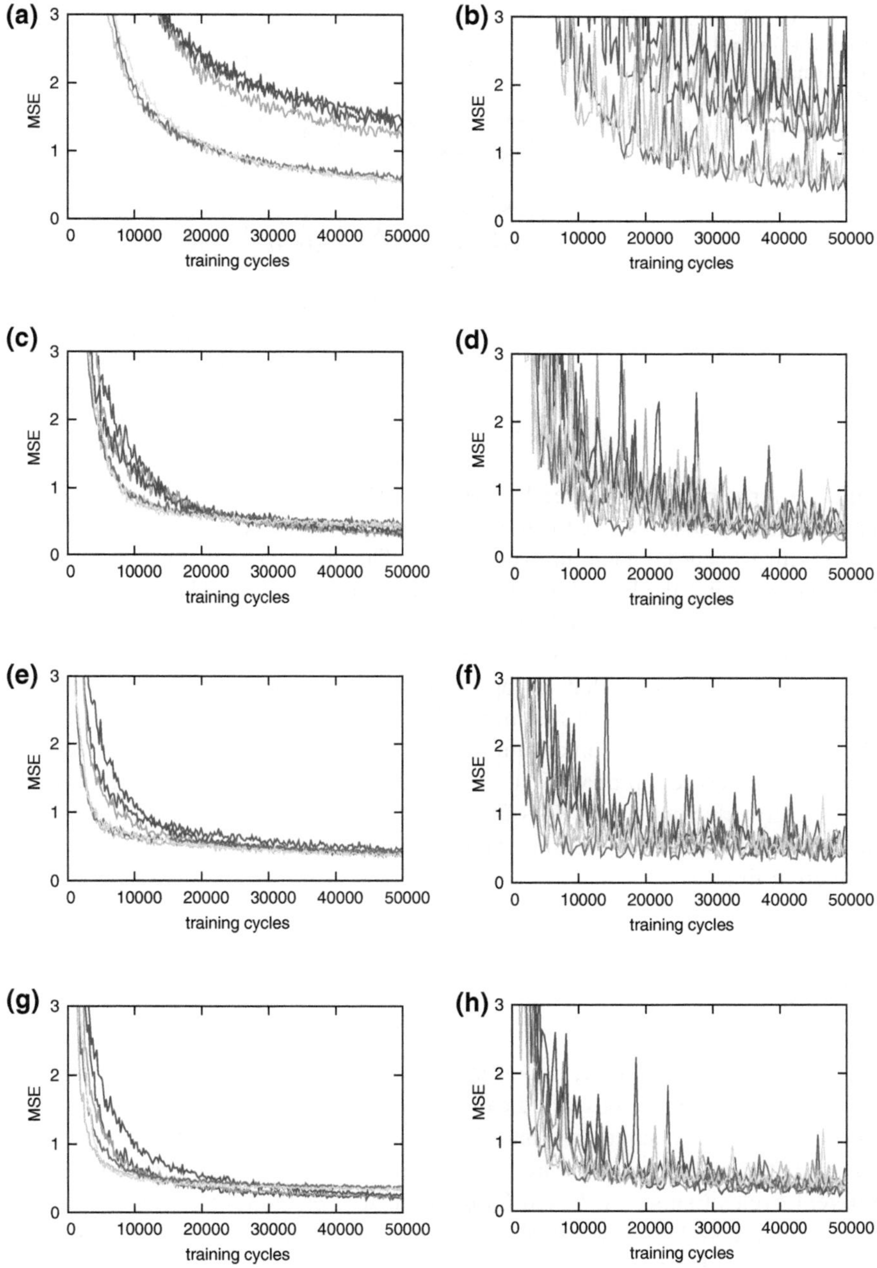

Fig. 3 Error graphs for different input formats and activation functions, graphs on the *left* display the learning *curves* of the same six training data sets and on the *right* side, the *curves* for the corresponding validation sets are shown. **a** training, log, [0,1]. **b** validation, log, [0,1]. **c** training, tanh [0,1]. **d** validation, tanh [0,1]. **e** training, log, [−1,1]. **f** validation, log, [−1,1]. **g** training, tanh [−1,1]. **h** validation, tanh [−1,1]

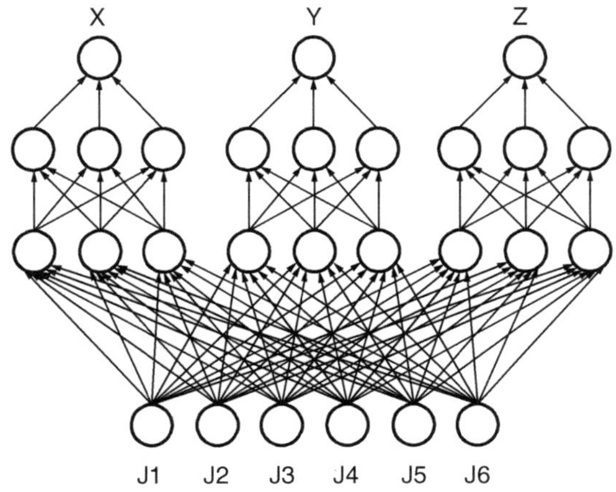

Fig. 4 Example of an ANN design for forward transformation with two hidden layers with three neurons each for all three subnets

principle for subnets with three by two hidden neurons. Derived from the examinations, the optimal layout for the considered test cases consist of three hidden layers with 15 neurons each for every subnet. The first two hidden layers as well as the input layer use the hyperbolic tangent activation function. The last hidden and output layers use the identity that is also the output function of all neurons. The input values are normalized to the interval [−1,1].

Experiments showed that to optimize the network for the dynamic case, adoption of the learning rules is more effective than altering the network layout. The examined learning algorithms were Backpropagation, Backpropagation-Momentum, Resilient propagation, and Quick propagation as implemented in the Neural Network Simulator SNNS [14]. Resilient propagation (cf. [15]) proved to have the best performance concerning residual error minimization as well as computing time.

7 Results

For evaluation of the initial coordinate system definition procedure, multiple independent runs of the described calibration were performed and evaluated. The positions of the measured grid and the reference position of the robot were not altered between the runs. Figure 5 shows the Euclidean distances among three different cycles (named A, B and C) before and after the proposed calibration of the coordinate system. The graph shows that the distance between the measured points of the different measuring cycles before the calibration is higher than the robot repeatability and the measurement uncertainty of the measurement system. This is attributed to the uncertainty of the static measurement process with the SpaceProbe. The similar curve progression of the B–C and A–C curve illustrates that measurement cycle A had a higher aberration than cycles B and C to each other. With increasing measurement index point numbers, the error rises because

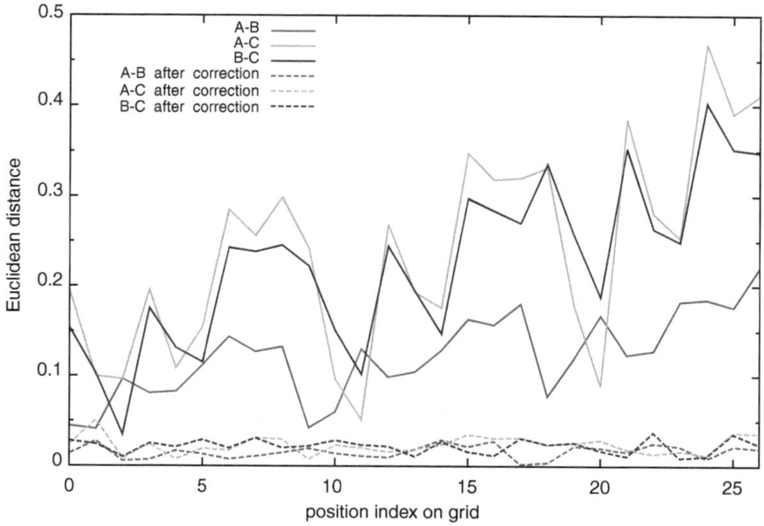

Fig. 5 Deviation of measured positions for three independent measurement cycles *A*, *B* and *C* before and after the coordinate system correction

they are at greater distance to the coordinate system center. In this case, rotation deviations have a stronger effect. The curves for the distances of the same three measurement cycles after the execution of the Levenberg–Marquardt method show only minor variance between the grid positions of the different cycles. These result from the robot repeatability error and measurement uncertainty. This clearly shows that the residual error of the suggested procedure for coordinate system definition is small enough (<0.05 mm) to evaluate robot precision.

The examined test cases of the performed experiments include circular and corner paths with different zone and speed values. Additionally these were executed with an extra payload of a 20 kg metal cylinder mounted to the robot flange. Figure 6 shows a projection of the corner path behavior into a 2-D illustration. It clearly shows the deviation of the actual robot movement from the ideal path geometry. It also shows the good approximation of the real robot movement by the ANN. The different zone values employed throughout the experiments determine the characteristics of the robot interpolation movement when it aims towards the next programmed target: A zone value >0 mm enables an interpolation motion towards the (n + 1)th target that starts before the n-th target is even reached. This approach prevents the problem that the robot would have otherwise nearly stopped at the corner target to maintain the exact geometric path definition.

The data presented in Figs. 7 and 8 was collected with a zone value of 0 and accordingly 5 mm. These parameters resulted in a velocity of ca. 70 mm/s and accordingly ca. 280 mm/s near the corner point. The considerable higher velocity of the motion with the 5 mm zone value accounts for the slightly higher deviation between planned and measured robot path. Both these velocities are considerably

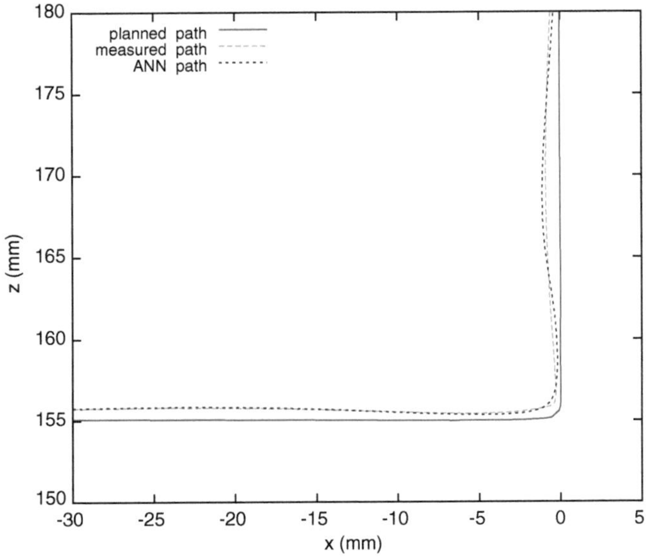

Fig. 6 Two-dimensional illustration of part of a movement path determined by planned, measured and trained position data for a robot movement with 1,000 mm/s speed, 0 mm zone value and no additional load

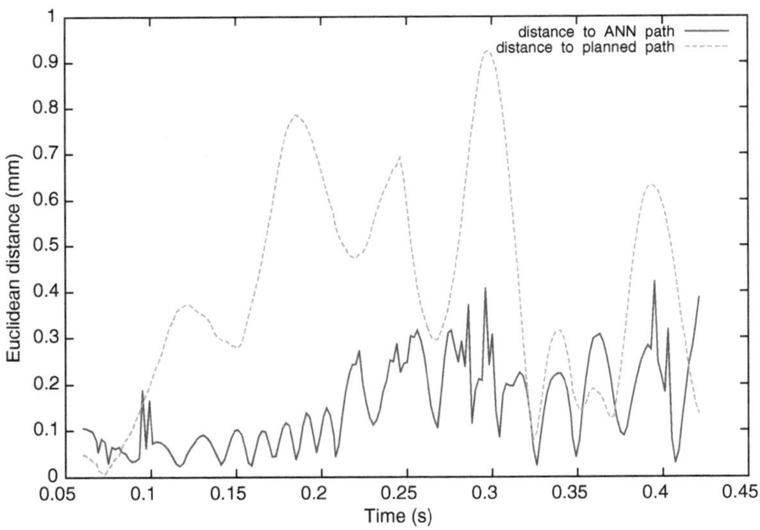

Fig. 7 ANN training result as error graphs for a corner path with 1,000 mm/s, zone value 0 mm and no additional load

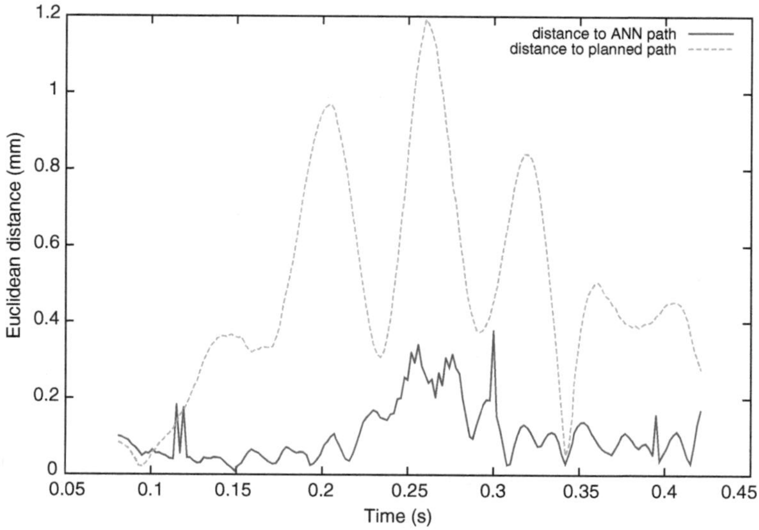

Fig. 8 ANN training result as error graphs for a corner path with 1,000 mm/s, zone value 5 mm and no additional load

smaller than the programmed speed of 1,000 mm/s. This desired speed was nevertheless reached in the linear path segments before and after the corner point. The high absolute distances between planned and measured path can be explained by this high velocity setting and the accordingly required high accelerations. Figures 7 and 8 show at the same time that the ANN was able to adjust to the erroneous robot movement behavior. The absolute distances between the planned and the ANN path are evidently smaller than the aforementioned deviations.

Another influencing factor is the payload attached to the robot. Especially the combination with high velocities, a payload equal or near to the maximum handling capacity represent the worst case scenario with respect to motion deviation errors. Even at a relatively low speed of 100 mm/s, the 20 kg load can increase the position error up to 0.6 mm as Fig. 9 shows. The diagram shows the two different segments of the corner path movement: The robot motion within the first segment runs parallel to the x–y-plane of the world coordinate system. Thus, the gravity takes effect perpendicular to the robot path. Within the second segment the gravity acts along the path resulting in a minor deviation from the movement path than in the first segment. As the second curve in Fig. 9 illustrates, the ANN adopted both segments with similar quality. Just near the corner point, the error increases, but generally keeps below the measured path error. Therefore, the ANN provides a more accurate approximation of the planned path.

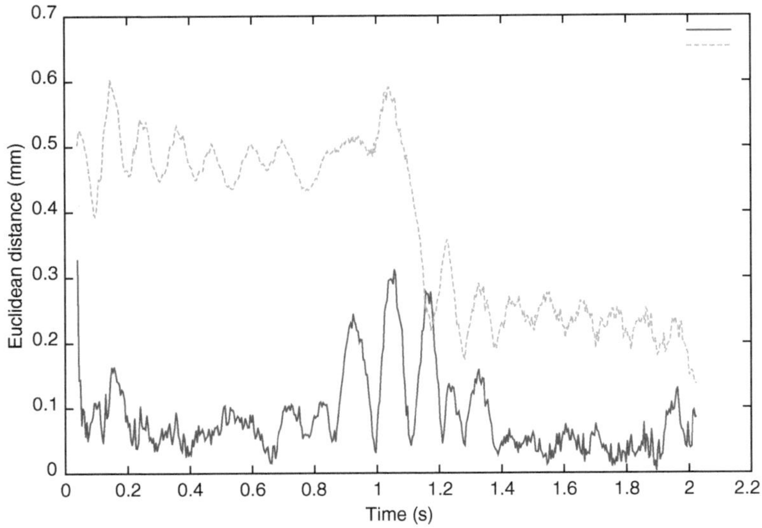

Fig. 9 ANN training result as *error* graphs for a corner path with 100 mm/s, zone value 20 mm and 20 kg load

8 Conclusion

Robot motion behavior is prone to inaccuracies from static and dynamic factors. Precise measurement of the robot movement opens the possibility to learn robot movement behavior. We presented a new way of calibrating robot coordinate systems to a vision based coordinate measurement machine. The residual error of the calibration technique could be minimized with the help of the Levenberg–Marquardt algorithm. We introduced the development of an artificial neural network to adopt robot movement behavior for individual path segments. Training the ANN with measured static and dynamic robot movement data could satisfyingly predict the robot motion behavior. It could be shown that the influence of high robot payloads also can be learned and predicted by the ANNs.

Acknowledgments This work has been funded by the Deutsche Forschungsgemeinschaft (German Research Foundation) as part of the Sonderforschungsbereich (collaborative research center) 708, project A4 "Efficient simulation of dynamic effects in surface oriented robot processing". We thank the Deutsche Forschungsgemeinschaft for the support of this work.

References

1. J.J. Craig, *Introduction to robotics: mechanics and control*, 3rd edn. (Addison-Wesley, Reading and Massachusetts, 2005)
2. J.M.S.T. Motta, R.S. Mcmaster, Modeling, optimizing and simulating robot calibration with accuracy improvement. J. Brazilian Soc. Mech. Sci. **21**, 384–401 (1999)

3. J. Moré, in *The Levenberg-Marquardt Algorithm: Implementation and Theory*, Numerical Analysis, volume 630 of Lecture Notes in Mathematics, ed. by G. Watson (Springer, Berlin/Heidelberg, 1978)
4. Levmar: Levenberg-Marquardt nonlinear least squares algorithms in C/C++. http://www.ics.forth.gr/lourakis/levmar/index.html. Accessed 02 Mar 2012
5. M. Dehghani, M. Ahmadi, A. Khayatian, M. Eghtesad, M. Farid, Neural network solution for forward kinematics problem of HEXA parallel robot. Am. Control Conf., 4214–4219 (2008)
6. A. Ghobakhloo, M. Eghtesad, Neural network solution for the forward kinematics problem of a redundant hydraulic shoulder. Ind. Electron. Soc. (IECON) 31st Annu. Conf. IEEE, 6 (2005)
7. H. Sadjadian, H.D. Taghirad, A. Fateh, Neural networks approaches for computing the forward kinematics of a redundant parallel manipulator. Int. J. Comput. Intell. **2**(1), 40–47 (2005)
8. L. Nguyen, R.V. Patel, K. Khorasani, Neural network architectures for the forward kinematics problem in robotics. JCNN Int. Jt. Conf. Neural Networks **3**, 393–399 (1990)
9. A.T. Hasan, A.M.S. Hamouda, N. Ismail, H.M.A.A. Al-Assadi, An artificial neural network strategy for the forward kinematics of robot control. Int. J. Intell. Syst. Technol. Appl. **2**, 41–49 (2007)
10. Y.I. Al-Mashhadany Inverse kinematics problem (IKP) of 6-DOF manipulator by locally recurrent neural networks (LRNNs). Int. Conf. Manage. Service Sci. (MASS), 1–5 (2010)
11. S. Manigpan, S. Kiattisin, A. Leelasantitham, A simulation of 6R industrial articulated robot arm using backpropagation neural network. Int Conf. Control Autom. Syst. (ICCAS), 823–826 (2010)
12. Y. Sahin, I. Eski, A QP artificial neural network inverse kinematic solution for accurate robot path control. J. Mech. Sci. Technol. **20**, 917–928 (2006)
13. T. Bhattacharjee, A. Bhattacharjee, A study of neural network based inverse kinematics solution for a planar three joint robot with obstacle avoidance. Assam Univ. J. Sci. Technol. **5**(2), (2010)
14. Stuttgart Neural Network Simulator (SNNS). http://www.ra.cs.uni-tuebingen.de/SNNS/. Accessed 02 Mar 2012
15. M. Riedmiller, H. Braun, A direct adaptive method for faster backpropagation learning: the RPROP algorithm. IEEE Int. Conf. Neural Networks **1**, 586–591 (1993)

Neural Network Control of Buoyancy-Driven Autonomous Underwater Glider

Khalid Isa and M. R. Arshad

Abstract This chapter presents a mathematical model and motion control analysis of a buoyancy-driven underwater glider. The glider mathematical model, which includes the presence of disturbance from the water currents, has been designed by using the Newton-Euler method. In order to predict and control the glider motion, a neural network control has been used as a model predictive control (MPC) as well as a gain tuning algorithm. The motion has been controlled by six control inputs: two forces of a sliding mass, a ballast pumping rate, and three velocities of water currents. The simulation results show the analysis of the motion control system for both neural network control approaches, and a comparison with the Linear Quadratic Regulator (LQR) controller is also included. The results show that the model is stable, and the neural network controller of MPC produced better control performance than the neural network gain tuner and the LQR, where the accuracy value of the MPC is 94.5 %.

"Based on Vertical motion simulation and analysis of USM underwater glider, Khalid Isa and Mohd Rizal Arshad which appeared in the Proceedings of the 5th International Conference on Automation, Robotics and Applications (ICARA 2011). © 2011 IEEE."

K. Isa (✉) · M. R. Arshad
Underwater Robotics Research Group (URRG), School of Electrical
and Electronic Engineering, Engineering Campus, Universiti Sains Malaysia (USM),
Nibong Tebal 14300 Pulau Pinang, Penang, Malaysia
e-mail: halid@uthm.edu.my

M. R. Arshad
e-mail: rizal@eng.usm.my

G. Sen Gupta et al. (eds.), *Recent Advances in Robotics and Automation*,
Studies in Computational Intelligence 480, DOI: 10.1007/978-3-642-37387-9_2,

1 Introduction

In recent decades, ocean resources have become international environmental, scientific and military issues. Therefore, an efficient underwater platform need to be developed for the exploitation of these resources, especially for applications such as oceanographic sampling [1–5] and related bathymetric purposes. The development of an underwater glider can fulfil these requirements because it is energy efficient and effective for long-term oceanographic sampling. The buoyancy-driven underwater glider concept was introduced in the late 1980s, which was formally proposed by oceanographer, Henry Stommel [1]. Currently, four underwater gliders are operational, which are known as the SLOCUM [2], Spray [3], Seaglider [4] and Deepglider [5]. The development of these gliders was inspired by the vision of Henry Stommel. These gliders were buoyancy propelled and had a similar design in terms of size, weight and configuration. However, the objective for each of these gliders is similar, which is to minimise energy consumption during the oceanographic exploration applications.

A typical underwater glider design consists of an ellipsoidal or cylindrical hull, wings, tail or rudder, internal electronic components and batteries. In addition, there are two internal actuators which consist of a ballast pump and a moving mass to control buoyancy and attitude respectively. These actuators enable the underwater glider to change buoyancy and weight to move vertically in the water column and use the hydrodynamic lift on the wings to create the forward motion. The hydrodynamic lift on the wings drives the glider forwards as it travels in a series of downward and upward glides. Due to the low power consumption, the buoyancy propulsion technique has the advantage of allowing the vehicles to undertake a mission of longer duration compared to that of conventional autonomous underwater vehicles (AUVs) [4, 6]. Although underwater gliders have demonstrated their greater efficiency, a change in vehicle configuration could further increase glider efficiency [7]. Due to this, research work on the underwater glider has experienced a substantial increase, and its related technologies are expanding rapidly.

Numerous underwater glider control techniques have been proposed by researchers, whether through simulation or actual experiment. Most existing gliders have used the PID and LQR controller to control the gliders attitude and motion [8–13]. In addition, the sliding mode control (SMC) has been used to control the underwater glider [14, 15], but the main constraint in SMC is the chattering effect, which can degrade the performance of the system, and make the system become unstable. Although these control systems have already demonstrated acceptable control results, they still have constraints in terms of control performance. The high nonlinearity of underwater vehicle dynamics and underwater disturbances are the main reasons that make it difficult to control the underwater glider [16, 17]. Due to this, it is highly desirable to design a controller that is robust and has an adaptive ability in order to deal with the constraints. Since the neural network is able to handle nonlinearity and is able to adapt to the changing conditions of the glider dynamics, we have chosen it as the controller for the motion control system of the buoyancy-driven USM underwater glider.

We have mathematically modelled the buoyancy-driven USM underwater glider and analysed the glider motion for the open-loop system in [18], and estimated the glider hydrodynamic coefficients and characteristics by using the Newtonian and Slender-body theory in [19]. In this work, the implementation of neural network control is implemented as the predictive control as well as the gain tuning algorithm. The simulations for both neural network approaches were performed by using Matlab. The motion characteristics for the glider were identified, and the neural network performance for both approaches were compared and analysed in the simulation results. In addition, we carried out a comparison of controller performance between the neural network controller and the LQR.

This chapter is organised as follows. In Sect. 2, the mathematical model and equations of motion for the USM buoyancy-driven underwater glider are presented. Section 3 describes the design of the neural network controller for the glider motion control system. Simulation results and discussion of analyses are presented in Sect. 4. Finally, a conclusion is given in Sect. 5.

2 Mathematical Model and Equation of Motion

In order to simulate and analyse the motion of the underwater glider, the dynamic parameters of the glider must first be modelled either using the analytical, experimental or computational method [20–22]. In this work, the glider model was formulated by using the analytical method, which is based on Newtonian and Slender-body theory. Although, there are several dynamic models of the underwater glider, e.g. Graver [23] and Mahmoudian [24], in these models the disturbance of water currents is neglected. Therefore, we have included disturbance from water currents in the dynamic model of the glider.

2.1 Kinematic Model

The USM underwater glider has a cylindrical hull with fixed wings and a tail. In this model, the hydrodynamic forces and moments on the glider are composed of the forces and moments acting on the cylindrical hull, wings, and the rudder. The components configuration and the parameter values of the glider are shown in Fig. 1 and Table 1, respectively. The parameter values were obtained from calculation and Solidworks.

In order to explain the kinematic model, Fig. 2 shows the frame of reference which was divided into a body-fixed frame (b-frame) and inertial frame (i-frame). The b-frame is a rotating frame with respect to the i-frame, which is considered as non-rotating frame in which the Newton's laws of motion applies. The centre of

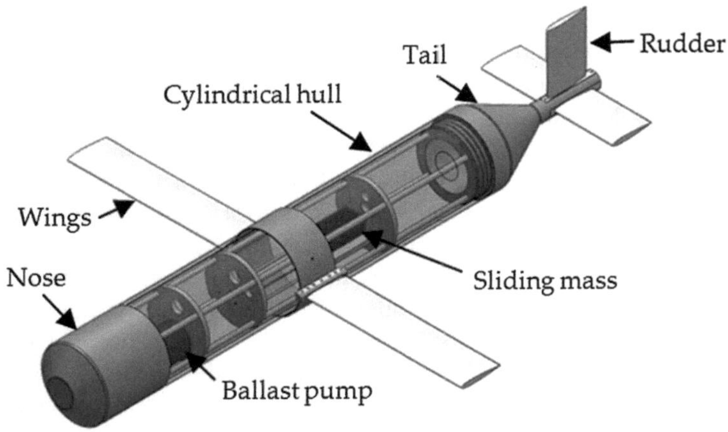

Fig. 1 Components configuration of the buoyancy-driven USM underwater glider

Table 1 Principal characteristics of the underwater glider

Characteristics	Value
Length, L	1.5 m
Body diameter, d	0.2 m
Wing length, W_L	0.4 m
Wing width, W_w	0.1 m
Rudder length, R_L	0.2 m
Rudder width, R_w	0.1 m
Hull mass, m_h	17.4583 kg
Ballast mass, m_b	0–0.850 kg
Sliding mass, m_p	2 kg
Water density, ρ	1025 kg/m^3

buoyancy (CB) is located at the centre of the cylindrical hull (origin of b-frame), and the centre of gravity (CG) is slightly offset from the CB in order to create a constant gravitational moment.

Referring to Fig. 2, the position of the glider with respect to the i-frame is denoted as $\eta = [b, \Theta]^T$, where $b = [x, y, z]^T$ and $\Theta = [\emptyset, \theta, \psi]^T$. The velocity of the glider is denoted as $V = [\vartheta, \omega]^T$ where $\vartheta = [u, v, w]^T$ and $\omega = [p, q, r]^T$. R is the rotational matrix that maps vector in b-frame into i-frame. Thus, the 6 DOF kinematic equations are defined as:

$$\dot{\eta} = \begin{bmatrix} R\Theta & 0_{3x3} \\ 0_{3x3} & T_{\Theta}\Theta \end{bmatrix} V, \tag{1}$$

where $T_{\Theta}\Theta$ is the transformation matrix of the Euler angle. In the longitudinal model, the glider travels in the vertical x-z plane with the state of r or yaw moment

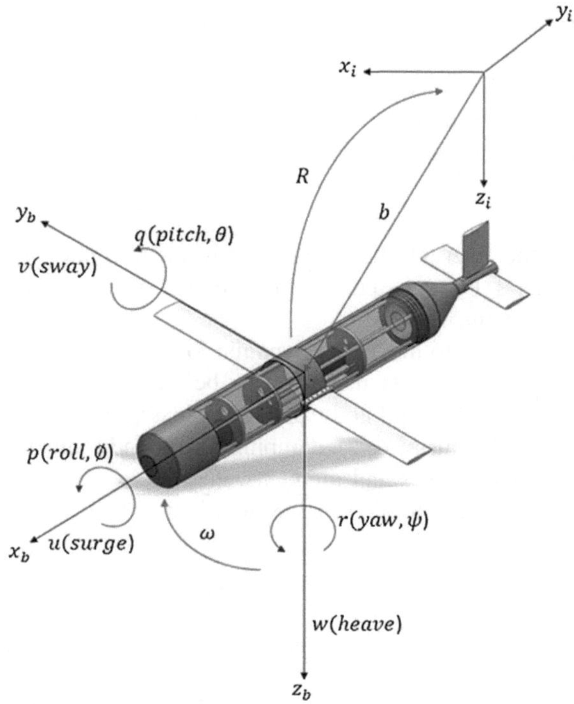

Fig. 2 Reference frame of the buoyancy-driven underwater glider

is zero or small. As a result, the hydrodynamic effects of side force are neglected in the model. In addition, other lateral states such as v, p, ψ and $\emptyset$ are also small.

2.2 Dynamic Model

The internal mass configuration of the glider is critical because the glider mass will change when travelling through the water column. There are three principal masses in the typical underwater glider: hull mass, m_h, internal moving point mass, m_p, and ballast point mass, m_b. In some models of glider, such as in [24], the internal moving point mass is divided into two: sliding moving mass, m_{px}, and rotating moving mass, m_{py}. However, for the USM underwater glider, only sliding moving mass is taken into account. Thus, the rigid body mass of the glider, m_{rb}, and net buoyancy, m_0, are defined as:

$$m_{rb} = m_h + m_b + m_p \text{ and } m_0 = m_{rb} - m \tag{2}$$

The variable m is the mass of fluid displaced. If m_{rb} is equal to m, the glider is neutrally buoyant. If m_0 is greater than zero, that means the glider tends to sink because it is heavy in water. If m_0 is lower than zero, the glider is buoyant and tends to rise. The position of m_p and m_b in the body frame is given by the vectors

r_p and r_b. Since the ballast mass is a variable mass with fixed position r_b, therefore; $\dot{r}_b = r_b = 0$

On the other hand, m_p has a fixed mass but variable position, r_p. The sliding mass moves along the *x-z* plane for controlling the pitch, creating the following moving mass velocity vector:

$$\dot{r}_p = [\dot{r}_{px}, 0, \dot{r}_{pz}] \tag{3}$$

The internal sliding mass movement creates a gravitational moment on the glider's body based on the amount of mass moved and the distance of its movement.

In order to model the dynamic forces and moments of the glider, there are four main components that need to be calculated. The components are the system inertia of the rigid body and added mass, *M*, Coriolis-centripetal of the rigid body and added mass, *C(V)*, damping forces and moments, *D(V)*, and gravitational and buoyancy forces and moments, $g(n)$.

According to [25], the system inertia matrix with respect to the origin is formulated as:

$$M = M_{RB} + M_A, \tag{4}$$

where M_{RB} is the rigid-body system inertia matrix and M_A is the added mass system inertia matrix.

In order to calculate the Coriolis forces and moments, $C(V)$, the velocity of water currents, V_c, is included in the equation as a disturbance. Thus, the Coriolis and centripetal force is defined as:

$$C(V) = C_{RB}(V)V + C_A(V_r)V_r, \tag{5}$$

where

$$V_r = V - V_c. \tag{6}$$

The C_{RB} and C_A represent the Coriolis of rigid body and added mass, respectively. On the other hand, the damping forces and moments, $D(V)$, act at a centre of dissipative force. The dissipative force is equal to the product of pressure and area, therefore the velocity of water currents must be taken into account as a disturbance. Thus, the damping force and moment is defined as:

$$D(V) = D + D_n(V_r), \tag{7}$$

where D is the linear damping matrix and $D_n(V_r)$ is the nonlinear damping matrix.

The gravitational and buoyancy forces, $g(n)$, are also called restoring forces. These forces occur due to weight, *W*, and buoyancy, *B*. The weight and buoyancy are defined as:

$$W = m_{rb}g \text{ and } B = \rho g \nabla, \tag{8}$$

where g is gravity, ρ is the water density and ∇ is the volume of the glider's body. It is necessary to design the underwater glider with positive buoyancy ($B > W$), so that the glider will surface automatically in the case of an emergency situation such as power failure.

2.3 Equations of Motion

As stated in (3), the velocity vector of sliding mass is denoted as $\dot{r}_p$. Therefore, the position of sliding mass is defined as $r_p = [r_{px}, r_{py}, r_{pz}]$ and the corresponding momentum is defined as $p_p = [p_{px}, p_{py}, p_{pz}]$. Let's say that the rate of change of p_p is equal to the net force acting on the sliding mass, therefore the force of sliding mass is defined as:

$$\overline{u} = [u_x, u_y, u_z] = p_p \times q + m_p g(R^T k). \tag{9}$$

The control input for the ballast pump is denoted as u_b, which is equal to the ballast mass rate, $\dot{m}_b$. In addition to the control input of the actuators, we also include the velocity and acceleration of the water currents, V_c, as the disturbance of the underwater glider plant. The current velocity and acceleration is defined as:

$$V_c = [V_{cx}, V_{cy}, V_{cz}, \dot{V}_{cx}, \dot{V}_{cy}, \dot{V}_{cz}] \tag{10}$$

Thus, the control input to the glider is denoted as:

$$u = [\overline{u}, V_c] = [u_x, 0, u_z, u_b, V_{cx}, V_{cy}, V_{cz}, 0, 0, 0] \tag{11}$$

According to the kinematic and dynamic model of the underwater glider, the 6 DOF nonlinear equations of motion for the generalized model of the underwater glider can be written as:

$$\dot{\eta} = [\dot{x}, \dot{y}, \dot{z}, \dot{\emptyset}, \dot{\theta}, \dot{\psi}] = J(\eta)V \tag{12}$$

$$\dot{V} = [\dot{u}, \dot{v}, \dot{w}, \dot{p}, \dot{q}, \dot{r}] = M^{-1}(-C(V) - D(V) - g(n)) \tag{13}$$

However, since the buoyancy-driven underwater glider glides vertically through the water column, we assumed that the motion in the horizontal plane is stable. Thus, several individual parameters in the equations of motion need to be rewritten and this can be done by referring to Graver [23]. In addition, the following differential equations of the sliding mass velocity and moment and the ballast mass rate also need to be included in the equations of motion.

$$\dot{r}_{px} = \frac{1}{m_p} p_{px} - u - r_{pz} q, \tag{14}$$

$$\dot{r}_{py} = \frac{1}{m_p} p_{py} - w + r_{px} q, \tag{15}$$

$$\dot{p}_{px} = u_x, \tag{16}$$

$$\dot{p}_{pz} = u_z, \tag{17}$$

$$\dot{m}_b = u_b. \tag{18}$$

3 Neural Network Controller Design

The neural network controller has been applied to the AUV because of its robustness and adaptability to the highly nonlinearity and dynamics environment of the vehicle. There are several research works that have implemented neural networks to control AUV [17, 26–29], but none of them are used to control the underwater glider. In this work, the neural networks is used for gain derivation and predictive control of the buoyancy-driven underwater glider motion control system.

3.1 Linearisation

In order to design the neural network controller, the nonlinear plant of the glider must be linearised. The linearisation is carried out about an operating point to obtain the state-space representation of the Multiple-Input-Multiple-Output (MIMO) system of the glider. In this work, the linearisation point is shown in Table 2.

This linearisation process produces the state-space that has 9 inputs, 17 states and 17 outputs. However, in order to control the glider motion and analyse it, only 6 inputs, 9 states and 9 outputs of interest were selected. Table 3 shows the selected states, inputs and outputs for the controller. The other states were not selected since they will not affect the dynamic of the underwater glider motion in the vertical plane.

Table 2 Linearisation operating point

Parameter	Value	
States	Position, η	$[0, 0, 0, 0, 0, 0]^T$
	Velocity, V	$[1, 0, 0, 0, 0, 0]^T$
	Sliding mass position, r_{px}, r_{pz} and forces, P_{px}, P_{pz} and ballast mass, m_b	$[0, 0, 0, 0, 0]^T$
Inputs	Sliding mass net force, u_x, u_z and ballast pumping rate, u_b	$[0, 0, 0]^T$
	Water current velocity and acceleration, V_c	$[0, 0, 0, 0, 0, 0]^T$

3.2 Neural Network Controller for Gain Derivation

The objective of designing the neural network controller is to derive the gain, K, for the glider motion control system. Therefore, we have designed the forward model of the neural network based on the mutlilayer perceptron (MLP) networks. Figure 3 shows the architecture of the forward model, where the model has one input layer, one hidden layer and one output layer. The input of the network is matrix B (desired control inputs) of the glider linear model and target output is matrix A (state outputs).

We used the sigmoid transfer functions in the hidden layer and output layer to estimate the output, and then used the states or outputs error to derive the gain for the controller. The state's error between the desired states and actual states is defined as:

$$E_s = \sum\nolimits_{i=1}^{n} A_{nn} - A_{ss}, \tag{19}$$

where A_{nn} is the matrix A, which obtained from the neural network and A_{ss} is the matrix A, which obtained from the state-space. Thus, the gain, K, is derived as:

$$K = -(E_s/B), \tag{20}$$

where B is the matrix B from the input layer.

In order to select the network parameters, which are weights and biases, we have trained the network by using the backpropagation training procedure. The backpropagation training algorithm for the MLP networks is an optimisation procedure, which is based on the gradient descent. Thus, a mean square error performance index can be minimised, and it is defined as:

$$w_{i,j}^{l}(k+1) = w_{i,j}^{l}(k) - \alpha s_i^l a_j^{l-1}, \tag{21}$$

$$b_i^l(k+1) = b_i^l(k) - \alpha s_i^l, \tag{22}$$

where α is the learning rate and s is the sensitivity of transfer function to change in the network input at layer, l.

Table 3 States, outputs and inputs of interest

Parameter	
States and outputs	Pitch angle, θ,
	Linear velocity of surge, u, and heave, w, and angular velocity of pitch, q
	Sliding mass position, r_{px}, r_{pz} and forces, P_{px}, P_{pz} and ballast mass, m_b
Inputs	Sliding mass net force, u_x, u_z and ballast pumping rate, u_b
	Water current velocity and acceleration, V_{cx}, V_{cy}, V_{cz}

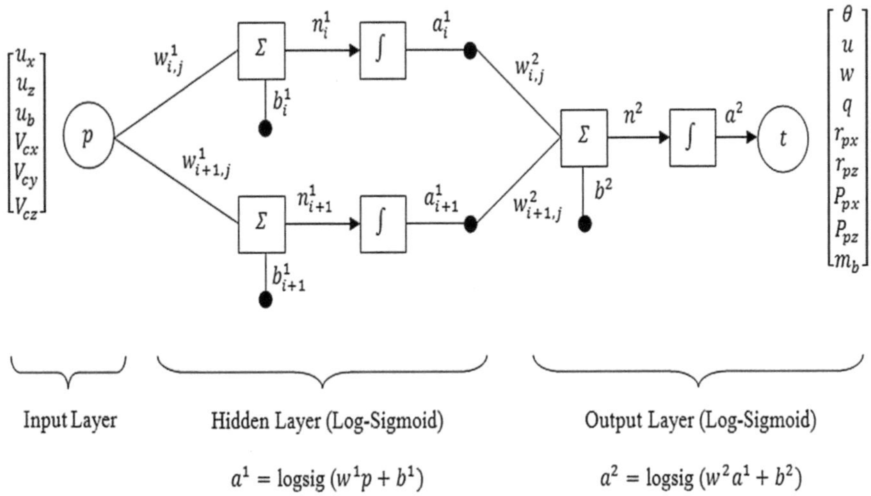

Fig. 3 The forward model of the neural network controller

3.3 *Neural Network Controller for Model Predictive Control (MPC)*

The design objective of the MPC controller is to map the desired control input as well as achieving the target output of the reference model. The MPC is chosen because of its ability to handle the MIMO system of nonlinear plant, with low rates of control update. The MPC architecture requires two neural network models of the glider plant. The models are a neural network plant model (forward model) and a neural network controller (inverse model). In addition, a performance function and optimisation procedure, which is used to evaluate system responses and to select the best control input, are also taken into account.

The neural network controller learns to produce the plant inputs, which were selected by the optimisation process, and the optimisation process can be replaced by the neural network controller when the training process is completed. Figure 4 shows the block diagram of the MPC for the underwater glider, and Fig. 5 shows the inverse model of the neural network controller.

The MPC system model of the underwater glider is given in the discrete time representation. The system outputs, y_n, controlled outputs, z_n, inputs, u_n and states, x_n are respectively defined as:

$$x_n = A_d x_n + B_d u_n, \tag{23}$$

$$y_n = C_d x_n, \tag{24}$$

$$z_n = C_z x_n, \tag{25}$$

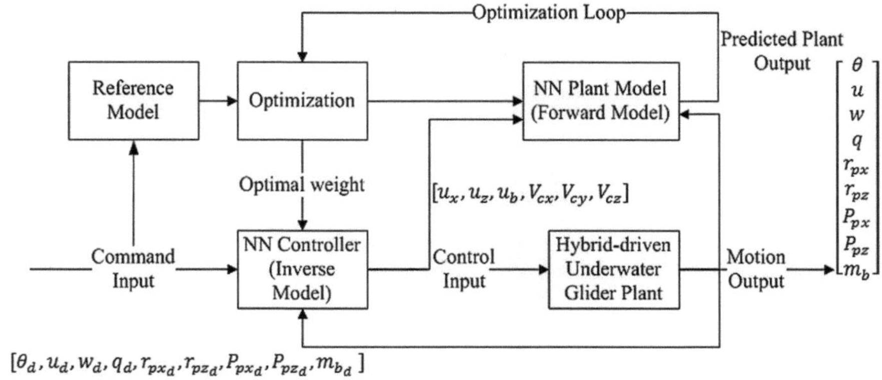

Fig. 4 Model predictive control of the underwater glider motion

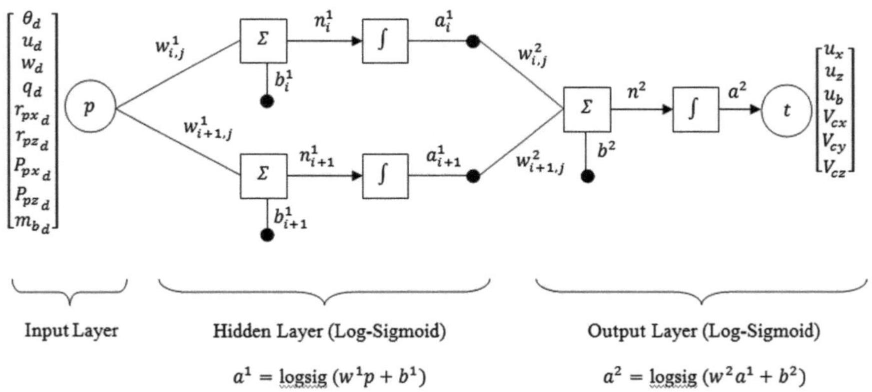

Fig. 5 The inverse model of the neural network controller

where *n* is the *n*th element of the system. In order to obtain the control input of the glider motion control system, the sampling time, control horizon and prediction horizon must be determined during the model prediction process. In this work, we have determined the value of sampling time, control horizon and prediction horizon as equal to 1, 15 and 5, respectively.

In addition, we have also specified the output constraints of the MPC control system in terms of the minimum and maximum value of the ballast mass, m_b, and sliding mass vector, r_{px}. These constraints are specified because the maximum value of ballast mass is 0.850 kg and the maximum value of sliding mass vector in *x*-direction is 0.40 m. Then, the MPC algorithm predicts the future system output at each sampling time for the predetermined prediction horizon. Finally, the control input at each sampling time is calculated based on the optimisation over the control horizon. Thus, there are three important steps that must be carried out

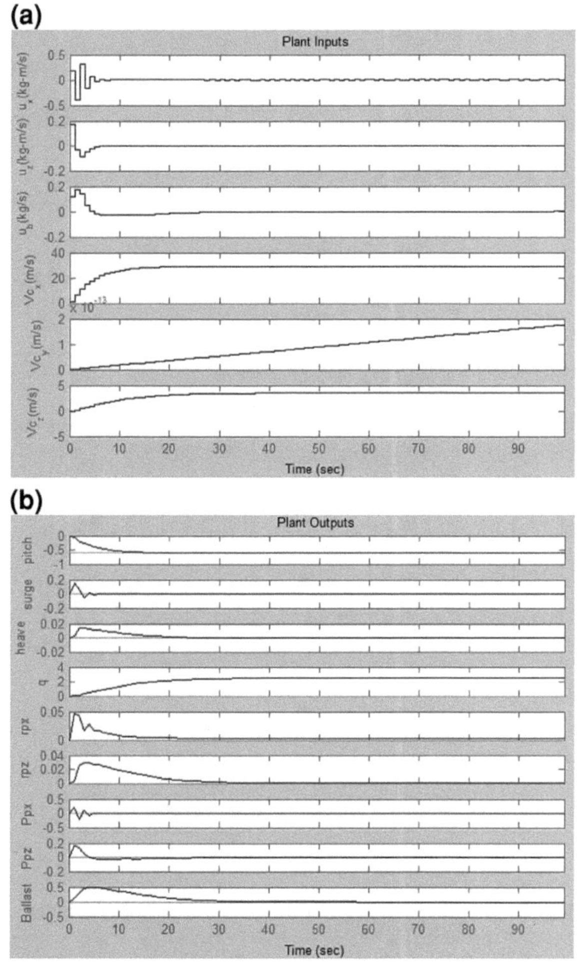

Fig. 6 **a** Plant inputs, and **b** plant outputs for the desired pitch angle, $\theta = -35°$

in the implementation of the MPC algorithm. The steps are prediction, optimisation and control.

4 Results and Discussion

In this section, we demonstrate the simulation results of the neural network control of the glider. The simulation was programmed using Matlab, and the results are presented in terms of the neural network controller for gain derivation and model predictive control. In addition, a comparison of the control performance between the neural network controller and LQR controller is presented.

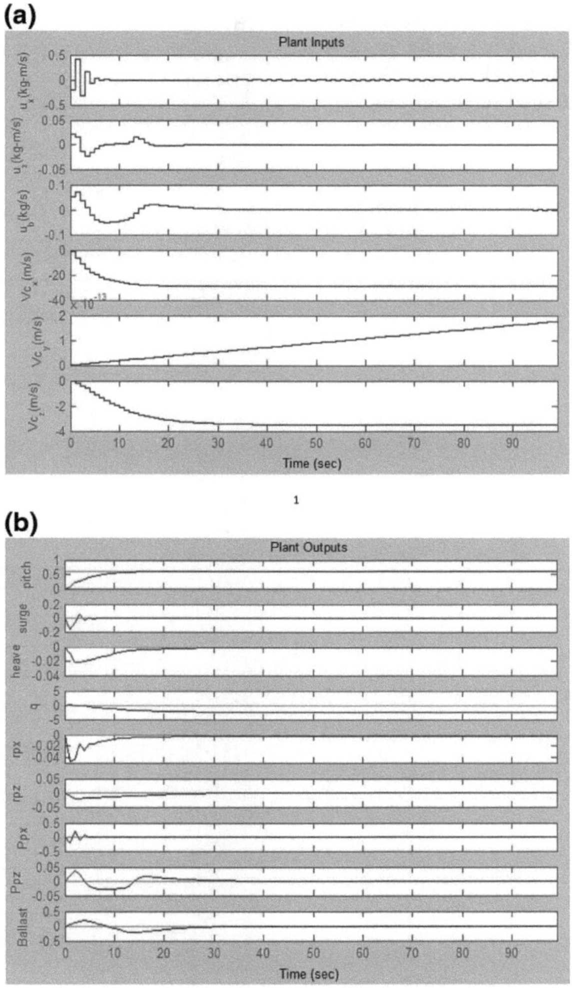

Fig. 7 **a** Plant inputs, and **b** plant outputs for the desired pitch angle, $\theta = 35°$

4.1 Model Predictive Control

We have simulated the model with a different value of desired outputs, where the range of the pitch angle, θ, was between $-45°$ to $45°$, and the range of surge velocity, u, was between 0 to 1 m/s. However, in this chapter, we demonstrate and compare the results of two desired pitch angles of downward and upward motion with different desired surge velocity. Thus, there are four sets of desired outputs which have been simulated in order to obtain the plant control inputs and the plant outputs. Figure 6a, b shows the graph of plant control inputs and actual plant outputs, respectively, which resulted from the simulation of the first set of desired outputs. In this first set of desired outputs, the value of the desired pitch angle, θ, was $-35°$ (-0.6109 rad) and the rest of the desired outputs were zero. According

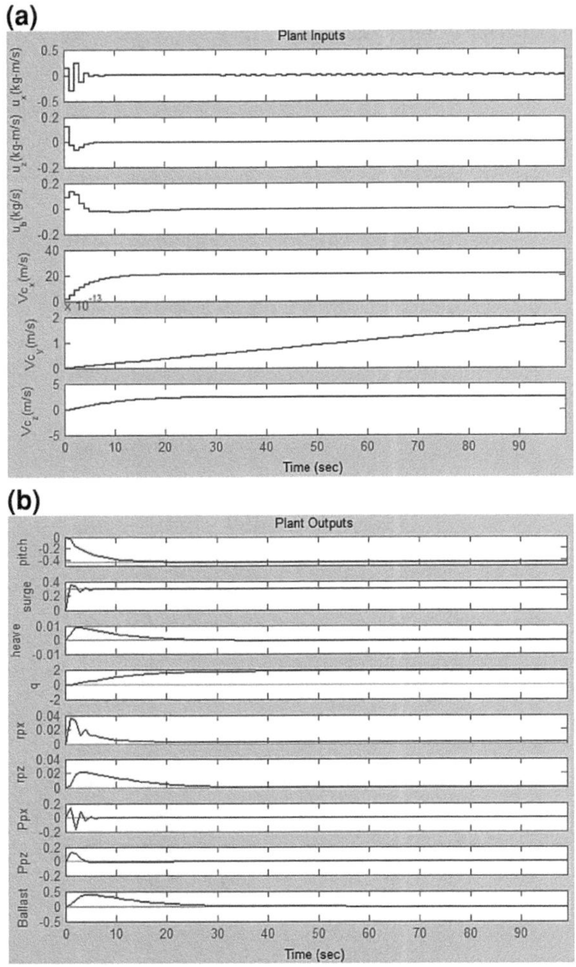

Fig. 8 **a** Plant inputs, and **b** plant outputs for the desired pitch angle, $\theta = -25^{\circ}$ and the desired surge velocity, $u = 0.3$ m/s

to the graph of plant output (Fig. 6b), the resulting or actual pitch angle converged to the desired angle, and the Fig. 6a shows the resulting control inputs, u = [0.1913, 0.1689, 0.1224, 1.1832, 0, −0.0596].

Figure 7a, b shows the plant inputs and outputs for the upward motion, where the desired pitch angle was 35° and the rest of the desired outputs were zero. The graphs show that the neural network controller was able to achieve the desired pitch angle and was able to predict the control inputs. The graphs of plant inputs show that the control input for the desired motion is u = [−0.1926, 0.0209, 0.0513, −1.2088, 0, −0.0270].

In the third simulation, the third set of desired outputs represent the downward motion, where the pitch angle was −25° and the surge velocity was 0.3 m/s. Figure 8a, b shows the resulting plant control inputs and actual plants outputs. The graphs show that the actual output for pitch angle and surge velocity are converged

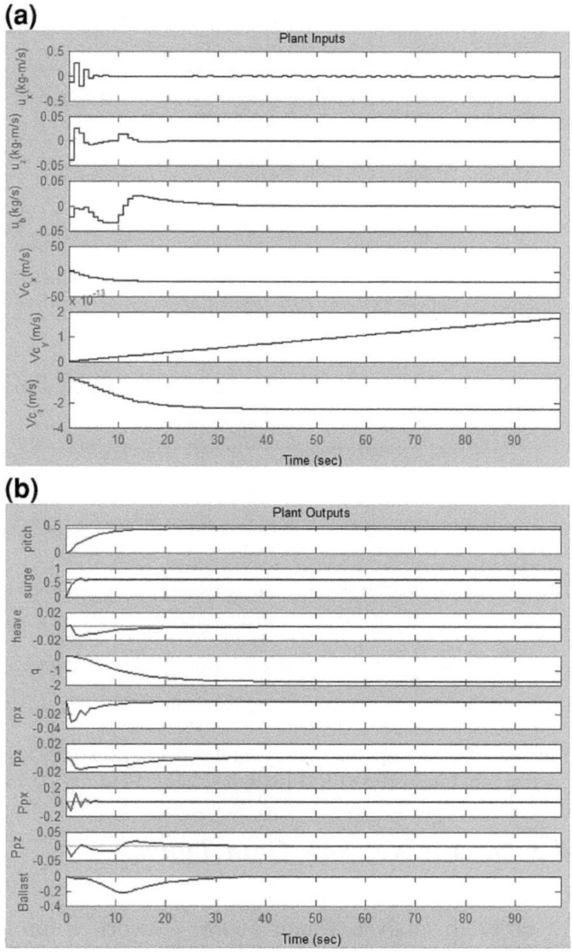

Fig. 9 **a** Plant inputs, and **b** plant outputs for the desired pitch angle, $\theta = 25°$ and the desired surge velocity, $u = 0.6$ m/s

to the desired pitch angle and surge velocity, and the predicted control input is u = [0.1443, 0.1258, 0.0889, 1.8294, 0, −0.0720].

Finally, the plant inputs and outputs for the fourth set of desired outputs, are shown in Fig. 9a, b. In the fourth set of desired outputs, the pitch angle was 25°, the surge velocity was 0.6 m/s and the rest of the desired outputs were zero. As a result, the graph shows that the actual output for pitch angle and surge velocity are converged to the desired output, and the predicted control input is u = [−0.1218, −0.0390, −0.0230, 1.1140, 0, −0.0457].

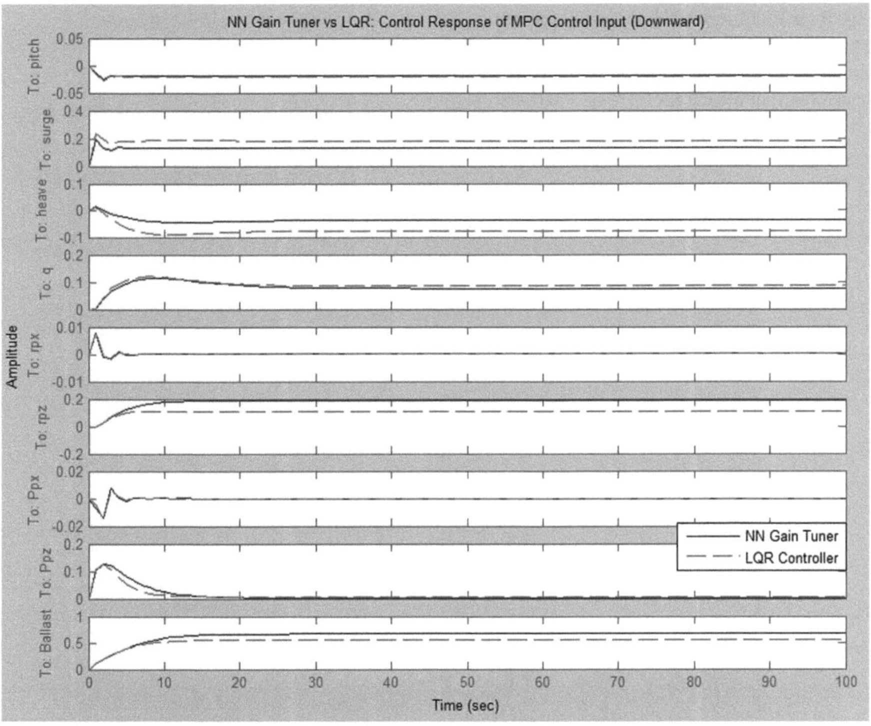

Fig. 10 Control response of the neural network gain tuner and the LQR controller based on MPC control input, u = [0.1913, 0.1689, 0.1224, 1.1832, 0, −0.0596]

4.2 Neural Network Gain Tuner

In order to analyse which neural network control approaches produced better performance, we have compared the actual plant output of MPC with the neural network gain tuner and the LQR controller. Therefore, we used the predicted control input that we obtained from the four sets of MPC simulation. Figure 10 shows the glider motion for the first control inputs, which resulted when the desired pitch angle, θ, was −35° (−0.6109 rad). The graphs show that both the neural network gain tuner and the LQR controller produced the pitch angle value of −0.0218 rad and −0.0266 rad respectively, which means the desired pitch angle is not achieved.

Figure 11 presents the resulting upward motion of the glider when the second set of plant inputs from the MPC is entered as the control input of the neural network gain tuner and the LQR controller. The graphs show that the desired angle of 35° is not achieved for both controllers. The neural network gain tuner shows the resulting value of the pitch angle is 1.4° (0.0247 rad) and the value of the pitch angle for LQR is 1.5° (0.0271 rad).

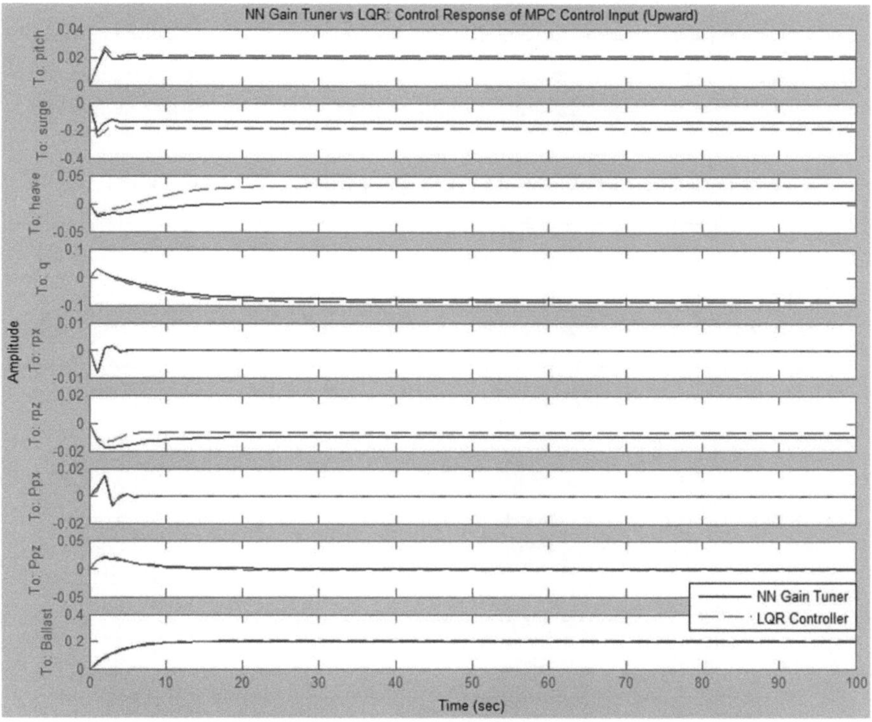

Fig. 11 Control response of the neural network gain tuner and the LQR controller based on MPC control input, u = [−0.1926, 0.0209, 0.0513, −1.2088, 0, −0.0270]

Figure 12 shows the downward motion of the glider for the third control inputs, which resulted when the desired pitch angle, θ, was −25° (−0.4363 rad) and the surge velocity was 0.3 m/s. The graphs show that both the neural network gain tuner and the LQR controller produced the pitch angle value of −0.0283 and −0.0323 rad, respectively, which means the desired pitch angle was not achieved. However, the surge velocity from the neural network gain tuner converged to the desired surge velocity, which is 0.3 m/s. On the other hand, the surge velocity from the LQR did not converge to the desired velocity, since the value is 0.44 m/s.

Finally, the resulting plant outputs from the neural network gain tuner and the LQR controller for the fourth control inputs of the MPC, are shown in Fig. 13. The graphs show that neither the desired pitch angle nor the desired surge velocity are achieved by either controller, where the error rate of pitch angle for neural network gain tuner is 98.1 % and the error rate for LQR is 96.8 %. On the other hand, the error rate of surge velocity for the neural network gain tuner is 23.5 % and the error rate for LQR is 41.3 %.

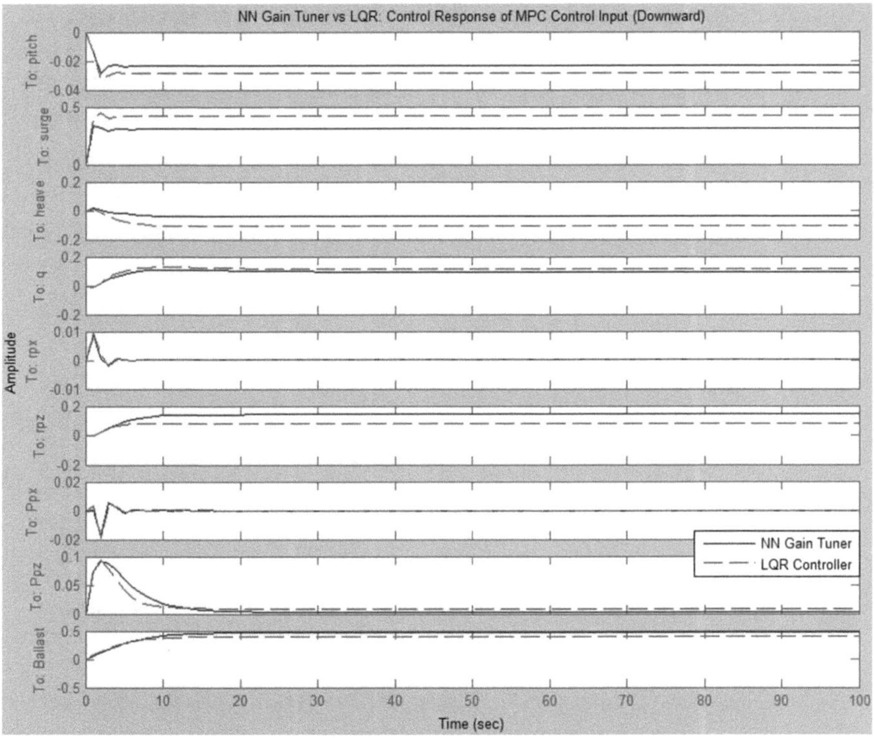

Fig. 12 Control response of the neural network gain tuner and the LQR controller based the MPC control input, u = [0.1443, 0.1258, 0.0889, 1.8294, 0, −0.0720]

5 Conclusion

This chapter presents the neural network motion control system of the buoyancy-driven underwater glider in the vertical plane, which is based on the linearised model of the nonlinear underwater glider plant. In addition, the presence of water currents as a disturbance is also included in the glider model. In this work, the neural network controller is used for the MPC and gain tuning or derivation algorithm. In order to predict the plant control inputs and plant outputs, four sets of desired outputs are used in the MPC. The desired outputs consist of 9 states of the glider linearised model. The resulting plant control inputs were then used on the neural network controller for gain derivation and the LQR controller. Therefore, the controllers' performance can be compared and analysed. According to the analysis, we found that the performance of the MPC is better than the neural network gain tuner and the LQR controller because of the efficient optimisation factor in the MPC. All the simulation results show that the glider model is stable, and the MPC able to compensate for disturbance and able to predict the control inputs and motion outputs, with a high accuracy rate of 94.5 %.

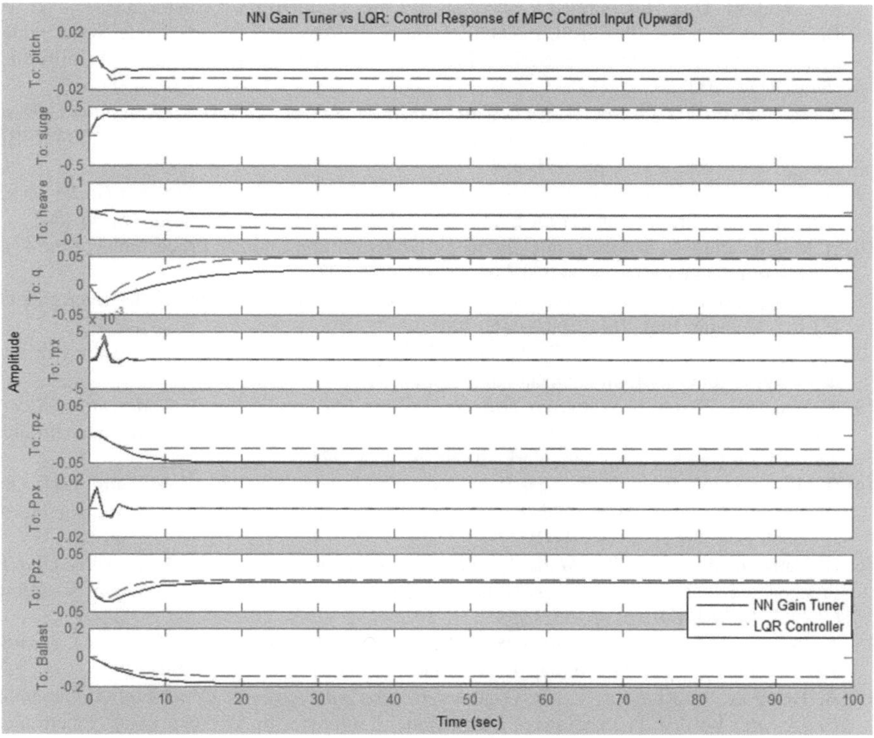

Fig. 13 Control response of the neural network gain tuner and the LQR controller based on the MPC control input, u = [−0.1218, −0.0390, −0.0230, 1.1140, 0, −0.0457]

Acknowledgments The author would like to thank the Malaysia Ministry of Higher Education (MOHE), ERGS-203/PELECT/6730045, Universiti Sains Malaysia (USM) and Universiti Tun Hussein Onn Malaysia (UTHM) for supporting the research.

References

1. H. Stommel, The Slocum mission. Oceanography **2**, 22–25 (1989).
2. D.C. Webb, P.J. Simonetti, C.P. Jones, SLOCUM: An underwater glider propelled by environment energy. IEEE J. Oceanic Eng. **26**(4), 447–452 (2001).
3. J. Sherman, R.E. Davis, W.B. Owens, J. Valdes, The autonomous underwater glider "spray". IEEE J. Oceanic Eng. **26**(4), 437–446 (2001).
4. C.C. Eriksen, T.J. Osse, R.D. Light, T.Wen, T.W. Lehman, P.L. Sabin, J.W. Ballard, A.M. Chiodi, Seaglider: A long range autonomous underwater vehicle for oceanographic research. IEEE J. Oceanic Eng. **26**(4), 424–436 (2001).
5. T.J. Osse, C.C. Eriksen, in *The Deepglider: A full Ocean Depth Glider for Oceanographic Research*. Proceedings of IEEE Oceans 2007 (IEEE Press, New York, 2007), pp. 1–12
6. D.L. Rudnick, R.E. Davis, C.C. Eriksen, D.M. Fratantoni, M.J. Perry, Underwater gliders for ocean research. Mar. Tech Soc. J. **38**(1), 48–59 (2004).

7. S.A. Jenkins, D.E. Humphreys, J. Sherman, J. Osse, C. Jones, N. Leonard, J.G. Graver, R. Bachmayer, T. Clem, P. Caroll, P. Davis, J. Berry, P. Wosley, J. Wasyl, Under water glider system study. Tech. report 53, Scripps Institute of Oceanography, University of California, (San Diego, 2003).
8. R. Bachmayer, J.G. Graver, N.E. Leonard, in *Glider control: A Close Look into the Current Glider Controller Structure and Future Developments*, Proceedings of OCEANS 2003 (IEEE Press, New York, 2003), pp. 951–954
9. D. C. Seo, G. Jo, H. S. Choi, in *Pitching Control Simulations of an Underwater Glider Using CFD Analysis*. Proceedings of OCEANS 2008 (IEEE Press, New York, 2008), pp. 1–5
10. N. Mahmoudian, C. Woolsey, in *Underwater Glider Motion Control*, Proceedings of the 47th IEEE Conference on Decision and Control, pp. 552–557
11. N.E. Leonard, J.G. Graver, Model-based feedback control of autonomous underwater gliders. IEEE J. Oceanic Eng. **26**(4), 633–645 (2001)
12. K. Lei, Z. Yuwen, Y. Hui, C. Zhikun, MATLAB-based simulation of buoyancy-driven underwater glider motion. J. Ocean Univ. China **7**(1), 133–188 (2008)
13. Y. Wang, H. Zhang, S. Wang, in *Trajectory Control Strategies for the Underwater Glider*, Proceedings of International Conference on Measuring Technology and Mechatronics Automation (IEEE Press, New York, 2009), pp. 918–921
14. B.-H. Jun, J.-Y. Park, F.-Y. Lee, P.-M. Lee, C.-M. Lee, K. Kim, Y.-K. Lim, J.-H. Oh, Development of the AUV 'ISiMI' and free running test in an ocean engineering basin. J. Ocean Eng. **36**(1), 2–14 (2009)
15. H. Yang, J. Ma, in *Sliding Mode Tracking Control of an Autonomous Underwater Glider*. Proceedings of International Conference on Computer Application and System Modeling *(ICCASM 2010)* (2010), pp. 555–558
16. A. Budiyono, Advances in unmanned underwater vehicles technologies: Modeling, control and guidance perspectives. Indian J. Geo-Mar. Sci. **38**(3), 282–295 (2009)
17. A. Reza, A. A. Khayyat, K. G. Osgouie, in *Neural Networks Control of Autonomous Underwater Vehicle*, Proceedings of International Conference on Mechanical and Electronics Engineering (ICMEE 2010) (2010), vol. 2, pp. 117–121
18. K. Isa, M. R. Arshad, in *Vertical Motion Simulation and Analysis of USM Underwater Glider*, Proceedings of the 5th International Conference on Automation, Robotics and Applications, 2011, pp. 139–144
19. K. Isa, M. R. Arshad, in*Dynamic Modeling and Characteristics Estimation for USM Underwater Glider*, Proceedings of IEEE Control and System Graduate Research Colloquium (ICSGRC) Incorporating the International Conference on System Engineering and Technology (ICSET, 2011) (2011), pp. 12–17
20. W. Wei, C. M. Clark, in *Modeling and Simulation of the VideoRay Pro III Underwater Vehicle*, Proceedings of OCEANS 2006 (IEEE Press, New York, 2006), pp. 283–287
21. Y. Li, L. Jian-Cheng, S. Ming-Xue, Dynamics model of underwater robot motion control in 6 degrees of freedom. J. Harbin Inst. Technol. **12**(4), 456–459 (2005)
22. F. Song, P. Edgar An, A. Folleco, Modeling and simulation of autonomous underwater vehicles: Design and implementation. IEEE J. Oceanic Eng. **28**(2), 283–296 (2003)
23. J.G. Graver, Underwater Gliders: Dynamics, Control and Design, Dissertation, Princeton University, 2005
24. N. Mahmoudian, Efficient Motion Planning and Control for Underwater Gliders, Dissertation, Virginia Polytechnic Institute and State University, 2009
25. T.I. Fossen, *Marine Control Systems: Guidance, Navigation and Control of Ships, Rigs and Underwater Vehicles* (Marine Cybernatics, Trondheim, 2002)
26. W.J. Pepijn, T.A. Johansen, A.J. Sorensen, C. Flanagan, D. Toal, Neural networks augmented identification of underwater vehicle models. J. Control Eng. Pract. **15**, 715–725 (2007). ELSEVIER
27. W.J. Pepijn, C. Flanagan, D. Toal, Neural network control of underwater vehicles. J. Eng. Appl. Artif. Intell. **18**, 533–547 (2005). ELSEVIER

28. J.H. Li, P.M. Lee, A neural network adaptive controller design for free-pitch- angle diving behavior of an autonomous underwater vehicle. J. Robot. Auton. Syst. **52**, 132–147 (2005). ELSEVIER
29. K. Ishii, T. Ura, An adaptive neural-net controller system for an underwater vehicle. J. Control Eng. Pract. **8**, I77–I184 (2000). ELSEVIER

Autonomous Guided Car Using a Fuzzy Controller

Miguel A. Olivares-Mendez, Pascual Campoy, Ignacio Mellado-Bataller, Ivan Mondragon, Carol Martinez and Jose Luis Sanchez-Lopez

Abstract The goal of the work described in this chapter is to develop a visual line guided system for being used on-board an Autonomous Guided Vehicle (AGV) commercial car, controlling the steering and using just the visual information of a line painted below the car. In order to implement the control of the vehicle, a Fuzzy Logic controller has been implemented, that has to be robust against curvature changes and velocity changes. The only input information for the controller is the visual distance from the image center captured by a camera pointing downwards to

Based on A Visual AGV-Urban Car using Fuzzy Control, by Miguel A. Olivares-Mendez and Pascual Campoy and Ignacio Mellado and Ivan Mondragon and Carol Martinez which appeared in the Proceedings of the 5th International Conference on Automation, Robotics and Applications (ICARA 2011). 2011 IEEE.

M. A. Olivares-Mendez (✉) · P. Campoy · I. Mellado-Bataller · C. Martinez · J. L. Sanchez-Lopez
Computer Vision Group Universidad Politénica de Madrid, Jose Gutierrez Abascal, 2, Madrid, Spain
e-mail: miguelolivaresmendez@gmail.com
URL: http://www.vision4uav.eu/?q=researchline/agv1

P. Campoy
e-mail: pascual.campoy@upm.es

I. Mellado-Bataller
e-mail: ignacio.mellado@upm.es

C. Martinez
e-mail: carolviviana.martinez@upm.es

J. L. Sanchez-Lopez
e-mail: jl.sanchez@upm.es

I. Mondragon
Centro Tecnológico de Automatización Industrial Departamento de Ingeniería Industrial Facultad de Ingeniería Pontificia Universidad Javeriana, Carrera 7 No 40-69C.P. 110311 Bogotá, D.C, Colombia
e-mail: ivanmond@yahoo.es

G. Sen Gupta et al. (eds.), *Recent Advances in Robotics and Automation*, Studies in Computational Intelligence 480, DOI: 10.1007/978-3-642-37387-9_3,

the guiding line on the road, at a commercial frequency of 30 Hz. The good performance of the controller has successfully been demonstrated in a real environment at urban velocities. The presented results demonstrate the capability of the Fuzzy controller to follow a circuit in urban environments without previous information about the path or any other information from additional sensors.

1 Introduction

Autonomous mobility is a central problem in Robotics, and more precisely, the control of autonomous guided vehicles (AGV) is a very ambitious non-linear control problem. In addition, if it is focussed on car-like vehicles, the difficulty is increased by the complexity of all the dynamic components, being very difficult to obtain the actual vehicle dynamic model. Despite this complexity, some works have used linear control methods to control an AGV using a PID, like [1]. Nowadays there are more people applying non-linear control system to this kind of vehicles, being Fuzzy Logic the most used technique, as it is seen in the reported literature [2] and [3]. Furthermore, this technique has an important role in the research with car-like mobile robots, like in [4] and [5], in which a golf car is used. The use of real urban cars in robotics has a major sponsor: the US Department of Defence through its various agencies. Notable examples include the DEMO I, II and III projects [6] and the DARPA Grand Challenge [7]. Two of the most important researchers in this area are Sebastian Thrun and M. Montemerlo, who won the DARPA Grand Challenge in 2005 [8], and developed another model to win the second place at the 2008 DARPA Urban-Challenge [9]. The hard requirements of these competitions forced to researchers to use many sensors to measure the environment with high accuracy and, this is unattainable for a big part of the research community. To reduce costs, some works have been performed with just a visual sensor, for instance, surface classification like [10] or object detection like [11]. Computer vision has been also used for guidance tasks with mobile robots like [2], with a camera pointing forwards and [1] with a camera pointing downwards and a field of view of 1 m high.

In this chapter is presented an AGV -urban car (Citron C3)that is guided by a painted line on the ground, with one camera capturing the forthcoming 30 cm in front of the vehicle. To control the steering of the car, a Fuzzy controller has been developed. Excellent results were obtained by testing the system in a real circuit for more than 3 km. This chapter is organized as follows. Section 2 explains the system of the urban vehicle. Section 3 shows the visual hardware and the visual algorithm. The Fuzzy controller is described in detail in Sect. 4. Some of the successful experiments are presented in Sect. 5 with detailed plots. To finish with the conclusions in Sect. 6.

Fig. 1 Automated Citron C3 pluriel

2 Car System

The car that was used for this research is a commercial vehicle "Citron C3 Pluriel" (Fig. 1). To move the steering wheel a manipulation of the power-assisted steering motor was used. This assistance system consists of an electric DC motor attached to the steering rack trough a gear. This motor drives the steering to the action of the driver on the steering wheel. This action is measured through a torque sensor located in the steering bar. The signal from the sensor is received by a control/power until that sends a PWM signal to the motor, to assist the steering movement. This device allowed a fast automation since the mechanical and electrical elements were already installed in the car. For our purpose, the connections of the motor were cut, and it was attached to a MAXON ADS 50/10 servo amplifier, with 240 W of peak power at 12 V. This card is able to generate a PWM signal whose duty cycle is proportional to an analog ± 10 V input signal. This input signal is generated by an Advantech USB-4711A acquisition card that is connected to an onboard computer. The necesary information to feedback the control system is provided by an absolute. The encoder gives the angular positions at a rate of 100 Hz.

During the tests, a human driver controlled the speed of the vehicle manually. In order to measure the vehicle speed, a non-contact speed sensor L-CE Correvit

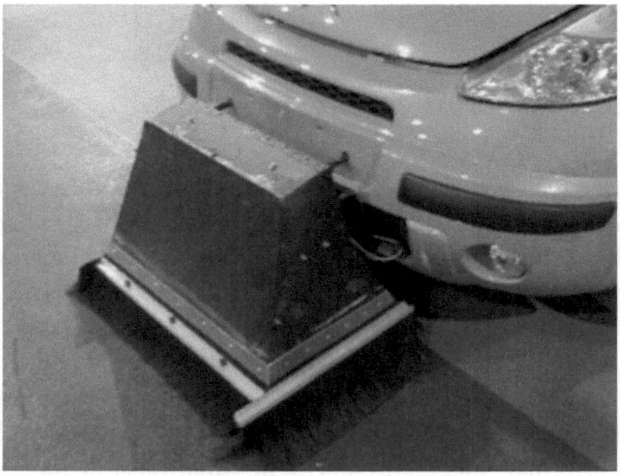

Fig. 2 Black metal box of the visual system

was installed. It sends the measured speed at a 100 Hz rate to the onboard computer.

The guiding line paint was produces with special pigments that appear blue-coloured when they are lit whit a bright ultraviolet light, while staying uncoloured under the normal light. In order to keep the illumination under control a special structure was designed and installed in front of the car. This structure is made up of a black metal box (Fig. 2), and it contains the camera the camera and the ultraviolet lamp. The restricted height (47 cm) of the box forced us to use a wide angle low distortion lens for the camera, in order to capture the whole scene at the bottom of the box, which is 60 × 40 cm. The box is isolated at its base from the daylight by rubber tabs and brushes. Despite this benefit, this isolation reduces the visual field to an area of 50 × 30 cm.

3 Visual System

The visual system is composed of all the hardware and software to extract the relative position between the car and the line.

3.1 Visual Hardware

For these tests, we use a laptop with a Core-Duo Centrino processor, running at 2.6 GHz, and 4 Gbytes of RAM. The operating system is Ubuntu 10.4. The camera is a Firewire Unibrain 520C, with a resolution of 320 × 240 pixels at 30 fps and a wide angle, low distortion lens attached. The whole set gives a field of view of 125° and a working distance of 10 mm. For the lighting, a special UV lamp with a wavelength of 365 nm (black light) is needed to excite the pigments of the line. To avoid black frames because of flickering, the operation frequency of the lamp is 25 kHz.

3.2 Visual Algorithm

For the detection of the line, a custom real-time computer vision algorithm was designed. The algorithm is able to detect the lines centroid and orientation under harsh conditions, such like a partially occluded and poorly painted line on a rough terrain, coping with non-compact line shapes. The line detection has been successfully tested at up to 30 kph.

On the front-end of the visual system, the camera captures the scene which is lit with UV light at 30 fps. First, a colour-based segmentation is performed on YUV space. Despite some other colour spaces were tested, YUV was found to be the best performer under different light conditions. A rectangular prism inside the YUV colour space is defined, so that only the pixel values inside this volume are considered to be part of the line. The result is a binary image where only the line pixels are set. This method proved to be robust detecting lines of different blue tones and brightness.

In the binary image, every 8-connected pixel group is marked as a blob. At the first step, to reduce the noise, blobs having an area outside a defined range are discarded. Then, for every survivor, centroid, dominant direction and maximal length are calculated, and those being too short are ignored. The remaining blobs are clustered according to proximity and parallelism, so each cluster becomes a candidate line. The centroid and dominant direction of each candidate line are calculated from the weighted sum of the features of its component blobs, being the weight of each blob proportional to its relative area. In this way, the algorithm is able to accurately detect lines that are fragmented because of ageing.

Finally, from the whole set of candidate lines, a detected line must be selected for the current frame. In order to do that, the distance between the centroids of every candidate line in the current frame and the detected line in the previous frame is measured. If the smallest distance is higher than a certain threshold, the detected line will be the leftmost or rightmost candidate line, depending on the user-defined criterion. Otherwise, the closest candidate line is taken as detected line. This mechanism avoids switching to fake lines when there are traces of old paintings along the circuit, even when it is deteriorated.

The algorithm outputs whether the line is detected or not and, if it is, it also outputs the error of the line in the x-axis from to the centre of the image and the direction of the line, expressed as an angle.

4 Fuzzy Controller

The steering control of the car includes two components. The first one is the Fuzzy controller and the other one is the integral of the error. The latter is added at the end of the control loop to the output of the controller, making a structure of $Fuzzy + I$, as it is shown in Fig. 3.

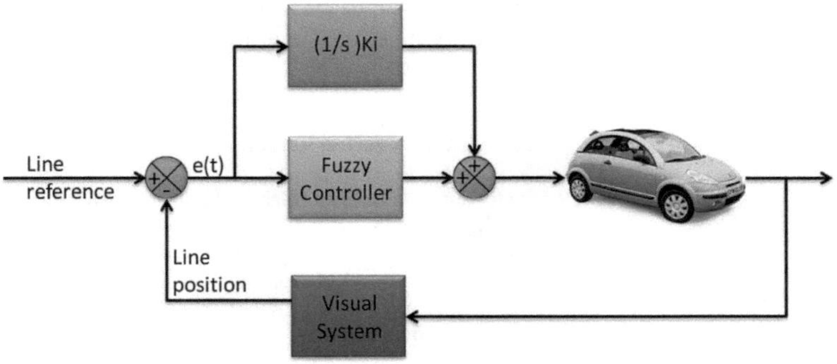

Fig. 3 Control loop of the visual servoing system.

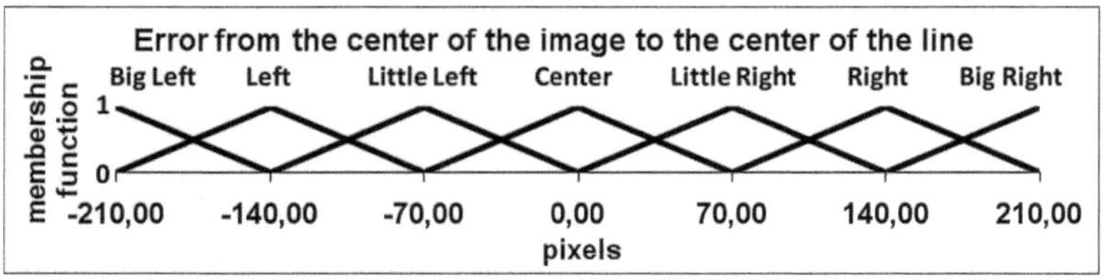

Fig. 4 First input variable of the Fuzzy controller: the error between the centre of the line and the centre of the image, in pixels.

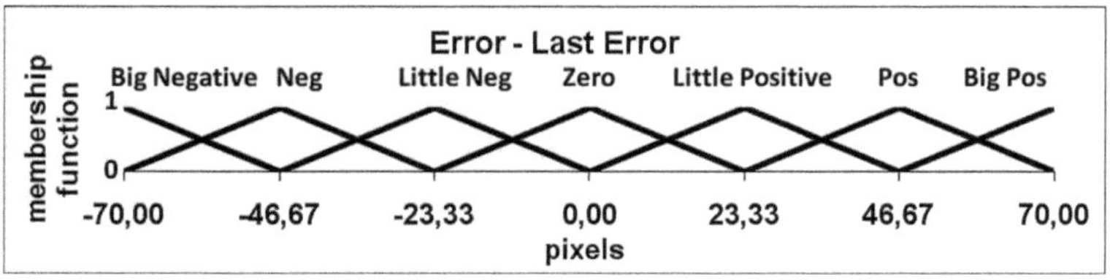

Fig. 5 Second input variable of the Fuzzy controller: the difference between the last error and the actual, in pixels.

The Fuzzy controller was implemented using the MOFS (*Miguel Olivares' Fuzzy Software*). This software was used previously to implement Fuzzy Controllers in other different platforms like a wheelchair [12] or in an unmanned helicopter, where it was applied to control a pan and tilt visual platform onboard the UAV [13] and for the autonomous landing of the aircraft [14]. With this software, it is possible to easily define a fuzzy controller with the required number of inputs and to select the type of membership functions, the defuzzification model and the inference operator. A more detailed explanation of this software can be found in [15].

The controller has two inputs and one output. All are fuzzyfied using triangular membership functions. The first input is defined as the error between the centre of the image and the centre of the line to follow (Fig. 4). The second input is the difference between current and previous error (Fig. 5). The output of the controller is the absolute turn of the steering wheel to correct this error, in degrees (Fig. 6). To obtain this output, 49 if-then rules were defined. The developed fuzzy system is a Mamdani type that use a height weight defuzzification model with the product inference model in Eq. 1.

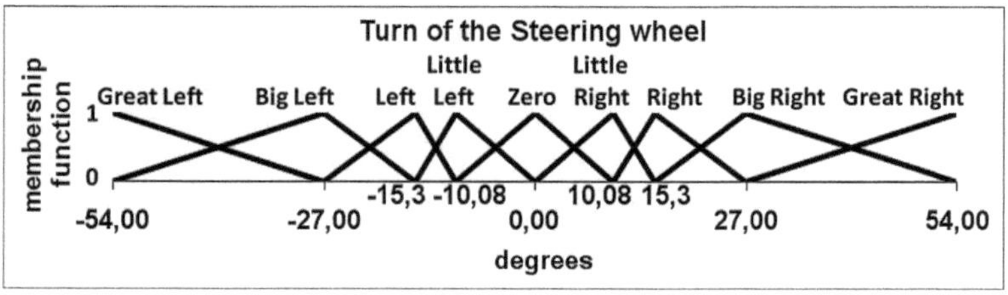

Fig. 6 Output variable of the Fuzzy controller: the steering wheel angle, in degrees.

$$y = \frac{\sum_{l=1}^{M} \bar{y}^{l} \prod_{i=1}^{N} \left(\mu_{x_i^l}(x_i)\right)}{\sum_{l=1}^{M} \prod_{i=1}^{N} \left(\mu_{x_i^l}(x_i)\right)} \quad (1)$$

where N and M represent the number of inputs variables and total number of rules respectively. $\mu_{x_i^l}$ denote the membership function of the lth rule for the ith input variable. $\bar{y}^l$ represent the output of the lth rule.

The calculation of the integrator value is shown in Eq. 2.

$$I_t = I_{t-1} + e \times \frac{1}{t} \times Ki \quad (2)$$

where e is the current error between the centre of the line and the centre of the image, t is the framerate, and Ki is a constant that appropriately weights the effect of the integrator, and for this case is equal to 0.6.

The initial idea of this work was to develop a controller for a circuit with short radius curves. In such conditions, the speed of the car can not be very high. Thus the actual velocity of the car is was not included in the Fuzzy controller, but it is taken into account multiplying the fuzzy output by $\frac{10}{v}$, being v the current velocity of the vehicle. The definition of the numerator value of this factor is based on the velocity, in kph, during a skilled human driving session, in which data was acquired to tune the rule base of the fuzzy controller. It is practically impossible for a human to drive faster than 10 kph while keeping the line in following error low enough to meet the requirements of the application. This is because the driver only sees 30 cm forward, and, at that speed, the contents of this area change completely every 0.108 s.

The driving session performed by the human at 10 kph output the necessary training data to modify the initial base of rules of the controller and the size of the fuzzy sets of its variables. For the definition of the fuzzy sets, a heuristic method was used based on the extraction of statistical measures from the training data. For the initial base of rules, we used a supervised learning algorithm, implemented in MOFS. This algorithm evaluates the situation (value of input variables) and looks for the rules that are involved in it (active rules). Then, according to the steering command given by the human driver, the weights of these rules are changed. Each time that the output of an active rule coincides with the human command, its weight will be increased. Otherwise, when the output differs from the human command, its weight will be decreased by a constant. Anytime the weight of a rule becomes negative the system sets the output of the rule to the one given by the human driver. Further details of the software are given at [15].

Table 1 shows the base of rules which enclose the knowledge of the controller.

Table 1 Base of rules of the Fuzzy controller

Error / Dot error	Big Left	Left	Little Left	Centre	Little Right	Right	Big Right
Big Negative	Great Left	Great Left	Big Left	Big Left	Left	Little Left	Zero
Negative	Great Left	Big Left	Big Left	Left	Little Left	Zero	Little Right
Little Negative	Big Left	Big Left	Left	Little Left	Zero	Little Right	Right
Zero	Big Left	Left	Little Left	Zero	Little Right	Right	Big Right
Little Positive	Left	Little Left	Zero	Little Right	Right	Big Right	Big Right
Positive	Little Left	Zero	Little Right	Right	Big Right	Big Right	Great Right
Big Positive	Zero	Little Right	Right	Big Right	Big Right	Great Right	Great Right

5 Experiments

To test the fuzzy controller, a closed loop line was painted with an oval shape, as shown in Fig. 7. The two curves are 20 and 11 m of radius and 72 and 34 m long, respectively. The stretches are 40 and 44 m long. The total length of the circuit is 190 m. First, we present system behaviour results after two different step perturbations were applied at different velocities and circuit curvatures. Subsequently, results for a continuous 18 laps test are presented. The total distance driven during the second test is 3.5 km.

5.1 Step Perturbation Test Series

In order to measure how good the fuzzy controller is, a set of step tests was made. The step value is 50 pixels, equivalent to more than 6 cm. This step was applied to the car at different velocities in straight lines and curves. Some of the results of these tests are shown after these lines.

Figure 8 shows the error measured when a +50 and −50 pixels step perturbation is applied to the system at 10 kph with a resulting RSME value of 7.166 cm. At it is shown, the system corrects the error in just 27 frames, which is about 1 s for an average rate of 28 frames per second during the test. The angle of the steering wheel versus the controller commands is shown in Fig. 9, in which a delay of 7–8 frames in the steering wheel action may be noticed. Ignoring this delay, the effective settling time would stay around 20 frames or 0.7 s.

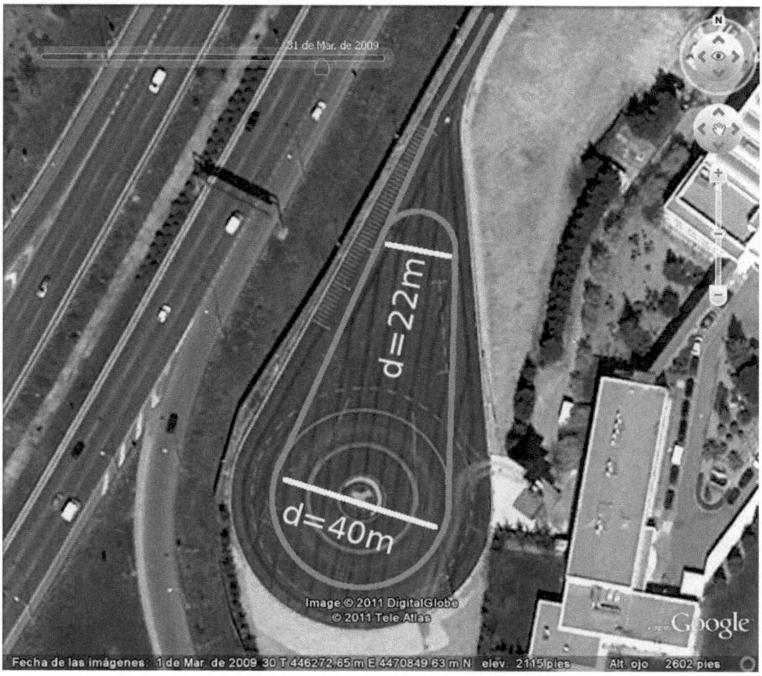

Fig. 7 Representation of the circuit on a Google Earth image.

Fig. 8 Representation of the error, in pixels, during the 50 pixels step test at 10 kph in a straight line. The measured RMSE is also shown at the *top*.

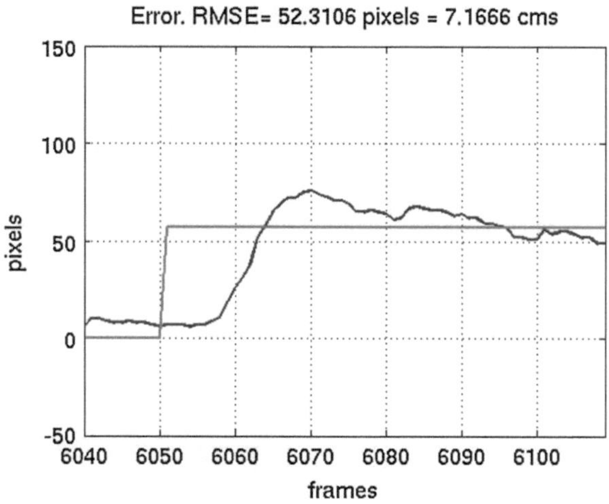

Figures 10 and 11 represent the results for a step perturbation test at 15 kph in a straight way. For this test the value of the RMSE is 7.0592 and the settling time is less than a second (25 frames).

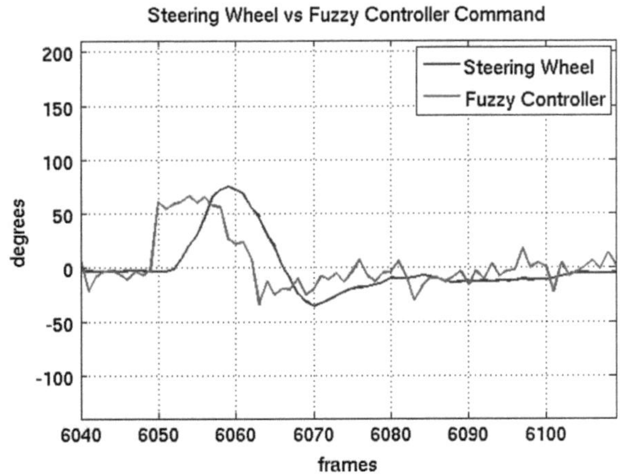

Fig. 9 Evolution of the steering wheel angle versus the controller commands during the 50 pixels step test at 10 kph in a *straight line.*

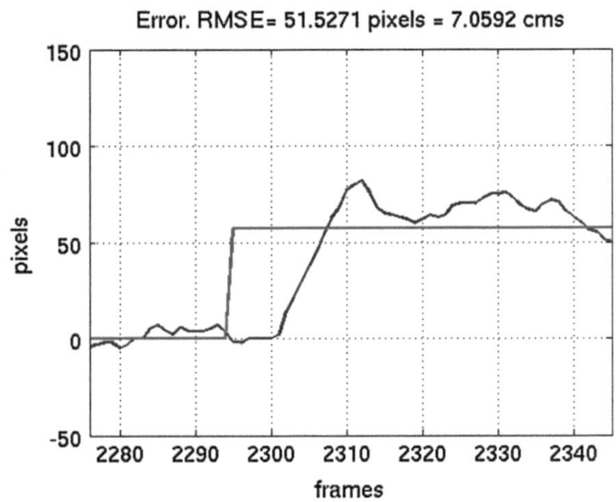

Fig. 10 Representation of the error in pixels during the 50 pixels step test in straight at 15 kph. The value of the RMSE of the test in this part of the circuit is 7.0592 cm

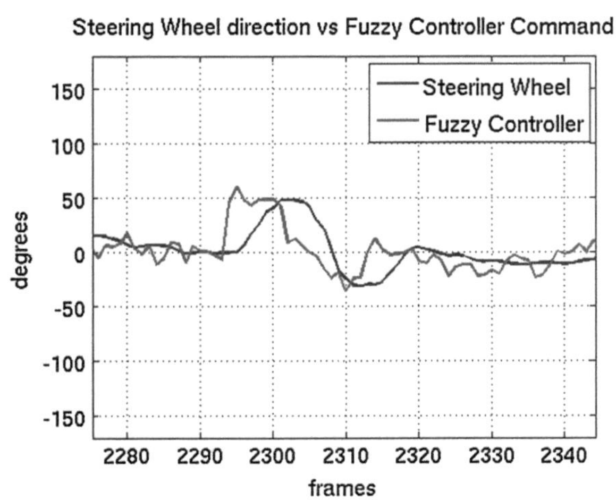

Fig. 11 Reperesentation of the movements of the steering wheel versus the value of the commands sent by the controller during the 50 pixels step test in straight at 15 kph

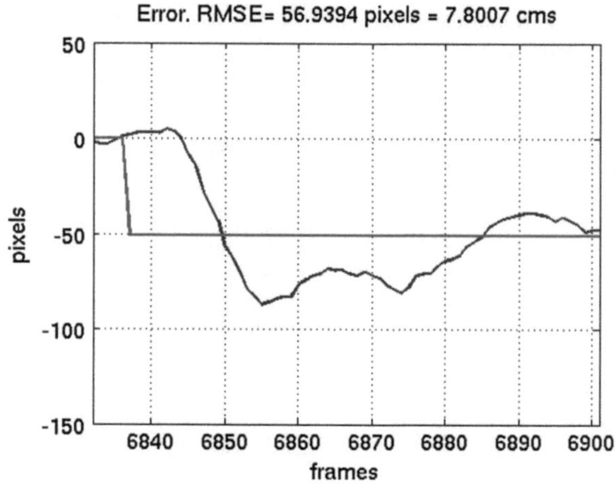

Fig. 12 Representation of the error in pixels during the 50 pixels step test in straight at 15 kph. The value of the RMSE of the test in this part of the circuit is 7.8007 cm

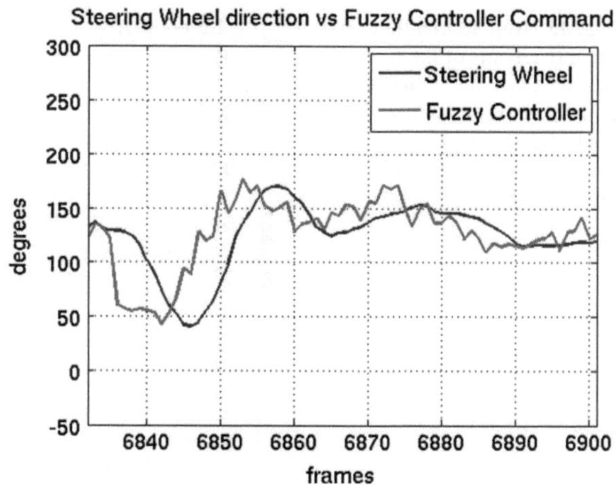

Fig. 13 Reperesentation of the movements of the steering wheel versus the value of the commands sent by the controller during the 50 pixels step test inside a curve at 10 kph

To test the robustness of the controller against step perturbations similar tests have been done when the vehicle was inside a curve. Figure 12 shows the step command and the evolution of the error at 10 kph. In this case the curve to the left and the step was done to the internal part of the curve (to the left).

The action of the controller and the response of the steering are shown in Fig. 13.

The test at 15 kph inside the curve has been done applying a perturbation in the direction against the curve, trying to move the car out of the curve. Figure 14 shows the evolution of this test comparing the step command and the error at each frame. As well as previous tests the Fig. 15 shows a comparison between the commands sent by the Fuzzy controller and the steering wheel position frame by frame.

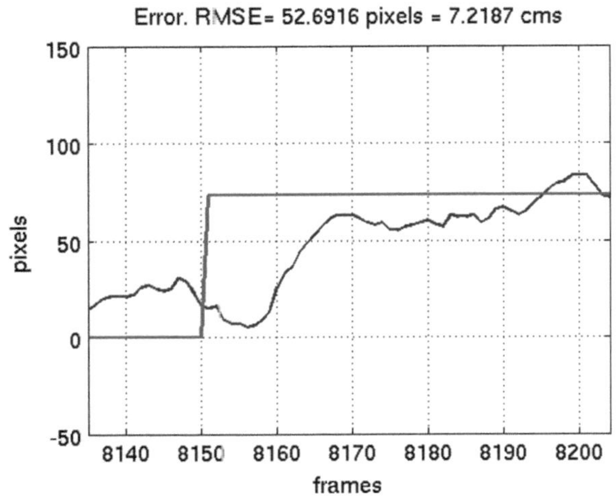

Fig. 14 Representation of the error in pixels during the 50 pixels step test inside a curve at 15 kph. The value of the RMSE of the test in this part of the circuit is 7.2187 cm

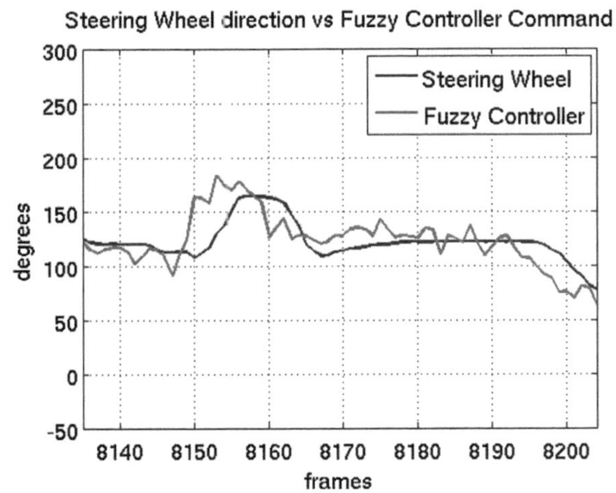

Fig. 15 Reperesentation of the movements of the steering wheel versus the value of the commands sent by the controller during the 50 pixels step test inside a curve at 15 kph

Also in this case the controller has a quick response which keep the vehicle over the line to follow. It must be taking into account that the lateral visual space of the system is just 25 cm per each side of the centre where is located the camera.

Finally the robustness of the controller has been tested by the application of a speed step command, by an emergency stop pressing the brake of the vehicle. Figure 16 shows the speed of the vehicle during this test. At the end of graph is shown the radical step from 10 to 0 kph. Figure 17 shows the evolution of the error during this test. To increase the complexity of the test the speed reduction has been done when the vehicle was inside a curve. At the beginning of the test the vehicle was located in a straight, gong inside a curve at the end of test. The transition from straight to curve is appreciable at the interval between the frames 280–420.

The response of the controller and the evolution of the steering wheel have been shown in Fig. 18.

Fig. 16 Representation of the speed of the vehicle during the test of an emergency stop

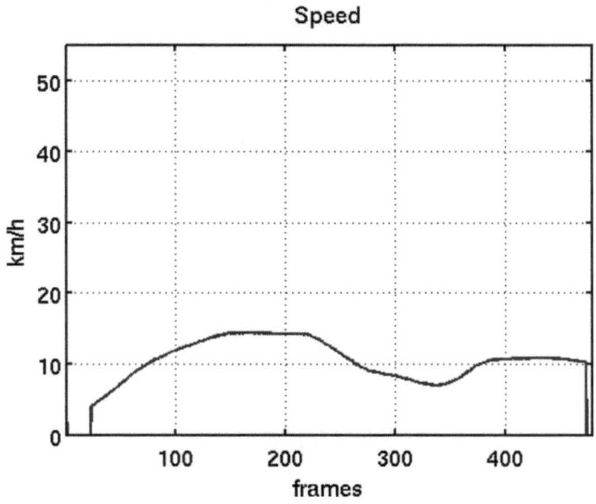

Fig. 17 Representation of the error in pixels during the test of an emergency stop

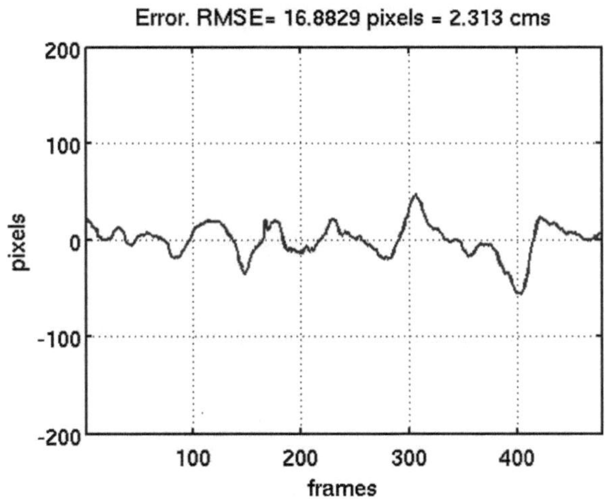

5.2 Continuous Driving Tests

In this tests, the car covered 18 laps of the circuit. In Fig. 19 the measured error during the whole test is shown. In this case, the RMSE was 5.0068 cm.

Figure 20 shows the comparison between the controller commands and the measured angle of the steering wheel. In the Figure, the changes between straight lines and curves may be noticed. In the straight lines, the steering wheel stays around zero degrees, while it turns between $-100°$ and $-150°$ in the first curve, and between -150 and -300 in the second one. It is more easily see in Fig. 21, in which the plot is scaled to show only one lap.

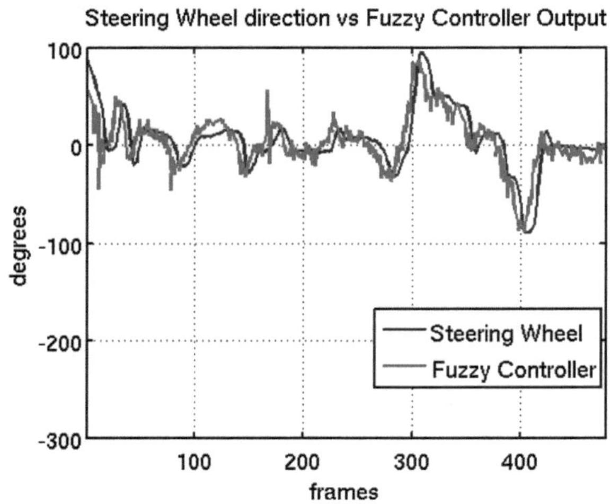

Fig. 18 Comparison between the Steering Wheel and the commands sent by the Fuzzy controller during the test of an emergency stop

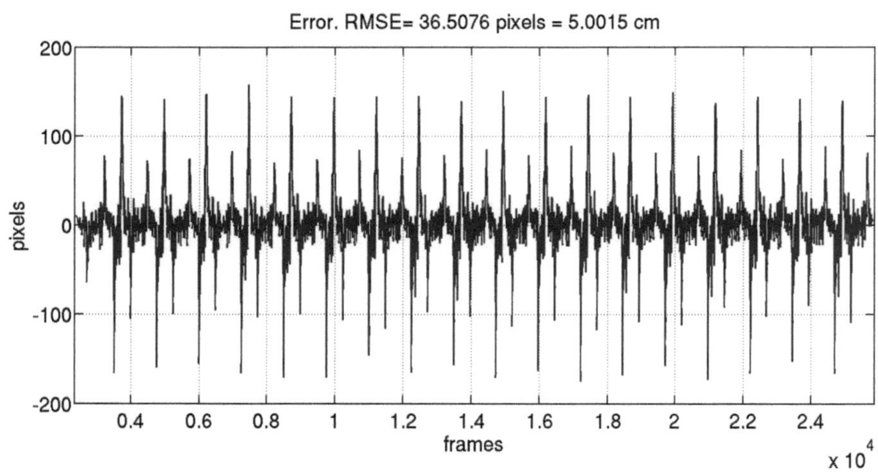

Fig. 19 Representation of the error in pixels during the 18 laps to the circuit. The value of RMSE for this test is 5.0015 cm

In Fig. 19 large error peak of even 170 pixels appear at every curvature change. However, they are decreased in a few frames by the controller. This errors appear because the circuit was not designed with clothoids. Therefore, curvature discontinuities happen when changing from straight line to curve and vice-versa. Figure shows a zoom of one of this instants in which a peak of -171 pixels occurs. The evolution of the error is plotted in Fig. 22a, while the output of the controller and the steering wheel angle are in Fig.

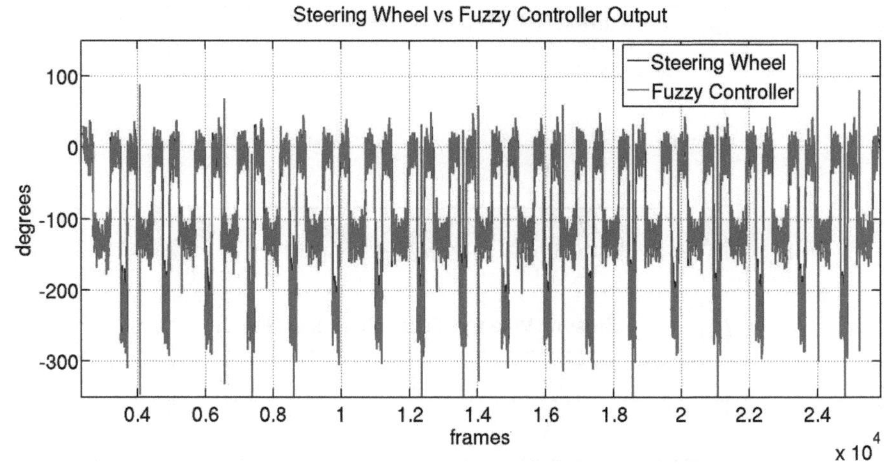

Fig. 20 Reperesentation of the movements of the steering wheel versus the value of the commands sent by the controller during the test of 18 laps to the circuit

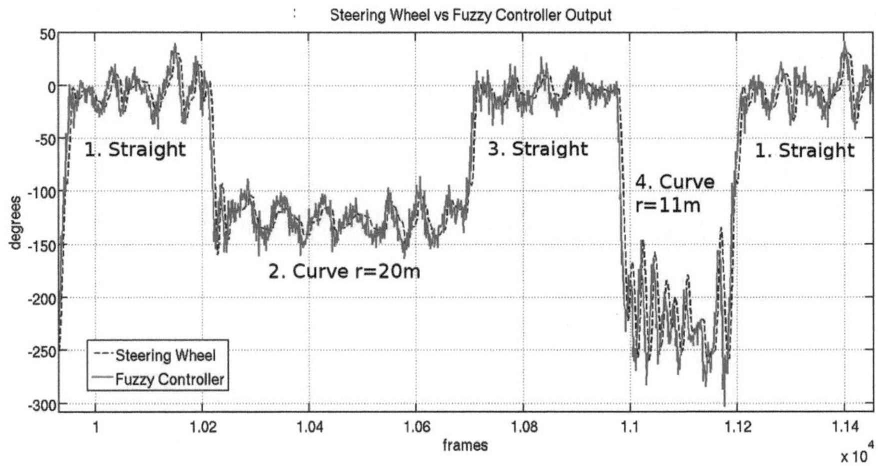

Fig. 21 Zoom to one lap of the circuit

The evolution of the vehicle speed is depicted in Fig. 23, which covers speeds between 12 and 13 kph.

Another test has been done to check the correct behavior of the controller when the speed is not so stable like in this test. In this case the speed was controller by a human changing its value during the test. Peak of 25 kph has been reach in some parts of the test as is shown in Fig. 24. The evolution of the error is shown in Fig. 25 with a RMSE of 5.8328. In this Figure also is appreciated the changes between straight and curve and vice-verse when the error peaks appears. These

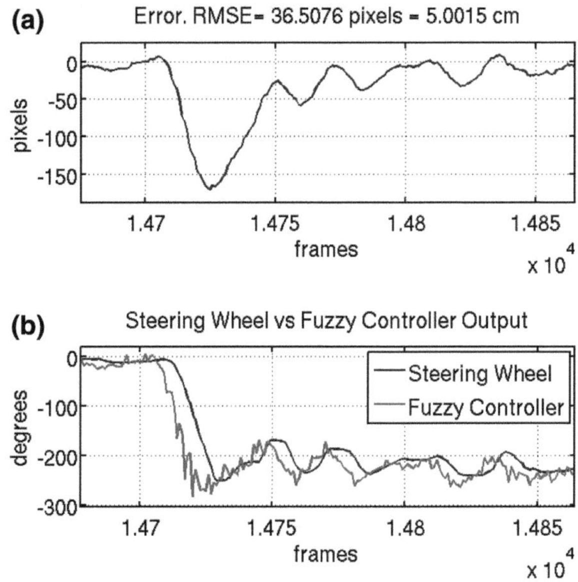

Fig. 22 Zoom of 170 pixels step at the beginning of the second curve

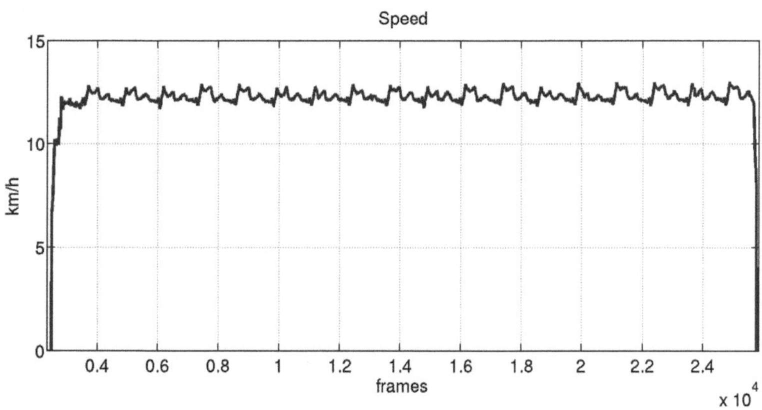

Fig. 23 Measure of the vehicle speed during the 18 laps test

transitions are appreciated too, in the variations of the steering wheel value in Fig. 26 as has been explained in the previous Fig. 21.

Finally Table 2 shows the results of these two test during 1 and 3.5 km.

In [16] is possible to see a video of some of these tests.

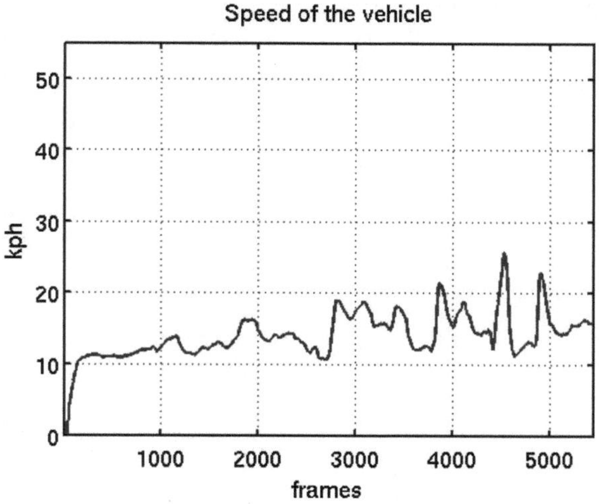

Fig. 24 Measure of the vehicle speed during the 4 laps test with speed variations

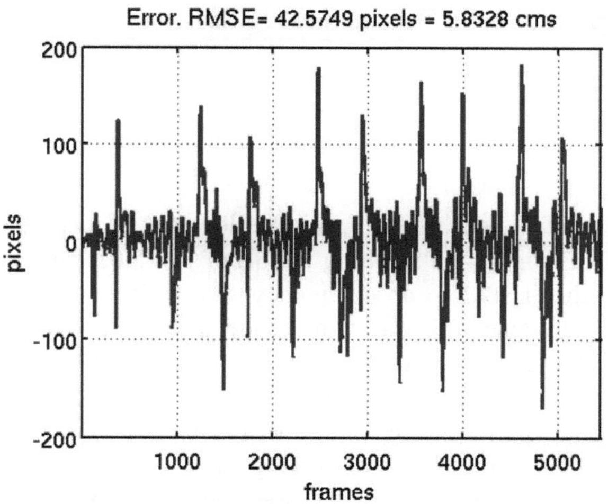

Fig. 25 Evolution of the error during the 4 laps test with speed variations. The RMSE during this test was 5.8328

6 Conclusion

This work presents a low-cost visual line-guided system for an urban-car controlled by a Fuzzy Logic controller. Strong results on real-world tests are presented in order to check the behavior of the controller. The quick response of the vehicle with step command tests and the excellent line-following behavior during long distance tests support the quality of the development control system. The controller reacts in about one second to a step perturbation of more than 6 cm in the visual system of the car. These tests have been done during straight and curve parts of the circuit. It must be taken into account that the steering wheel of the car has a

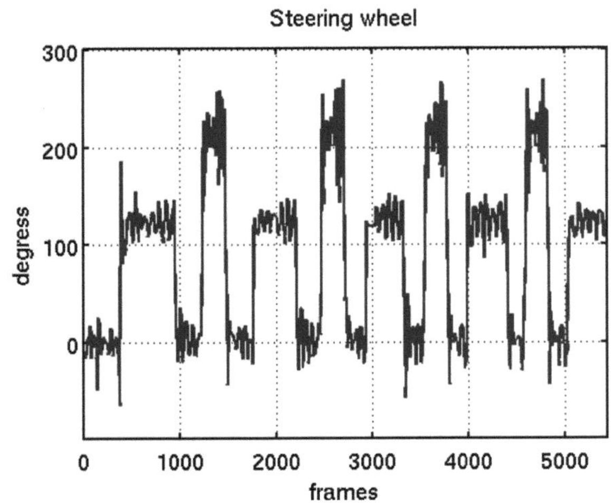

Fig. 26 Measure of the steering wheel of the vehicle during the 4 laps test with speed variations

Table 2 Results obtained with long duration tests

Number of laps	Kms	Speed (kph)	RMSE
18	3.5	13	5.0068
4	1	variable	5.8328

reaction delay of 0.7 s. Another long test of more than 3 km (18 laps of the circuit) was run with a resulting RMSE of just 5.01 cm. In order to check the robustness other tests like an emergency stop inside a curve and a long test of 1 km has been done with variations of the speed of the vehicle, reaching peaks of 25 kph. The successful results show the excellent behavior of the controller, despite the reduced perception span given by the camera, which puts some limitations on the AGV performance, being the maximum attainable velocity the most important.

Currently, we are working on a custom visual signaling system to acquire information about the position of the car and the forthcoming features of the path. This update will give some anticipation capabilities that will enable higher velocities and a better controller behavior. On the other hand, we intend to join all the four inputs inside the fuzzy controller, including the error, its derivative, its integral and the car velocity.

Acknowledgments The work reported in this chapter is the product of several research stages at the Computer Vision Group Universidad Politécnica de Madrid. The authors would like to thank the company SIEMENS España S.A. that has made possible the research described in this chapter through several contracts, and the INSIA-UPM Institute and the people at LABIE for their support and the provided facilities.

References

1. J.W. Lee, S.U. Choi, C.H. Lee, Y.J. Lee, K.S. Lee, in *A Study for Agv Steering Control and Identification Using Vision System*. Proceedings of IEEE international symposium on industrial electronics, 2001 (ISIE 2001), vol 3, pp. 1575–1578 (2001)
2. W.S. Wijesoma, K.R.S. Kodagoda, in *Design of Stable Fuzzy Controllers for an Agv*. Intelligent robots and systems (IROS 2000). Proceedings 2000 IEEE/RSJ international conference, vol 2, pp. 1044–1049 (2000)
3. Y. Dianyong, X. Hui, in *Application of Fuzzy Control Method to Agv*. Proceedings of robotics, intelligent systems and signal processings 2003 IEEE international conference, vol 2, pp. 768–772 (2003)
4. Q. Yuan, Y. Zhang, H. Wu, H. Wang, in *Fuzzy Control Research in the Courses of Smart Car*. International conference on machine vision and human-machine interface (MVHI), pp. 764–767 (2010)
5. K.R.S. Kodagoda, W.S. Wijesoma, E.K. Teoh, Fuzzy speed and steering control of an agv. IEEE Trans. Control Sys. Technol. **10**(1), 112–120 (2002)
6. C.M. Shoemaker, J.A. Bornstein, in *The Demo iii ugv Program: A Testbed for Autonomous Navigation Research*. Intelligent control (ISIC). Proceedings of held jointly with IEEE international symposium on computational intelligence in robotics and automation (CIRA), Intelligent systems and semiotics (ISAS), pp. 644–651 (1998)
7. DARPA, Darpa Gran Challenge (2011), http://www.darpagrandchallenge.com
8. S. Thrun, M. Montemerlo, H. Dahlkamp, D. Stavens, A. Aron, J. Diebel, P. Fong, J. Gale, M. Halpenny, G. Hoffmann, K. Lau, C. Oakley, M. Palatucci, V. Pratt, P. Stang, S. Strohband, C. Dupont, L.-E. Jendrossek, C. Koelen, C. Markey, C. Rummel, J. van Niekerk, E. Jensen, P. Alessandrini, G. Bradski, B. Davies, S. Ettinger, A. Kaehler, A. Nefian, P. Mahoney, Winning the darpa grand challenge. J. Field Robot. (2006) (accepted for publication)
9. M. Montemerlo, J. Becker, S. Bhat, H. Dahlkamp, D. Dolgov, S. Ettinger, D. Haehnel, T. Hilden, G. Hoffmann, B. Huhnke, D. Johnston, S. Klumpp, D. Langer, A. Levandowski, J. Levinson, J. Marcil, D. Orenstein, J. Paefgen, I. Penny, A. Petrovskaya, M. Pflueger, G. Stanek, D. Stavens, A. Vogt, S. Thrun, Junior: The stanford entry in the urban challenge. J. Field Robot. (2008)
10. L. Lu, C. Ordonez, E.G. Collins, E.M. DuPont, in *Terrain Surface Classification for Autonomous Ground Vehicles Using a 2d Laser Stripe-Based Structured Light Sensor*. IEEE/RSJ international conference on intelligent robots and systems 2009 (IROS 2009), pp. 2174–2181 (2009)
11. R. Manduchi, A. Castano, A. Talukder, L. Matthies, Obstacle detection and terrain classification for autonomous off-road navigation. Auton. Robots **18**, 81–102 (2004)
12. M.A.O. Mendez, J.A.F. Madrigal, in *Fuzzy Logic User Adaptive Navigation Control System for Mobile Robots in Unknown Environments*. IEEE international symposium on Intelligent signal processing 2007 (WISP 2007), pp. 1–6 (2007)
13. M. Olivares, P. Campoy, C. Martinez, I. Mondragon, A pan-tilt camera fuzzy vision controller on an unmanned aerial vehicle. Intell. Robot Sys. (2009)
14. M.A. Olivares-Mendez, I.F. Mondragon, P. Campoy, C. Martinez. in *Fuzzy Controller for uav-Landing Task Using 3d-Position Visual Estimation*. IEEE international conference on fuzzy systems (FUZZ), pp. 1–8 (2010)
15. I. Mondragn, M. Olivares-Mendez, P. Campoy, C. Martnez, L. Mejias, Unmanned aerial vehicles uavs attitude, height, motion estimation and control using visual systems. Auton. Robots **29**, 17–34 (2010). doi:10.1007/s10514-010-9183-2
16. Computer Vision Group, Polytechnic University of Madrid. Agv test video http://www.vision4uav.eu/?q=researchline/agv1 2010

On-Line Adaptive Fuzzy Modeling and Control for Autonomous Underwater Vehicle

O. Hassanein, Sreenatha G. Anavatti and Tapabrata Ray

Abstract Autonomous Underwater Vehicles (AUVs) have gained importance over the years as specialized tools for performing various underwater missions in military and civilian operations. The autonomous control of AUV poses serious challenges due to the AUVs' dynamics. Its dynamics are highly nonlinear and time varying and the hydrodynamic coefficients of vehicles are difficult to estimate accurately because of the variations of these coefficients with different navigation conditions and external disturbances. This work is going to deal with the system identification of AUV dynamics to obtain the coupled nonlinear dynamic model of AUV as a black box. This black box has an input-output relationship based upon on-line adaptive fuzzy technique to overcome the uncertain external disturbance and the difficulties of modeling the hydrodynamic forces of the AUVs instead of using the mathematical model with hydrodynamic parameters estimation. The fuzzy model's parameters are adapted using the back propagation algorithm. Fuzzy control system is applied to guide and control the AUV with mathematical model, fuzzy model and adaptive fuzzy model. The simulation results show that the performance of the AUV with the fuzzy control with adaptive fuzzy model is having better dynamic performance as compared to others model even in the presence of noise and parameter variations.

Based on "Fuzzy Modeling and Control for Autonomous Underwater Vehicle", by "O. Hassanein, Sreenatha G. Anavatti and Tapabrata Ray" which appeared in the Proceedings of the 5th International Conference on Automation, Robotics and Applications (ICARA 2011).

O. Hassanein (✉) · S. G. Anavatti · T. Ray
School of Engineering and Information Technology, UNSW@ADFA, Canberra, Australia
e-mail: o.hassan@adfa.edu.au

S. G. Anavatti
e-mail: s.anavatti@adfa.edu.au

T. Ray
e-mail: t.ray@adfa.edu.au

G. Sen Gupta et al. (eds.), *Recent Advances in Robotics and Automation*,
Studies in Computational Intelligence 480, DOI: 10.1007/978-3-642-37387-9_4,

1 Introduction

The autonomous underwater vehicle (AUV) is an underwater system that contains its own power and is controlled by an onboard computer. Many names are given to these vehicles according to their configuration such as Remotely Operated Vehicles (ROVs), Unmanned Underwater Vehicles (UUVs), and submersible devices. The fundamental task for these devices is fairly well defined as the vehicle is able to follow a predefined trajectory to reach the final destination. The main advantage of the AUV is that it does not need a human operator. Therefore, is capable of doing operations that are too dangerous for humans.

Applications of AUVs can be divided into three groups [1]. These three groups are maritime (marine) security, oceanography, and submerged structure inspection and maintenance. According to [2] the two most significant technological challenges in AUV's design are power and autonomy. Power sources limit the mission time of the vehicle and autonomy limits the degree to which an AUV can be left unattended by human operators.

As presented in [3], the fuzzy system is used as a system identifier for nonlinear dynamic system. They show that the fuzzy system can be viewed as a three-layer feed forward network. The neurofuzzy modeling techniques are presented in [4]. The system identification is a powerful approach for that technique and was demonstrated by its application to the identification of an Ocean Voyager AUV. Reference [5] proposed a fuzzy logic control that uses Genetic Algorithm for membership function optimization.

Reference [6] is concerned with practical System Identification (SI) method in order to obtain a model of AUV using input-output data obtained from test trails. The autopilot deployed was an LQG controller for testing the modeling in simulation environment and tested on real system used ROV.

Reference [7] presented a simple model identification method for UUV and applied this method to the underwater robot GARBI. The system identification was aimed at decoupling the different degrees of freedom in low speed vehicles. Least Square techniques were used to estimate the UUV dynamics.

Australian Defence Force Academy (ADFA) AUV has been developed and built in UNSW@ADFA. Figure 1 shows ADFA AUV model that has one electrical thruster power for forward motion. Two sets of electrical pumps used for maneuvering in horizontal plane and vertical plane. The middle box is used for carrying the sensors, battery and the electronic accessories. The hydrodynamic coefficients of this AUV are calculated under certain conditions using CFD method to derive the exact mathematical model.

Fig. 1 Australian defence force academy AUV

Fig. 2 Six degrees of freedom of an AUV

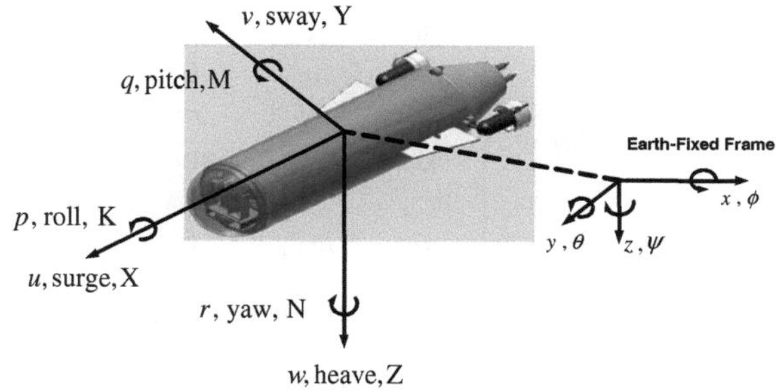

The general dynamic equation describing the mathematical model of the ADFA AUV is provided in Sect. 2. The design details of the system identification using fuzzy system are discussed in Sect. 3. The design details of adaptive fuzzy model are presented in Sect. 4. Moreover, fuzzy control system and numerical simulation results are presented in Sects. 5 and 6 respectively. Finally, the chapter is concluded in Sect. 7.

2 AUV's Mathematical Model

The hydrodynamic forces per unit mass acting on each axis will be denoted by the uppercase letters X, Y and Z. u, v and w represent the forward, lateral and vertical speeds along x, y and z axes respectively. Similarly, the hydrodynamic moments on AUV will be denoted by L, M and N acting around x, y and z axis respectively. The angular rates will be denoted by p, q and r components.

Dynamics of AUVs, including hydrodynamic parameters uncertainties, are highly nonlinear, coupled and time varying. According to [8], the six DOF nonlinear equations of motion of the vehicle are defined with respect to two coordinate systems as shown in Fig. 2. The equations of motion for underwater vehicle can be written as follows [9]:

$$M\ddot{q} + C(\dot{q})\dot{q} + D(\dot{q})\dot{q} + G(q) = \tau \tag{1}$$

where M is a 6×6 inertia matrix as a sum of the rigid body inertia matrix, M_R and the hydrodynamic virtual inertia (added mass) M_A. $C(\dot{q})$ is a 6×6 Coriolis and centripetal matrix including rigid body terms $C_{RB}(\dot{q})$ and terms $C_A(\dot{q})$ due to added mass $D(\dot{q})$ is a 6×6 damping matrix including terms due to drag forces. $G(q)$ is a 6×1 vector containing the restoring terms formed by the vehicle's buoyancy and gravitational terms. τ is a 6×1 vector including the control forces and moments.

The rigid body inertia matrix M_{RB}, can be represented as;

$$M_{RB} = \begin{bmatrix} m & 0 & 0 & 0 & mz_g & -my_g \\ 0 & m & 0 & -mz_g & 0 & mx_g \\ 0 & 0 & m & my_g & -mx_g & 0 \\ 0 & -mz_g & my_g & I_x & -I_{xy} & -I_{xz} \\ mz_g & 0 & -mx_g & -I_{yx} & I_y & -I_{yz} \\ -my_g & mx_g & 0 & -I_{zx} & -I_{zy} & I_z \end{bmatrix} \tag{2}$$

The added mass, M_A, can be represented as;

$$M_A = \begin{bmatrix} X_{\dot{u}} & X_{\dot{v}} & X_{\dot{w}} & X_{\dot{p}} & X_{\dot{q}} & X_{\dot{r}} \\ Y_{\dot{u}} & Y_{\dot{v}} & Y_{\dot{w}} & Y_{\dot{p}} & Y_{\dot{q}} & Y_{\dot{r}} \\ Z_{\dot{u}} & Z_{\dot{v}} & Z_{\dot{w}} & Z_{\dot{p}} & Z_{\dot{q}} & Z_{\dot{r}} \\ K_{\dot{u}} & K_{\dot{v}} & K_{\dot{w}} & K_{\dot{p}} & K_{\dot{q}} & K_{\dot{r}} \\ M_{\dot{u}} & M_{\dot{v}} & M_{\dot{w}} & M_{\dot{p}} & M_{\dot{q}} & M_{\dot{r}} \\ N_{\dot{u}} & N_{\dot{v}} & N_{\dot{w}} & N_{\dot{p}} & N_{\dot{q}} & N_{\dot{r}} \end{bmatrix} \tag{3}$$

The rigid body Coriolis and centripetal matrix, $C_{RB}(\dot{q})$ is given by;

$$C_{RB} = \left[\begin{matrix} 0 & 0 & 0 \\ 0 & 0 & 0 \\ 0 & 0 & 0 \\ -m(y_g q + z_g r) & m(y_g p + w) & m(z_g p - v) \\ m(x_g q - w) & -m(z_g r + x_g p) & m(z_g q + u) \\ m(x_g r + v) & m(x_g r - u) & -m(x_g p + y_g q) \end{matrix} \right. \\ \left. \begin{matrix} m(y_g q + z_g r) & -m(x_g q - w) & -m(x_g r + v) \\ -m(y_g p + w) & m(z_g r + x_g p) & -m(y_g r - u) \\ -m(z_g p - v) & -m(z_g q + u) & m(x_g p + y_g q) \\ 0 & -I_{yz} q - I_{xz} p + I_z r & I_{yz} r + I_{xy} p - I_y q \\ I_{yz} q + I_{xz} p - I_z r & 0 & -I_{xz} r - I_{xy} q + I_x p \\ -I_{yz} r - I_{xy} p + I_y q & I_{xz} r + I_{xy} q - I_x p & 0 \end{matrix} \right] \tag{4}$$

The added mass Coriolis and centripetal matrix, $C(\dot{q})$ is given by;

$$C_A = \begin{bmatrix} 0 & 0 & 0 & 0 & -a_3 & a_2 \\ 0 & 0 & 0 & a_3 & 0 & -a_1 \\ 0 & 0 & 0 & -a_2 & a_1 & 0 \\ 0 & -a_3 & a_2 & 0 & -b_3 & b_2 \\ a_3 & 0 & -a_1 & b_3 & 0 & -b_1 \\ -a_2 & a_1 & 0 & -b_2 & b_1 & 0 \end{bmatrix} \tag{5}$$

where

$$
\begin{aligned}
a_1 &= X_{\dot{u}}u + X_{\dot{v}}v + X_{\dot{w}}w + X_{\dot{p}}p + X_{\dot{q}}q + X_{\dot{r}}r \\
a_2 &= Y_{\dot{u}}u + Y_{\dot{v}}v + Y_{\dot{w}}w + Y_{\dot{p}}p + Y_{\dot{q}}q + Y_{\dot{r}}r \\
a_3 &= Z_{\dot{u}}u + Z_{\dot{v}}v + Z_{\dot{w}}w + Z_{\dot{p}}p + Z_{\dot{q}}q + Z_{\dot{r}}r \\
a_4 &= K_{\dot{u}}u + K_{\dot{v}}v + K_{\dot{w}}w + K_{\dot{p}}p + K_{\dot{q}}q + K_{\dot{r}}r \\
a_5 &= M_{\dot{u}}u + M_{\dot{v}}v + M_{\dot{w}}w + M_{\dot{p}}p + M_{\dot{q}}q + M_{\dot{r}}r \\
a_6 &= N_{\dot{u}}u + N_{\dot{v}}v + N_{\dot{w}}w + N_{\dot{p}}p + N_{\dot{q}}q + N_{\dot{r}}r
\end{aligned}
\tag{6}
$$

The damping term of AUV is separated into two different terms, a linear and quadratic term. The linear term and the quadratic term can be given as;

$$
\begin{aligned}
D_L(v) &= \begin{bmatrix} X_u & X_v & X_w & X_p & X_q & X_r \\ Y_u & Y_v & Y_w & Y_p & Y_q & Y_r \\ Z_u & Z_v & Z_w & Z_p & Z_q & Z_r \\ K_u & K_v & K_w & K_p & K_q & K_r \\ M_u & M_v & M_w & M_p & M_q & M_r \\ N_u & N_v & N_w & N_p & N_q & N_r \end{bmatrix} \\
D_Q(v) &= \begin{bmatrix} X_{|u|u} & X_{|v|v} & X_{|w|w} & X_{|p|p} & X_{|q|q} & X_{|r|r} \\ Y_{|u|u} & Y_{|v|v} & Y_{|w|w} & Y_{|p|p} & Y_{|q|q} & X_{|r|r} \\ Z_{|u|u} & Z_{|v|v} & Z_{|w|w} & Z_{|p|p} & Z_{|q|q} & Z_{|r|r} \\ K_{|u|u} & K_{|v|v} & K_{|w|w} & K_{|p|p} & K_{|q|q} & K_{|r|r} \\ M_{|u|u} & M_{|v|v} & M_{|w|w} & M_{|p|p} & M_{|q|q} & M_{|r|r} \\ N_{|u|u} & N_{|v|v} & N_{|w|w} & N_{|p|p} & N_{|q|q} & N_{|r|r} \end{bmatrix}
\end{aligned}
\tag{7}
$$

The restoring force and moment vector in the body fixed coordinate system is;

$$
gRB(\eta) = \begin{bmatrix} (\mathrm{W}-\mathrm{B})\sin\theta \\ -(\mathrm{W}-\mathrm{B})\cos\theta\sin\phi \\ -(\mathrm{W}-\mathrm{B})\cos\theta\cos\phi \\ -(\mathrm{y_G W}-\mathrm{y_B B})\cos\theta\cos\phi + (\mathrm{z_G W}-\mathrm{z_B B})\cos\theta\sin\phi \\ (\mathrm{z_G W}-\mathrm{z_B B})\sin\theta + (\mathrm{x_G W}-\mathrm{z_B B})\cos\theta\cos\phi \\ -(\mathrm{x_G W}-\mathrm{x_B B})\cos\theta\sin\phi - (\mathrm{y_G W}-\mathrm{y_B B})\sin\theta \end{bmatrix}
\tag{8}
$$

where *W* and *B* represent the weight and buoyancy of the vehicle respectively. The dynamic model of thrusters and pumps have been included in the present study.

3 Fuzzy Modeling of AUV

Fuzzy systems are known for their capabilities to approximate any nonlinear dynamic system [10]. The main idea of fuzzy control is to build a model of a human control expert who is capable of controlling the plant without thinking in mathematical model terms [11]. The nature of fuzzy logic offers a control solution

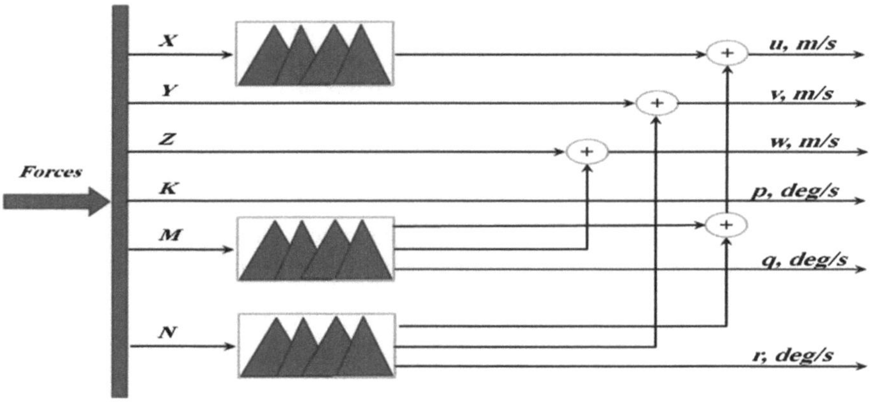

Fig. 3 The configuration of the fuzzy modeling for the AUV

when a mathematical model is not well known or not known at all and has good dynamic performance.

Fuzzy Modeling is the method of describing the characteristics of a system using fuzzy rules, and it can express complex non-linear dynamic systems by linguistic if-then rules [12] as the following

$$R^{(l)}: \text{IF } (x_1 \text{is } F_1^l \text{ and} \ldots\ldots\ldots \text{and } x_n \text{ is } F_n^l) \text{THEN y is } G^l \tag{9}$$

where $x = (x_1, \ldots\ldots, x_n)^T \in U$ and $y \in R$ the inputs and output the fuzzy system, respectively, F_1^l and G_i are labels of fuzzy sets in U, and R, respectively, and $l = 1, 2, \ldots, M$.

The AUV's fuzzy modeling is constructed based upon the input-output data that has been characterized from the open loop system results of the AUV's mathematical model. The input data is considered as the force generated by thruster or pump which moves the AUV in a certain direction, and the output data is considered as the resulted linear or angular velocity of the vehicle taking into account the coupling effect of the other DOF in that direction.

Figure 3 shows the configuration of the proposed fuzzy model which consists of three fuzzy systems that represent surge, pitch and yaw motions.

For surge, pitch and yaw fuzzy models, the input is the force required for the thruster to produce the desired motion of the AUV corresponding to desired pose. The output is the linear velocity in x-direction, *u*. in forward motion and there are three outputs for this fuzzy model.

The first output of the fuzzy model in pitch and yaw motions is the linear velocity in x-direction and this output represents the coupling effect on forward direction, which means the effect of the pitch or yaw motion on forward motion. The second output represents the coupling effect on heave and sway directions respectively. The third output is the angular velocity about y and z-axes respectively. The membership functions for input and outputs of fuzzy model are shown in Figs. 4 and 5.

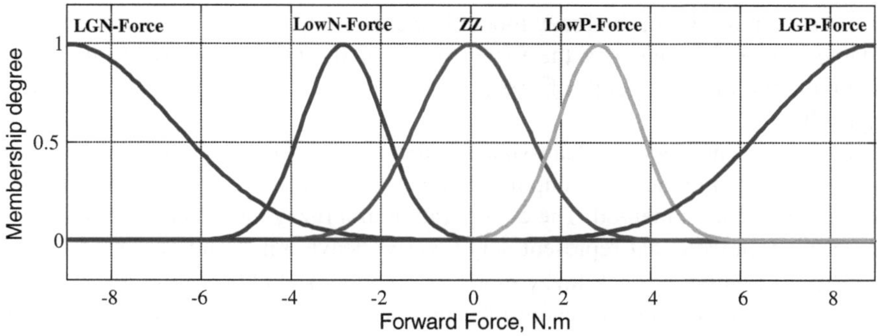

Fig. 4 Forward force membership function

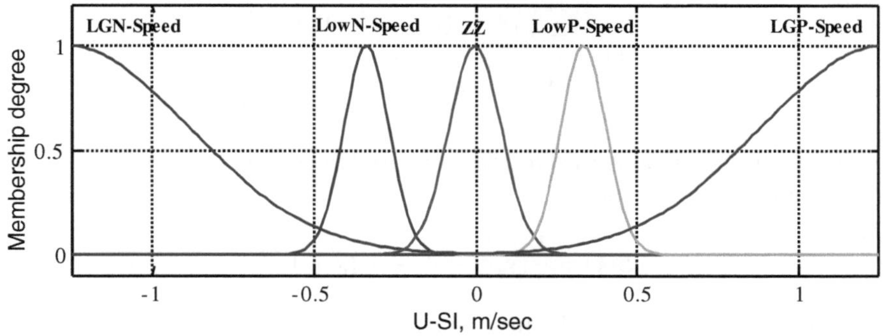

Fig. 5 Linear velocity in x-direction membership function

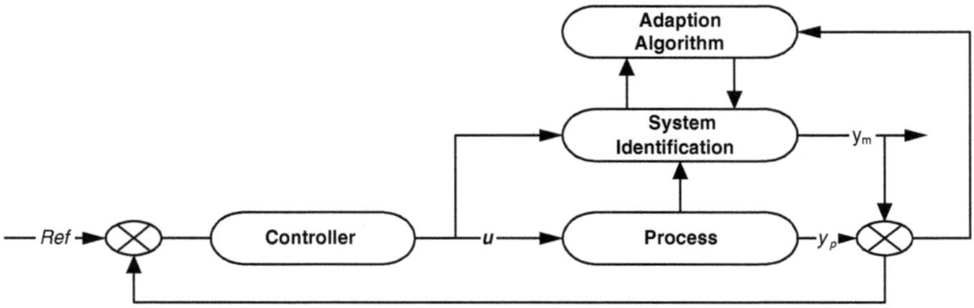

Fig. 6 Tracking of error function between the process and fuzzy model

4 Adaptive Fuzzy Model

An adaptive fuzzy model is a fuzzy model with a training algorithm where the model is synthesized from a bundle of fuzzy If-Then rules. Each fuzzy membership is characterized by certain parameters. The training algorithm adjusts these parameters based on numerical inputs and outputs data.

As shown in Fig. 6, the fuzzy model is placed in parallel with the process to be adapted. It aims to identify the fuzzy model of the process online by using the input–output measurements of the process based on back propagation training algorithm.

In the present work, the Gaussian membership functions are used in the fuzzy model. Fuzzy inference based upon a product sum is considered with the centre of area deffuzification method. The configuration of a fuzzy model system consists of three fuzzy models that represent surge, sway, heave, pitch and yaw motions.

The final output of each fuzzy model is given by Eq. (11) with c_i^l and σ_i^l which representing the centre and width of Gaussian memberships for input variable x_i for the rule i, l and n being the number of rules of fuzzy model and number of inputs respectively.

$$y_m(k+1) = \frac{\sum_{l=1}^{M} B^l \left[\prod_{i=1}^{n} exp\left(-0.5\left(\frac{x_i - c_i^l}{\sigma_i^l}\right)^2\right)\right]}{\sum_{l=1}^{M} \left[\prod_{i=1}^{n} exp\left(-0.5\left(\frac{x_i - c_i^l}{\sigma_i^l}\right)^2\right)\right]} \tag{10}$$

The gradient method is used to adapt the fuzzy model parameters based on the following objective function;

$$E(k) = \frac{1}{2}\left(y_m(k+1) - y_f(k+1)\right)^2 \tag{11}$$

where *E(k)* is error between the fuzzy model and the actual plant outputs.

If *Z(k)* represents the parameter to be adapted at iteration *k* in the fuzzy model, the training algorithm seeks to minimize the value of the objective function by [13];

$$z(k+1) = z(k) - \alpha \frac{\partial E}{\partial Z} \tag{12}$$

To train B^l;

$$B^l(k+1) = B^l(k) - \alpha \frac{y_m - y_f}{b} z^l \tag{13}$$

To train c_i^l;

$$c_i^l(k+1) = c_i^l(k) - \alpha \frac{y_m - y_f}{b} (B^l - y_m) z^l \frac{2(x_i - x^l(k))^1}{\sigma_i^{l2}(k)} \tag{14}$$

To train σ_i^l;

$$\sigma_i^l(k+1) = \sigma^l(k) - \alpha \frac{y_m - y_f}{b} (B^l - y_m) z^l \frac{2(x_i - x^l(k))^2}{\sigma_i^{l3}(k)} \tag{15}$$

where

$$b = \sum_{l=1}^{M} z^l,\ z^l = \prod_{i=1}^{n} \exp\left(-0.5\left(\frac{x_i - c_i^l}{\sigma_i^l}\right)^2\right) \tag{16}$$

The learning rate α in Eq. (13) [14] has a significant effect on the stability and convergence of the system. The proper choice of the learning rate is therefore very important.

4.1 Convergence of Fuzzy Controller

Each learning rate parameter of the weight, the mean, and the variance, α, has a significant effect on the convergence. A higher learning rate may enhance the convergence rate but can reduce the stability of the system. A smaller value of the learning rate guarantees the stability of the system but slows the convergence. To ensure a quick and stable convergence of fuzzy model parameters a convergence analysis of the learning rate will be considered according to the following theorem.

Theorem [14, 15]: *Let α be the learning rate for the parameters of fuzzy controller and g_{max} be defined as $g_{max} := \max_k \| g(k) \|$ where $g(k) = \delta y(k)/\delta z(k)$ and $\|.\|$ is the usual Euclidean norm in $\Re^n$ and let $= \delta y_m/\delta u$. Then the convergence is guaranteed if α is chosen as*;

$$0 < \alpha < \frac{2}{S^2 g_{max}^2} \tag{17}$$

5 AUV Fuzzy Control System

The inputs to the fuzzy controller are the error and error difference. The fuzzy controller output can be represented as;

$$u(t) = f(e(t), \Delta e(t) \tag{18}$$

where *u(t)* is control signal, *e(t)* is the error between the reference and the process output and Δ*e(t)* is the error difference. Using the fuzzy inference based upon product sum gravity at given input *(e(t),Δe(t))* and the Gaussian membership functions for all fuzzy sets, the final output of the fuzzy controller is given by;

$$u(t) = \frac{\sum_{j=1}^{h} B_j \left[\exp\left(-0.5\left(\frac{e(t)-ce_j}{\sigma e_j}\right)^2\right)\right]\left[\exp\left(-0.5\left(\frac{\Delta e(t)-c\Delta e_j}{\sigma \Delta e_j}\right)^2\right)\right]}{\sum_{j=1}^{h} \left[\exp\left(-0.5\left(\frac{e(t)-ce_j}{\sigma e_j}\right)^2\right)\right]\left[\exp\left(-0.5\left(\frac{\Delta e(t)-c\Delta e_j}{\sigma \Delta e_j}\right)^2\right)\right]} \tag{19}$$

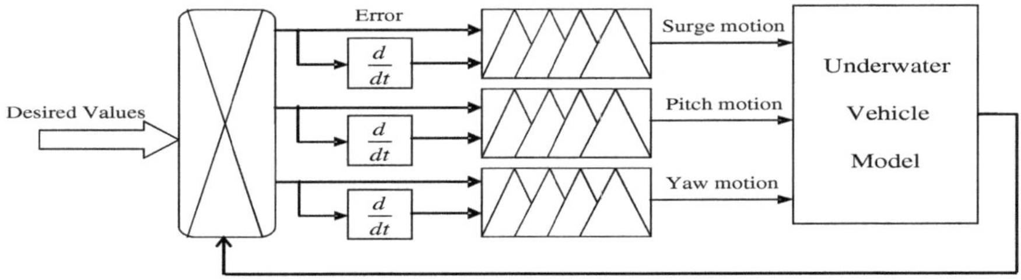

Fig. 7 The structure of the fuzzy control system for the AUV

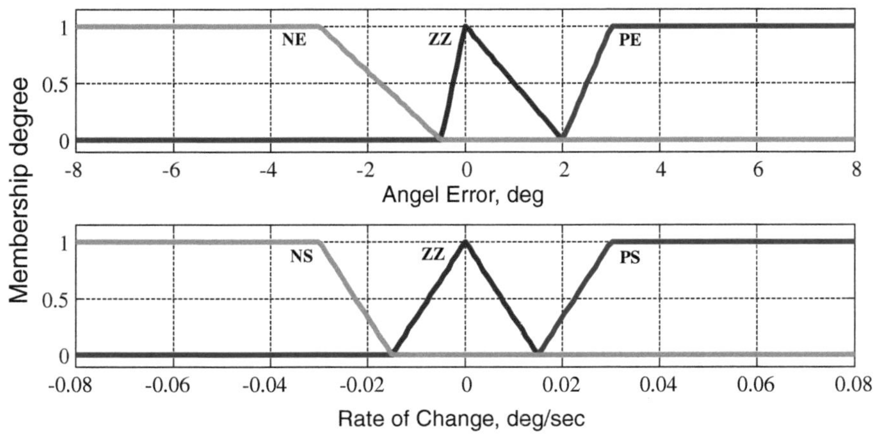

Fig. 8 Memberships functions for the error and the rate of change of the error

where $ce_j, \sigma e_j, c\Delta e_j$ and $\sigma\Delta e_j$ represent the centre and width of Gaussian membership for the error and error difference respectively for the rule j and k is the number of rules of fuzzy controller.

Generally, the depth of AUV can be maintained by controlling the pitch and forward motion. Moving right and left of AUV can be achieved by controlling the yaw angle. The pitch and yaw angles are controlled by controlling the operating time of the pump responsible for moving the AUV in that direction.

As shown in Fig. 7 there are three fuzzy controllers in surge, pitch and yaw motions. Controlling of AUV in surge, pitch and yaw motions are sufficient to determine the position and orientation of the AUV in three dimensions.

For surge motion, the inputs to fuzzy controller are the position error and its error difference. The output of the fuzzy controller is the force required by the thruster. The membership functions for this controller are shown in Fig. 8.

For pitch and yaw motions, the inputs to fuzzy controller are angle error and its error difference. The output of the fuzzy controller is the time required to operate the pump to rotate the AUV in the required direction. The membership functions are the same for surge fuzzy control with different values.

6 Simulation Results

A MATLAB program is conducted to simulate the dynamics of the AUV by using Rung-Kutta fifth order method with tolerance 0.00001. The AUV is required to move with fuzzy controller which applied on both mathematical model and adaptive fuzzy model with some disturbance applied on the system velocities during the simulation.

6.1 *Open Loop Accuracy Evaluation of AUV Model*

The most widely used method for measuring performance and the accuracy indicators of the AUV system is the root mean square error (RMSE). The root mean square error is defined as:

$$RMSE = \left(\frac{1}{n} \sum_{i=1}^{n} d_i^2 \right)^{1/2} \quad (20)$$

where n is the number of data pairs and d_i is the difference between *ith* desired and *ith* measured values. The RMSE is said to provide information on the short-term performance of a model by allowing a term by term comparison of the actual difference between the desired value and the measured value. The smaller the value is, the better the model's performance.

Table 1 shows the performance of the open loop mathematical model, fuzzy model and adaptive fuzzy model for the AUV and same operating conditions have been applied on all models. It is clear that there is a significant improvement in the performance of the AUV with the adaptive fuzzy model rather than other models.

6.2 *Results for Adaptive Fuzzy Model with Fuzzy Controller*

The learning rates for the fuzzy model are initially set to one to train the model parameters. Then the convergence conditions were verified at every sampling time.

Table 1 The RMSE values for AUV mathematical model, fuzzy model and adaptive fuzzy model in each direction

Direction	Mathematical model	Fuzzy model	Adaptive fuzzy model
X	38.1652	7.6558	7.0802 e−004
Y	1.2808	0.2748	5.5305 e−004
Z	2.5209	0.8275	2.4197 e−004
Pitch	1.0074	0.869	0.0161
Yaw	0.7188	0.5161	0.0024

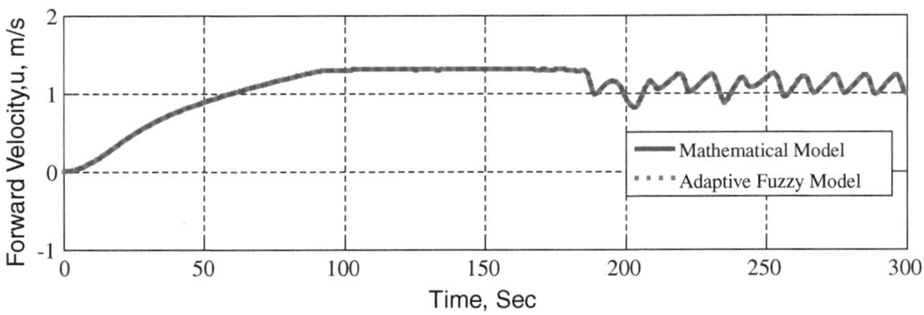

Fig. 9 The forward velocity of the AUV with mathematical model and system identification

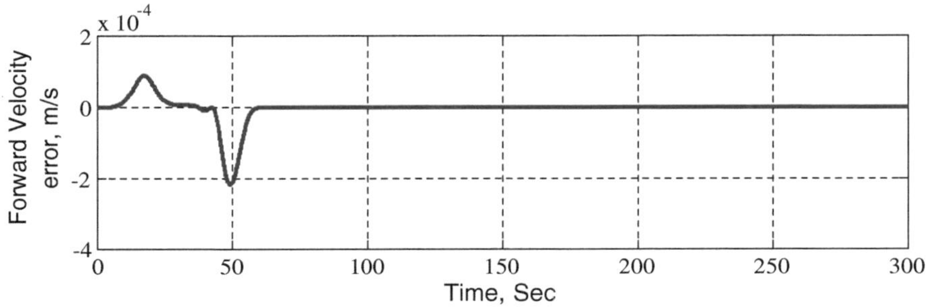

Fig. 10 The error between mathematical model and system identification in u

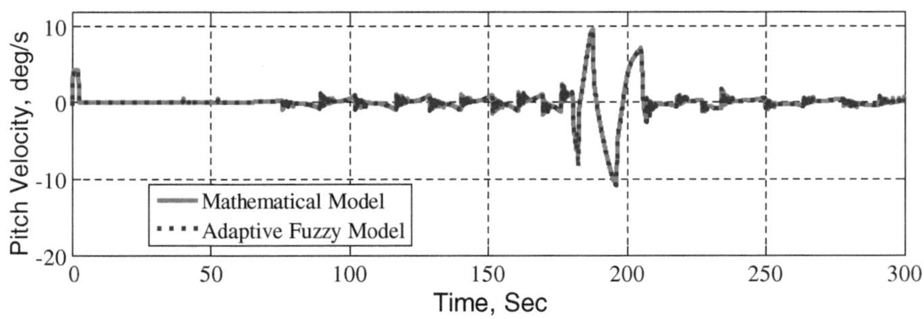

Fig. 11 The pitch velocity of the AUV with mathematical model and system identification

In the present work, it was found that the value was always within the limit of convergence for the fuzzy controller.

Figures 9 and 10 show the linear velocity on forward direction, u, and the error between the mathematical and adaptive fuzzy model in the forward velocity respectively. It is seen that the adaptive fuzzy model successfully tracks the plant and the maximum error between them is about 0.0002 m/s.

Figures 11 and 12 show the angular velocity on pitch motion, q, and the error between the mathematical and adaptive fuzzy model in that motion, respectively.

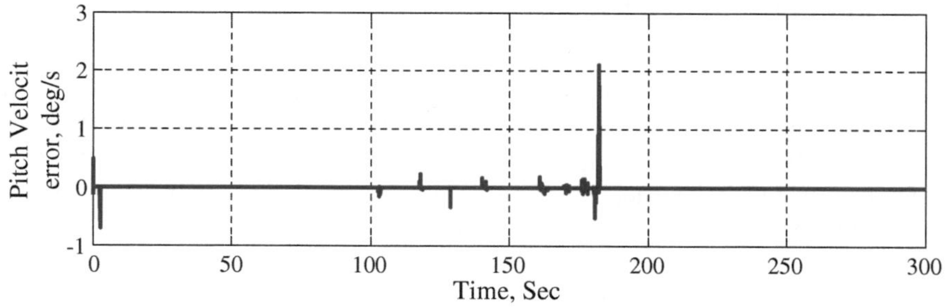

Fig. 12 The error between mathematical model and system identification in q

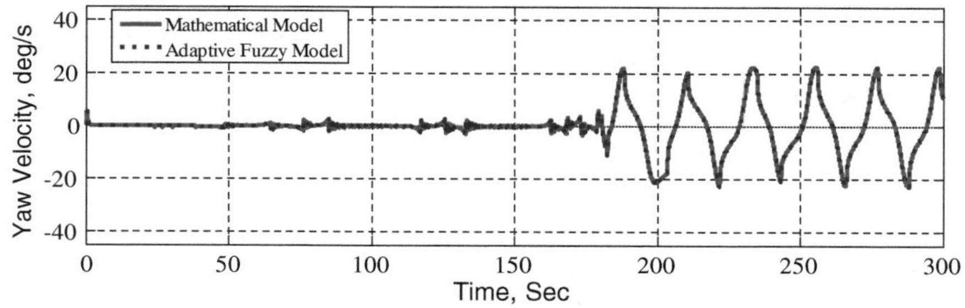

Fig. 13 The yaw velocity of the AUV with mathematical model and system identification

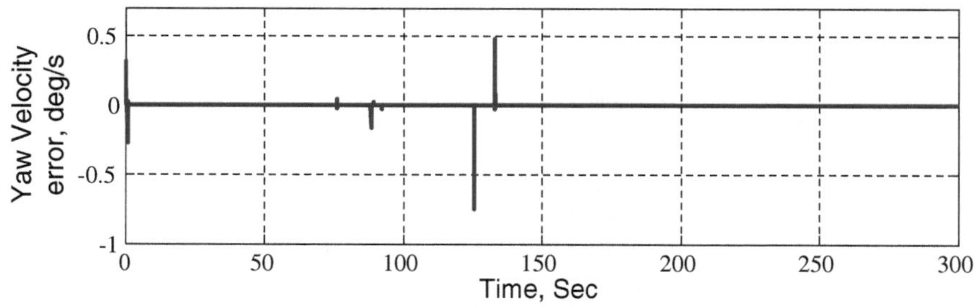

Fig. 14 The error between mathematical model and system identification in r

It is clearly seen that the behavior of the adaptive fuzzy model is same as the mathematical model. The system identification has the capability to track the plant successfully whatever the change in the operating condition.

Figures 13 and 14 show the angular velocity on yaw motion, *r*, and the error between the mathematical and adaptive fuzzy model in that motion, respectively. It is seen that the system identification is very good in tracking the plant behavior successfully regardless of the operating condition or disturbance.

7 Concluding Remarks

The chapter presents the numerical simulation results of the adaptive fuzzy modeling as system identification and mathematical modeling of the AUV using fuzzy controllers. Fuzzy modeling is used to identify the model of the AUV using input–output data. The back propagation as a training algorithm for the fuzzy system proves the fast convergence of the fuzzy system and the fuzzy identifier successfully achieved a similar performance of the process.

References

1. K.M. Tan, T. Liddy, A. Anvar, T.-F. Lu, *The Advancement of an Autonomous Underwater Vehicle (AUV) Technology*. Industrial Electronics and Applications, 2008. ICIEA 2008, 3–5 June 2008, pp. 336–341
2. S. Holtzhausen, *Design of an Autonomous*. Master Thesis, KwaZulu-Natal: University of KwaZulu-Natal (2010)
3. L.-X. Wang, J. M. Mendel, in *Back-Propagation Fuzzy System as Nonlinear Dynamic System Identifiers*. Fuzzy Systems, IEEE International Conference (1992), pp. 1409–1418
4. K.M. Bossley, M. Brown, C.J. Harris, Neurofuzzy identification of an autonomous underwater vehicle. Int. J. Syst. Sci. **30**(9), 901–913 (1999)
5. J. Guo, S.H. Huang, *Control of an Autonomous Underwater Vehicle Testbed Using Fuzzy Logic and Genetic Algorithms*. Autonomous Underwater Vehicle Technology, 1996. AUV '96, Proceedings of the 1996 Symposium (IEEE, Monterey, 1996), pp. 485–489
6. W. Naeem, in *Model Predictive Control of an Autonomous Underwater Vehicle*. UKACC Conference control (Sheffield, IFAC, 2002), pp. 19–23
7. Citeseerx, http://citeseerx.ist.psu.edu/viewdoc/download?doi=10.1.1.18.4264&rep=rep1&type=pdf
8. J. Avadi-Moghaddama, A. Bagherib, An adaptive neuro-fuzzy sliding mode based genetic algorithm control system for under water remotely operated vehicle. Expert Syst. Appl. 647–660 (2009)
9. T.I. Fossen, *Guidance and Control of Ocean Vehicles* (Wiley, New York, 1994)
10. T. Asokan, M. Singaperumal, *Autonomous Underwater Robotic Vehicle (AUV) Project Indian Institute of Technology*, Madras, Jan 2008
11. L. X. Wang, Fuzzy *System are Universal Approximators*, IEEE Proceeding of International Conference on Fuzzy Systems, pp. 1163–1170, 1992
12. T. Takagi, M. Sugeno, Fuzzy identification of systems and its applications to modelling and control. IEEE Trans. Syst. Man Cyb. 15, 116–132
13. L.-X. Wang, *Adaptive Fuzzy System and Control, Design and Stability Analysis* (Prentice Hall, New Jersey, 1994)
14. A. S. Shaaban, G. A. Sreenatha, Y. C. Jin, in *Indirect Adaptive Fuzzy Control of Unmanned Aerial Vehicle*. Proceedings of the 17th World Congress, the International Federation of Automatic Control (Seoul, Korea, 2008), pp. 13229–13243
15. Y.C. Chen, C.C. Teng, A model reference control structure using a fuzzy neural network. J. Fuzzy Sets Syst. **73**, 291–312 (1995)

Process Rescheduling and Path Planning Using Automation Agents

Munir Merdan, Wilfried Lepuschitz, Benjamin Groessing and Markus Helbok

Abstract The process industry faces a permanently changing environment, where sudden component failures can significantly influence the system performance if not treated in an appropriate amount of time. Moreover, current market trends have to be met such as short production times, a low price as well as a broad spectrum of product and process varieties. Distributed intelligent control systems based on agent technologies are seen as a promising approach to handle the dynamics in large complex systems. In this chapter, we present a multi-agent system architecture capable to answer to the major requirements in the process domain. The architecture is based on agents with diverse responsibilities as well as tasks and separates the control software of agents controlling hardware components into two levels, the high level control and the low level control. Our system architecture has also the ability to flexibly reschedule allocated jobs in the case of resource breakdowns in order to minimize downtimes. This goes hand in hand with a dynamic path finding algorithm to enhance the flexibility of transport tasks. The system is currently tested and evaluated in the Odo Struger Laboratory at the Automation and Control Institute.

Based on "Advanced Process Automation Using Automation Agents", by Munir Merdan, Wilfried Lepuschitz, and Emilian Axinia which appeared in the Proceedings of the 5th International Conference on Automation, Robotics and Applications (ICARA 2011) © 2011 IEEE.

M. Merdan (✉) · W. Lepuschitz · B. Groessing
Vienna University of Technology, Automation and Control Institute, Vienna, Austria
e-mail: merdan@acin.tuwien.ac.at

W. Lepuschitz
e-mail: lepuschitz@acin.tuwien.ac.at

B. Groessing
e-mail: groessing@acin.tuwien.ac.at

M. Helbok
COPA-DATA GmbH, Karolingerstrasse 7b 5020 Salzburg, Austria
e-mail: markush@copadata.com

G. Sen Gupta et al. (eds.), *Recent Advances in Robotics and Automation*,
Studies in Computational Intelligence 480, DOI: 10.1007/978-3-642-37387-9_5,

1 Introduction

The process industry faces a permanently changing environment and has to meet current market trends such as short production times, a low price as well as a broad spectrum of product and process varieties. Due to these reasons, the importance of automation in the process industry has increased dramatically in recent years. It has become a vital force in the entire chemical, oil, gas and biotechnology industries [1]. The proper function of an automation system is critical for the operation of the majority of plants today, since it is the automation system that performs control and other advanced functions including optimization, scheduling, planning, monitoring, etc. [2]. Traditional approaches for solving scheduling problems encounter great difficulties when applied in real situations, because these scheduling methods use simplified theoretical models and are centralized in the sense that all computations are carried out in a central computing unit [3]. Besides, to make appropriate decisions, an automation system relies on exchanging different kinds of data such as process measurements, diagnostic data or historical data. Nevertheless, due to different standards in the domain, information is represented in different syntaxes as well as semantics and on different levels of abstraction. Considering the ongoing information and knowledge explosion, intelligent process control systems are needed that include features for knowledge and information management to effectively manage and access information for efficient decision making [1, 4]. In addition, deviations from the original schedule and information about equipment breakdowns provided by the control and monitoring system will eventually trigger rescheduling mechanisms and in some cases require a reconfiguration of the system. The automation system should be able to change quickly and cost-effectively from its current configuration to another configuration without being taken off-line [5]. Hence, online modification of the system configuration is a key requirement for future process automation systems [2].

The introduction of artificial intelligence techniques is seen as a promising trend in the process industry [1]. The application of a multi-agent system (MAS) is recognized as a convenient way to handle the dynamics in large complex systems reducing the complexity, increasing flexibility and enhancing fault tolerance [6]. Agents cooperate and coordinate their actions in order to achieve their own as well as the system's aims, which are beyond the capabilities possessed by an individual agent. Moreover, by being able to use parallel computation and to apply diverse strategies and methods for solving their simple local problems, agents can significantly improve the efficiency and increase the performance of the entire system [7]. A further advantage of these approaches is their better reactivity and adaptability in highly dynamic environments. Most of the scheduling models for chemical processes assume that all required data is certain and therefore do not include uncertainty in the calculation. However, it can be shown that a schedule generated by a deterministic model based on nominal values of the parameters may be infeasible upon the occurrence of uncertainty [8]. Besides, the presence of real-time information permanently forces the revision of pre-established schedules and the consideration of new conditions and constraints related to certain environments. Under such circumstances a fast reaction is required in terms of identifying alternative resources and respective time slots to continue production. Besides, the ability to flexibly reschedule allocated jobs is needed in the case of resource breakdowns in order to minimize downtimes. Furthermore, these rescheduling actions should be performed autonomously and with a modest computational effort. Despite a resource failure, the system should be able to autonomously continue its activities (as may be the case in a degraded mode) while the broken resource is repaired. Furthermore, the reallocation of jobs awarded to a broken resource is required, so these jobs are performed by other resources and do not unnecessarily block other jobs. Effective resource reallocation processes minimize the amount of overcapacity needed to cope with unpredictable events and consequently reduce the related inventory costs needed for these overcapacities. Moreover, the maximal usage of potential process redundancies and flexible routing capabilities is required in order to minimize the influence of a failure [9].

In this chapter, we introduce a new architecture of automation agents to answer to the requirements in the process domain mentioned as mentioned above. The architecture separates the control software into two levels, the high level control (HLC) and the low level control (LLC), related to specific concerns based on their ability to execute particular activities, e.g. to perform failure detection and recovery. This combination enables efficient process scheduling, monitoring and diagnosis tasks, where the information from diverse data sources is combined in a flexible manner to make conclusions about the overall state of the process [10, 11].

The chapter is structured as follows. The subsequent section introduces a motivation example. Section 3 presents the architecture of the MAS. In Sect. 4 we describe system rescheduling and path planning and Sect. 5 concludes this chapter.

2 State of the Art

Conventional control approaches based on hierarchical architectures are limited in dealing with emerging requirements due to their inflexible structures and operating rules [12]. Agent-based control systems have the capability to respond quickly and correctly to dynamic changes in the production environment, and differ from the conventional approaches with their inherent capabilities to adapt to emergence without external intervention [13]. The approaches mentioned in this section show some developments in this field.

The management of chemical batch plants is based on collecting and processing huge amounts of data, which is subsequently exploited. This data is a valuable source of information for decision-making, scheduling, process control, fault analysis, etc. A common representation of knowledge is needed to merge the mass of heterogeneous process information. There are many standards for information representation in the field, but the agents need to be able to address both low level and high level information in the same context [10]. One approach for the integration of information systems is the usage of shared ontologies. Ontologies can be used to describe the semantics of the information sources and make the contents explicit, thereby enabling the integration of existing information repositories [4]. In this context, Sesen et al. [14] developed an ontological framework for automated regulatory compliance in pharmaceutical manufacturing, Batres et al. [15] provided a brief overview of an upper ontology based on ISO 15926, Venkatasubramanian et al. [4] introduced an ontological informatics infrastructure for pharmaceutical product development and manufacturing, and Muñoz et al. [16] presented a batch process ontology. We use these ontologies as the roots for a description of a process automation system and for a definition of the ontology-based world model of each agent types. The usage of ontologies for knowledge representation and high level reasoning is a major step ahead in the area of agent-based control solutions [17].

Most of the earlier developed scheduling methods have difficulties in solving actual industrial problems such as changed schedules resulting from changed production orders, due to the complexity of real-life manufacturing constraints [18]. Hamaguchi et al. [5] presented an approach for solving scheduling problems. Seilonen et al. [19] introduced an approach suitable also for handling failures. To incorporate uncertainty in the process industries, it is necessary to investigate a new unified framework for planning and scheduling under uncertainty [20]. Besides, it is important to define an overall framework for specifying the system as well as component behavior in the occurrence of exceptions [21]. This goes hand in hand with the requirement of integrating a dynamic path finding method for determining appropriate material paths between the tanks. In comparison to hard-coded paths as often applied currently in industrial practice, dynamic path finding with agent technology can greatly enhance a production system's flexibility during runtime. But also extensions to the system can be applied without having to reprogram any hard-coded paths.

3 System Architecture

In order to reduce its complexity, the control of a process automation system can be organized in several "hierarchically" ordered layers as it is also done in current industrial solutions. We specified four layers: Management, Planning, Scheduling and Executive Layer. The Management Layer is responsible for the entire system functionality. It is also concerned with the communication with the external environment and provides solutions for complex problems related to the global environment. It accepts orders on a routine basis. The Planning Layer links process planning with product design. It is basically concerned with the sequencing of process steps as well as with the identification of product types and quantities to be produced. The Scheduling Layer is concerned with the synchronization of production needs with available resource capacities. This layer is responsible for negotiating with the resources and the task allocation between resources. The Executive Layer is related to the process system's equipment. On this layer, the production tasks are executed considering the resources' constraints and abilities, their performances are measured and if a failure or disruption is diagnosed, the higher layers are informed. The architecture of the MAS is based on the specified layered structure with a particular agent type for each layer (see Fig. 1). In the following the different types of agents and their general activities are described.

The Task Agent determines the equipment for performing the operations of the recipe by searching for Automation Agents with suitable equipment in the list of the Directory Facilitator Agent. When appropriate Automation Agents are found, the tasks for the first operation of the recipe are created including transport tasks.

Work Agents receive tasks from Task Agents and manage Automation Agents. The tasks can either be transport or production tasks. After receiving a transport task with the associated source and destination information, the Work Agent generates a path using the path finding algorithm as described in Sect. 4. Ultimately, the Work Agent sends orders to Automation Agents to have them make state changes on hardware components. These changes can be for instance opening a valve or activating a pump. Newly created or changed routes due to system modifications are automatically considered by this agent and the system.

Automation Agents, key components of our architecture, are those agents incorporating some kind of physical representation. They control manufacturing resources (such as a valve or a pump) providing particular processes or services. Each Automation Agent manages its local scheduling and negotiates with the Task

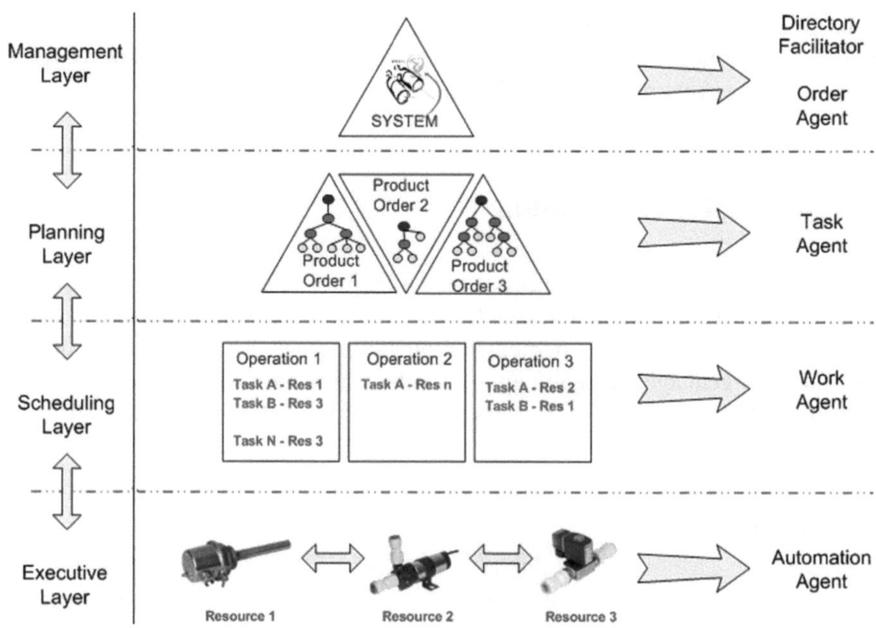

Fig. 1 Architecture of the developed MAS

Agent about supply and free timeslots in which the particular operations can be performed. One of the main features of the Automation Agent architecture is the distinction between the HLC and the LLC within each agent [22].

The HLC layer is in charge of higher level diagnostics, of the coordination with other agents and of adaptation based on the representation of the world. It incorporates an ontology-based world model that provides an explicit representation of the agent's immediate environment and supports reasoning about its state. The application of the ontology enables the systematic integration of different functionalities for the controlled process (i.e. in particular planning, scheduling and plant/unit control). Also this provides a common language for a better communication and creates a modular and internally consistent standard that can help reduce engineering costs [16].

The LLC is responsible for managing the physical component by using a limited set of reactive behaviors and requires means to access the physical I/Os of the controllers to gather sensor data as well as to issue commands to actuators. We base our LLC on the standard IEC 61499 [23], which is well-suited for distributed applications due to its event-driven execution model and offers a framework for the integration of run-time control and basic diagnostic applications. Our architecture was successfully used for controlling physical components that integrate "on the fly" reconfiguration abilities of the LLC [22], for monitoring and diagnostic tasks [24], as well as to enhance reconfigurability, robustness and fault tolerance of a transportation system [25].

Finally, the Directory Facilitator Agent (DF Agent) manages lists of Automation Agents and their services. Automation Agents register themselves in this list to be found by the Task Agent for the execution of tasks. In this context it is important to notice that Order Agent, Task Agent and Work Agent are designed likewise to the Automation Agents' HLC.

4 Rescheduling and Routing

The overall economic effectiveness of the entire manufacturing system is strongly coupled with the production schedule, since it needs to synchronize the entire system activities to achieve particular production goals. The application of distributed intelligent agent technology is considered as a promising approach and has been applied in manufacturing process planning and scheduling as well as the shop floor control domain [26].

4.1 Rescheduling

We use an event-driven rescheduling policy with rescheduling actions triggered upon the recognition of an exceptional event which could cause a significant disruption of the system [27]. Rescheduling is started once the related Automation

Agent notices a failure of its component and informs all related Work Agents about this issue. Each Work Agent reschedules assigned jobs that are already scheduled to the failed resource and excludes this component from participation in subsequent negotiations for future jobs until it is repaired. In order to maximize the overall system throughput as well as to minimize the flow time and make span, each Work Agent balances the jobs that need to be rescheduled between alternative components using the Contract Net Protocol [28]. Firstly, the Work Agent sequences all jobs originating from the agenda of the failed component according to their urgency, asks the DF agent for alternative resources for the particular job and sends a call for proposal to them. After the Work Agent has received the proposals for a particular job from the Automation Agents, it compares the proposals and awards the job to the best suited component. These steps are repeated until all jobs from the agenda of the failed component are rescheduled.

4.2 Dynamic Path Planning

Considering the complexity of process systems and their dynamic nature, where particular anomalies such as a component breakdown or an overload can cause that parts of the systems are not in function, the process of choosing the best route at such a specific moment can be difficult. Besides, when a component (pump, valve, etc.) breaks down, the related pipes are usually blocked and cannot be used until the component is repaired. If the system is inflexible and therefore not able to adequately reroute the medium using alternative routes, additional expenses are caused that again reduce the profitability of the production process. The route planning and best path algorithms have been highly researched topics in computer science for many years [29]. Based on Dijkstra's algorithm [30], we implemented a simple, but very efficient Shortest Path Algorithm, which is used by the Work agents for calculating an appropriate path in the case of a transport task. The laboratory process plant located in the Odo Struger Laboratory (see Fig. 2) is represented within the ontology.

The process system ontology is represented by an XML schema (Fig. 3) and a compatibility matrix. The main components such as tanks, valves and pumps are specified in the schema with their name, usage status and various other attributes providing further information for routing algorithms, business logic and monitoring applications. Constraints regarding media flow directions can be defined at any connection point between a pipe and a component. Combining this information with a media compatibility matrix enables advanced routing concepts and determines, if pipes need to be cleaned for routing incompatible media. Hence, the constraints of system components (e.g., pipe currently used or dirty as well as the compatibility between mediums) are also incorporated in the ontology and regulated through appropriate agent behaviors. The Work Agent is able to use the representation of the process plant and applies Dijkstra's algorithm to find the shortest usable path for performing the transport task.

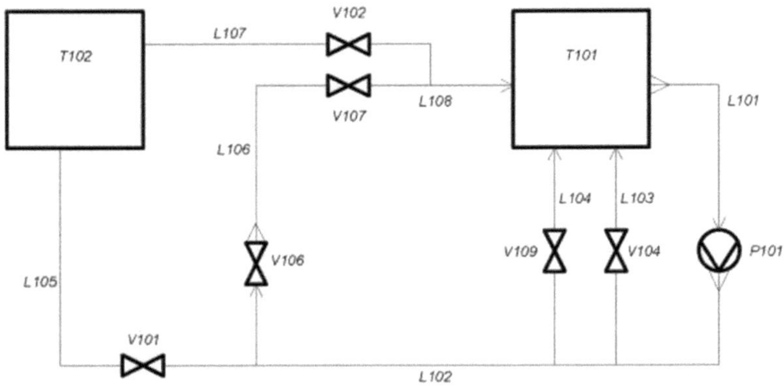

Fig. 2 Pipe and instrumentation diagram of the laboratory process plant

```xml
<ontology>
  <components>
    <tank name="T101" medium="opc:T101.MEDIUM" transferState="opc:T101.TRANS_STATE" functionalState="opc:T101.FUNC_STATE" capacity="10" currentLevel="opc:T101.LEVEL" unit="" />
    ...
    <valve name="V109" medium="opc:V109.MEDIUM" transferState="opc:V109.TRANS_STATE" functionalState="opc:V109.FUNC_STATE" serviceState="opc:V109.SERV_STATE" routing="opc:V109.ROUTING" unit="" />
    <pump name="P101" medium="opc:P101.MEDIUM" transferState="opc:P101.TRANS_STATE" functionalState="opc:P101.FUNC_STATE" serviceState="opc:P101.SERV_STATE" routing="opc:P101.ROUTING" unit="" />
  </components>
  <pipes>
    <pipe name="L101" medium="opc:L101.MEDIUM" transferState="opc:L101.TRANS_STATE" length="45" diameter="1.5" routing="opc:L101.ROUTING" unit="">
      <link direction="in">T101</link>
      <link direction="out">P101</link>
    </pipe>
    ...
    <pipe name="L107" medium="opc:L107.MEDIUM" transferState="opc:L107.TRANS_STATE" length="30" diameter="1.5" routing="opc:L107.ROUTING" unit="">
      <link>T102</link>
      <link>V102</link>
    </pipe>
    <pipe name="L108" medium="opc:L108.MEDIUM" transferState="opc:L108.TRANS_STATE" length="45" diameter="1.5" routing="opc:L108.ROUTING" unit="">
      <link>V102</link>
      <link>V107</link>
      <link direction="out">T101</link>
    </pipe>
  </pipes>
<compatibility file="./CM-compatibility.csv" />
</ontology>
```

Fig. 3 XML schema of the system ontology

5 Conclusion and Outlook

In this chapter we presented a system architecture that integrates various types of agents in a consistent architecture able to manage an entire batch process system. The agents cooperate and coordinate their activities in order to accomplish particular orders. As a core of our architecture, we integrated a concept of Automation Agents that offers the opportunity of a new approach focusing on the definition of system components with every component having its own intelligence. Considering that deviations from an original production schedule happen more and more often in the present dynamic production environment, we integrated an appropriate rescheduling mechanism able to trigger rescheduling and in some cases the reconfiguration of the system. Additionally, this architecture enables a fast reaction on component breakdowns or traffic jams and, by identifying alternative routes and the consideration of new conditions and constraints, an adequate reconfiguration of the system as well as the rerouting of material.

The presented MAS architecture is currently evaluated on an existing laboratory process plant located in the Odo Struger Laboratory at the Automation and Control Institute. In our ongoing work, we are measuring and evaluating the performance of the presented architecture on this real system and will compare the results with similar approaches performed with other architectures. Furthermore, we tend to integrate our architecture with the real batch management software zenon[1] developed by COPA-DATA.

Acknowledgments This work has been supported by the Austrian Research Promotion Agency (FFG) project PrOnto (829576) under the BRIDGE program.

References

1. S. Jämsä-Jounela, Future trends in process automation. Annu. Rev. Control **31**(2), 211–220 (2007)
2. T. Samad, P. McLaughlin, J. Lu, System architecture for process automation: review and trends. J. Process Control **17**(3), 191–201 (2007)
3. R. Brennan, Toward real-time distributed intelligent control: a survey of research themes and applications systems. IEEE Trans. Man Cybern. Part C Appl. Rev. **37**(5), 744–765 (2007)
4. V. Venkatasubramanian, C. Zhao, G. Joglekar, A. Jain, L. Hailemariam, P. Suresh, P. Akkisetty, K. Morris, G. Reklaitis, Ontological informatics infrastructure for pharmaceutical product development and manufacturing. Comput. Chem. Eng. **30**, 1482–1496 (2006)
5. T. Hamaguchi, T. Hattori, M. Sakamoto, H. Eguchi, Y. Hashimoto, T. Itoh, in *Multi-Agent Structure for Batch Process Control*. Proceedings of the IEEE international conference on control applications, vol. 2 (2004), pp. 1090–1095
6. N. Jennings, S. Bussmann, Agent-based control systems: why are they suited to engineering complex systems? Control Syst. Mag. IEEE **23**(3), 61–73 (2003)
7. J. Reaidy, P. Massotte, D. Diep, Comparison of negotiation protocols in dynamic agent-based manufacturing systems. Int. J. Prod. Econ. **99**, 117–130 (2006)
8. C.A. Floudas, X. Lin, Continuous-time versus discrete-time approaches for scheduling of chemical processes: a review. Comput. Chem. Eng. **28**, 2109–2129 (2004)
9. H. Aytug, M.A. Lawley, K. McKay, S. Mohan, R. Uzsoy, Executing production schedules in the face of uncertainties: a review and some future directions. Eur. J. Oper. Res. **161**, 86–110 (2005)
10. T. Pirttioja, A. Pakonen, I. Seilonen, A. Halme, K. Koskinen, in *Multi-Agent Based Information Access Services for Condition Monitoring in Process Automation*. Proceedings of the 3rd IEEE international conference on industrial informatics (INDIN '05), (2005), pp. 240–245
11. B. Sahovic, Automation Agents with a Reflective World Model for Batch Process Automation, Master Thesis, Technische Universität Wien (2010), Online: http://media.obvsg.at/AC07808679
12. N. Chokshi, D. McFarlane, *A Distributed Coordination Approach to Reconfigurable Process Control* (Springer, London, 2008)
13. R. Srinivasan, Artificial intelligence methodologies for agile refining: an overview. Knowl. Inf. Syst. **12**, 129–145 (2007)

[1] http://www.copadata.com/

14. M.B. Sesen, P. Suresh, R. Banares-Alcantara, V. Venkatasubramanian, An ontological framework for automated regulatory compliance in pharmaceutical manufacturing. Comput. Chem. Eng. **34**, 1155–1169 (2010)
15. R. Batres, M. West, D. Leal, D. Price, Y. Naka, An upper ontology based on ISO 15926. Comput. Aided Chem. Eng. **20**, 1543–1548 (2005)
16. E. Muñoz, A. Espuña, L. Puigjaner, Towards an ontological infrastructure for chemical batch process management. Comput. Chem. Eng. **34**(5), 668–682 (2010)
17. M. Obitko, V. Mařík, in *Ontologies for Multi-Agent Systems in Manufacturing Domain*. Proceedings of the 13th international workshop on database and expert systems applications (DEXA '02), (2002)
18. X. Hong, S. Jiancheng, in *Multi-Agent Based Scheduling for Batch Process*. Proceedings of the 8th international conference on electronic measurement and instruments (ICEMI '07), (2007), pp. 2-464–2-467
19. I. Seilonen, K. Koskinen, T. Pirttioja, P. Appelqvist, A. Halme, in *Reactive and Deliberative Control and Cooperation in Multi-Agent System Based Process Automation*. Proceedings of the IEEE international symposium on computational intelligence in robotics and automation (CIRA '05) (2005), pp. 469–474
20. Z. Li, M. Ierapetritou, Process scheduling under uncertainty: review and challenges. Comput. Chem. Eng. **32**(4–5), 715–727 (2008)
21. A. Bonfill, A. Espuña, L. Puigjaner, Proactive approach to address the uncertainty in short-term scheduling. Comput. Chem. Eng. **32**, 1689–1706 (2008)
22. W. Lepuschitz, A. Zoitl, M. Vallée, M. Merdan, Towards self reconfiguration of manufacturing systems using automation agents. IEEE Trans. Syst. Man Cybern. Part C Appl. Rev. **41**(1), 52–69 (2011)
23. A. Zoitl, Real-Time Execution for IEC 61499, ISA (2008)
24. M. Merdan, M. Vallée, W. Lepuschitz, A. Zoitl, Monitoring and diagnostics of industrial systems using automation agents. Int. J. Prod. Res. **49**(5), 1497–1509 (2011)
25. M. Vallée, M. Merdan, W. Lepuschitz, G. Koppensteiner, in *Decentralized Reconfiguration of a Flexible Transportation System*, Industrial Informatics, IEEE Trans. (2011 in press)
26. W. Shen, Q. Hao, H.J. Yoon, D.H. Norrie, Applications of agent-based systems in intelligent manufacturing: an updated review. Adv. Eng. Inform. **20**, 415–431 (2006)
27. D. Ouelhadj, S. Petrovic, A survey of dynamic scheduling in manufacturing systems. J. Sched. **12**, 417–431 (2009)
28. R. Smith, The contract net protocol: high-level communication and control in a distributed problem solver (1981)
29. S. Russell, P. Norvig, *Artificial Intelligence: A Modern Approach* (Prentice Hall/Pearson Education, Englewood Cliffs, 2003)
30. E.W. Dijkstra, A note on two problems in connexion with graphs. Numer. Math. **1**, 269271 (1959)

Model Based Path Planning Module

Sobers L. X. Francis, Sreenatha G. Anavatti and Matthew Garratt

Abstract Autonomous Ground Vehicles (AGVs) have been the topical area in the recent past in which the researchers are working to enhance the intelligence of the vehicles for different applications. This chapter provides the importance of kinematic and dynamic constraints in a path planning module for AGVs. The kinematic and dynamic models are taken into account to model the actual mobile robot including the torques provided by the motors and the restriction to the robot mobility induced by the constraints. Result discusses the importance of the model in the path planning module. The model of the mobile robot is compared with the behaviour of a Pioneer 3DX robot.

1 Introduction

Path planning has been considered as the main challenge in the field of autonomous mobile systems. Several researchers are currently looking at this problem in the presence of uncertainties like dynamic obstacles, partially known

"Based on Dynamic Model of Autonomous Ground Vehicle for the Path Planning Module, by Sobers L.X. Francis et al., which appeared in the Proceedings of the 5th International Conference on Automation, Robotics and Applications (ICARA 2011). © 2011 IEEE."

S. L. X. Francis (✉) · S. G. Anavatti · M. Garratt
SEIT, UNSW@ADFA, Canberra 2600, Australia
e-mail: S.francis@adfa.edu.au

S. G. Anavatti
e-mail: S.anavatti@adfa.edu.au

M. Garratt
e-mail: m.garratt@adfa.edu.au

G. Sen Gupta et al. (eds.), *Recent Advances in Robotics and Automation*,
Studies in Computational Intelligence 480, DOI: 10.1007/978-3-642-37387-9_6,

environments, etc. A number of algorithms for path planning have been developed over the years. One of the reasons for the continued work in this field can be summarised in the words of Miller [1], as

> The simplifying assumptions made about the modules associated with the module actually being designed are too simple. Implementing a module in a real robot system makes it difficult to make oversimplifications.

In general, it is difficult to come up with a very good model for a mobile robot without any assumptions. In order to develop an effective path planning technique, knowledge about the robot dynamics is very important. Mathematical path planning techniques assume that the robot is a point mass and can move in any direction and able to reach the goal flawlessly [2]. Most researchers use a simplified kinematic constraint to plan a feasible path. For real world scenarios, these assumptions are often invalid.

Many path planning techniques have been discussed in the literature. However, the resulting paths by the search algorithms do not tend to be smooth and hence do not include kinematic constraints. Some well-known methods [3] have been used to generate the smooth path but they have a drawback of discontinuity of curvature at the joint nodes connecting the lines and arc [4]. Clothoid curve [5], Bezier curve [6] are used by the researchers to model the smooth curves to get the continuous curve, which meets the curvature constraint.

The planned path for a real robot should satisfy the practical conditions imposed by: (a) Speed and acceleration limitation which is caused by the actuator (DC motor). (b) Dynamics of the vehicle in reacting to the actuator forces and (c) Position and velocity [7]. The kinematic model of a mobile robot is essentially the description of the admissible instantaneous motions in respect of the constraints. On the other hand, the dynamic model accounts for the reaction forces and describes the relationship between the motions and the generalized forces acting on the robot. The trajectories have to be generated in respect of the kinematic constraints imposed by the wheels. For the control of rotary motion, a common actuator is the DC motor [8]. Such a device is constrained in input voltage and currents, which, without load, is equivalent to velocity and acceleration.

This chapter proposes a model which replicates the actual Pioneer robot's behaviour. First the collision free path consists of a series of coordinates, generated for the path planning module. These sequences of points are generated by suitable path planning technique from the literature. These set of points are formed as a continuous curve. To form a smooth movement, the curve fitting is used as a part of path planning module. The path planning algorithm takes into account the model of the robot. Results are discussed to show the variations in the paths due to the incorporation of the dynamics.

The chapter is organized as follows: Sect. 2 describes the grid based path planning technique and smoothing by cubic spline, Sect. 3 shows the modelling of Pioneer mobile robots, and Sect. 4 gives an overview on models implementation in MATLAB. Chapter ends with conclusion and future work on this topic.

2 Path Planning

2.1 Robot Position

The vector of joint variables, q provides a representation of a configuration space $\wp = \Re^2$. The robot is denoted by the notation ρ and $\rho\ (q)$ represents the subset of the workspace that is occupied by the robot in the Cartesian space. The posture coordinates of the robot is completely specified by the three variables $q = x,\ y,\ \theta$, where x, y represent the position in the 2D space and θ represents the orientation with respect to the OX axis. The path planning problem is to find a path from an initial posture q_{start} to a particular target posture q_{goal} in the configuration space, such that the robot does not collide with any obstacle as it traverses [9]. In the path planning literature, a collision free path from start to final location is represented as a continuous map, $(\Im)$.

$$\Im : [start, goal] \Rightarrow \wp_{collisionfreepath}$$
$$\Im(start) = q_{start} : \Im(goal) = q_{goal}$$

The collision free path is determined by the D^* lite search algorithm [10]. After the path is calculated, cubic spline interpolations are used to create a smooth path using kinematic constraints.

2.2 Cubic Splines

Considering the points generated by the path planning algorithms in x -direction, q_{x}, the ith segment can be constructed [11] as in (1).

$$\mathrm{X_i(t) = a_i(t-1)^2(2t+1) + b_i t^2(3-2t) + c_i t(t-1)^2 - d_i t^2(1-t)} \qquad (1)$$

where, X_i is the parameter connecting the points q_i and q_{i+1} in the x-direction, $0 \le \mathrm{t} \le 1$ is the relative time of motion in each segment $i = 1, 2, \ldots, N-1$ and N is the total number of points. The smoothened continuous curves are generated through a series of unique cubic polynomials which are fitted between each of these points, with the stipulation that the curve obtained be continuous and appear smooth.

3 Dynamic Modelling

For a mobile robot, the wheel configuration may restrict the controllable degree of freedom (DOF) to two: typically the robot can turn on the spot or move forward but it cannot directly move side-wards. The vehicle is said to be under actuated or

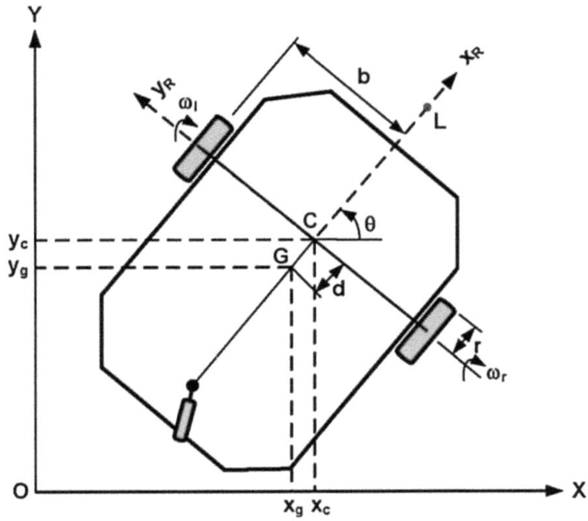

Fig. 1 Structure of mobile robot model

non-holonomic. In all physical systems actuators are subject to saturation as they can only deliver a certain amount of force. The inputs of the kinematic models are analogous to velocities while the inputs of the dynamic models are either torques or acceleration [7]. The structure of the robot model is given in Fig. 1.

Here, r is the radius of the wheel, d is the distance between the centre of mass and the centre of wheel axis, b is the distance from the centre of the robot frame to driving wheel. The x_c, y_c are the Cartesian coordinates of the point C, which is the origin of mobile robot body axis and located at the intersection of the X_R-axis and Y_R-axis. The point L is the look-ahead point located on the X_R-axis of the mobile robot.

3.1 Kinematic Model

Let x_g, y_g be the coordinates of the centre of the mass of the robot platform and θ be the angle between the heading direction and the OX-axis. The wheel angular velocities are dependant on linear and angular velocities of the robot as in (2). The notation v represents the linear velocity and $\omega = \dot{\theta}$ denotes the angular velocity of the robot [12].

$$\begin{bmatrix} v \\ w \end{bmatrix} = \begin{bmatrix} \frac{r}{2} & \frac{r}{2} \\ \frac{r}{2b} & -\frac{r}{2b} \end{bmatrix} \begin{bmatrix} w_r \\ w_l \end{bmatrix} \tag{2}$$

The kinematic model of the mobile robot [13] with two conventional fixed wheels on the same axis and one rear conventional off-centered orientable wheel (caster wheel) can be written as in (3).

$$\begin{bmatrix} \dot{x}_g \\ \dot{y}_g \\ \dot{\theta} \end{bmatrix} = \begin{bmatrix} \frac{r(b\cos\theta - d\sin\theta)}{2b} & \frac{r(b\cos\theta + d\sin\theta)}{2b} \\ \frac{r(d\cos\theta + b\sin\theta)}{2b} & \frac{r(-d\cos\theta + b\sin\theta)}{2b} \\ \frac{r}{2b} & -\frac{r}{2b} \end{bmatrix} \begin{bmatrix} w_r \\ w_l \end{bmatrix} \tag{3}$$

3.2 Dynamic Model

The dynamic equations of the motion of a mobile robot can be deduced by Lagrangian formulation [11] as in (4).

$$M(q)\ddot{q} + C(q,\dot{q}) + F(q,\dot{q}) = B(q)\tau + A^T(q)\lambda \tag{4}$$

where, $M(q)$ is a positive symmetric inertia matrix. $C(q,\dot{q})$ represents the Centrifugal and Coriolis force and torques. $B(q)$ is the matrix of motor torques applied to the robot. $A(q)$ is the constraints matrix associated with the non-holonomic system. λ is associated with the independent kinematic constraints.

Due to the constraints, there exists a Jacobian matrix $S(q)$ that satisfies the following relation (5).

$$\dot{q} = S(q)\eta \tag{5}$$

where, $\eta = [v, \omega]^{\tau}$.

The constraint can be written in the Pfaffian form [14] as in (6).

$$S(q)^T A(q)^T = 0 \tag{6}$$

Practically, the mobile robot is restricted to its maximum steering angle and maximum speed. The translation and rotational speeds must be within their limits. The turning rate is limited by the speed of the mobile robot. Due to the constraints imposed by the dynamics of the vehicle, it is necessary to pre-analyse the path that is planned, to ensure it is traversable by the mobile robot dynamics.

3.3 Actuator Model

The mobile robot is driven by two differential DC motors. The mathematical model characterises the behaviour of DC motor controlled by the armature voltage. The dynamic model relating the voltage v to the torque τ is obtained as in (7). The load torque is obtained after the gear box which has a gears reduction ratio.

The speed of the wheel $\dot{\Phi}$ is related with the rotational force, torque applied to the wheels.

$$\tau = g_r \mu \left\{ K_T \left(\frac{V_a - K_E g_r \dot{\Phi}_i}{R_a} \right) - \tau_{fr} \right\} \tag{7}$$

Most mobile robots have low level PID velocity controllers to track the input reference velocities and do not allow the motor voltage to be driven directly [8]. The PD velocity controller used in this model to track the reference velocities is described by the following Eq. (8).

$$\begin{bmatrix} v_v \\ v_\omega \end{bmatrix} = \begin{bmatrix} k_{p_1}(v_{ref} - v) - k_{D_1}(\dot{v}_{ref} - \dot{v}) \\ k_{p_2}(\omega_{ref} - \omega) - k_{D_2}(\dot{\omega}_{ref} - \dot{\omega}) \end{bmatrix} \tag{8}$$

3.4 Pioneer Mobile Robot

The Pioneer 3*DX* has high-speed, high torque, reversible-DC motors, each equipped with encoders for position and speed and advanced dead-reckoning. The mobile robot parameters [15] can be obtained from the mobile robot manufacturer data as shown in Table 1. Some of the parameters are estimated.

The Pioneer servers use a common PID system with the wheel-encoder feedback to adjust a Pulse-Width-Modulated (PWM) signal at the motor drives to control the power to the motors. The motor-duty cycle is 50 ms; pulse-width is proportional 0–500 for 0–100 % of the duty cycle. Number of encoder ticks per revolution equals 500. The swing radius of the mobile robot is 0.26 m.

Here, M_c is the mass of the robot chassis. M_w is the mass of a single wheel and its rotor. I_c is the moment of inertia of the robot chassis about the vertical axis. I_w is the moment of inertia of the wheel about the wheel diameter. I_m is the moment of inertia of the wheel about the wheel axis.

Table 1 Pioneer robot parameters

Parameter	Value	Units
Mc	15.764	kg
Mw	0.35	kg
r	0.0976	m
b	0.1689	m
d	0.07	m
Ic	15.625	kg m^2
Iw	0.0047	kg m^2
Im	0.4879	kg m^2

4 Experiment

The physical model of the Pioneer mobile robot is incorporated in the simulation model as seen in Fig. 2, which is modelled from the actual motor parameters and mobile robot parameters. The inputs of the model are the desired position (q).

The model is assigned to the following constraints similar to the Pioneer 3DX mobile robot.

- Max. Linear velocity = 400 mm/s
- Max. Angular velocity = 300 mm/s
- Max. Acceleration = 300 mm/s/s
- Max. Deceleration = 300 mm/s/s.

The PD controller which is used in the model to track the velocity, has the gain value of

$$k_{p_1} = k_{p_2} = 10; \; k_{D_1} = k_{D_2} = 10$$

For the experiment, the heuristic based search algorithm, D^* lite algorithm is used to get the collision free path from the start to goal position. These set of points which represents the x, y position in the continuous map ($\Im$), are smoothed to create a continuous path using cubic spline interpolation.

$$\Im : [start, goal] \Rightarrow \wp_{collisionfreepath}$$

The model is tested in three different scenarios as shown in Figs. 3 and 6. In the Figure, the blue circle represents the start position q_{start}, the green rhombus represents the target goal position q_{goal}, the black square blocks and circles denote the obstacles where the robot can't traverse, the purple circles are the set of points which is obtained from the D^* lite algorithm. Cubic splines are used to create a smooth path from the points obtained above.

During path planning, generally, the robot is considered to be a point mass as shown. Due to the physical dimensions of the robot, the posture and the limitations on the achievable velocities/accelerations, the trajectory will get modified as shown in Fig. 4.

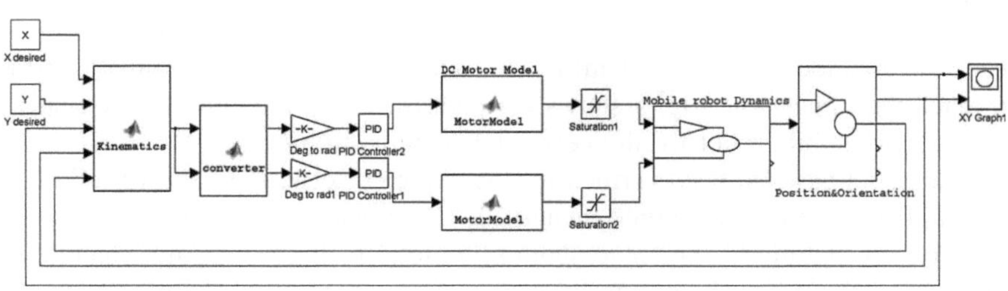

Fig. 2 Robot dynamic model in simulink

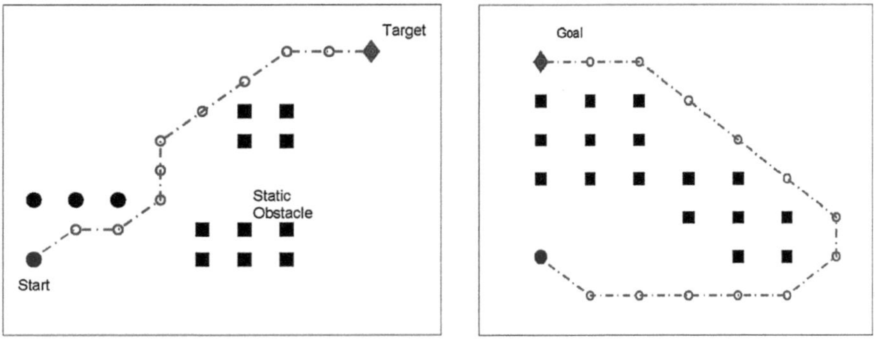

Fig. 3 Collision free paths in scenarios I and II

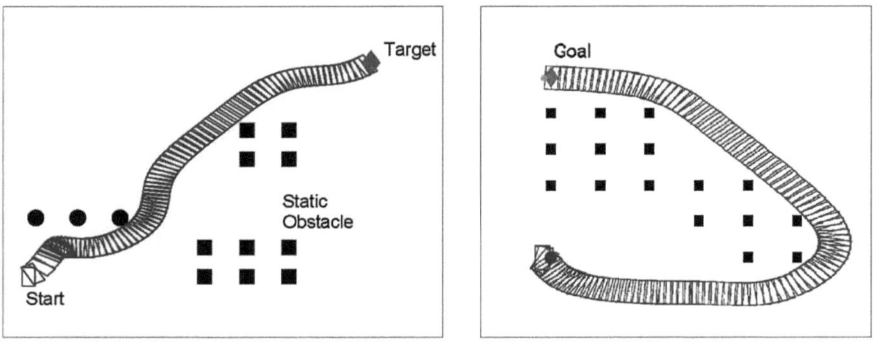

Fig. 4 Path traced by the mobile robot model

Since the mobile robot has its own dynamic constraints, the path which is traced by the robot, is shown in Fig. 4. This indicates the changes in the allowable path by taking the dynamic constraints into account. The simulated model behaviour is tested with the actual Pioneer robot for different conditions. The model exactly replicates the mobile robot path with the same time.

Initially the mobile robot is located along the *x-axis* at (0, 0) in the Scenario 1 as in Fig. 3. Due to the dynamics of the AGV, it takes a curvy path to reach its next point.

The time taken by the Pioneer 3DX mobile robot from start location (0, 0) to reach the goal location (2 and, 1 m) is 7.0 s and is the same as the simulation time to reach the target. The Fig. 5 shows the path traced by the point L, look-ahead point of the mobile robot for the generated smooth curves.

The model has also been verified for the cluttered scenario as shown in Fig. 6. The collision free path is initially planned by D * lite algorithm and then these set of points are optimised further to shorten the total path but does not change the topology of the path. The optimised path is represented by blue lines. Thus, the

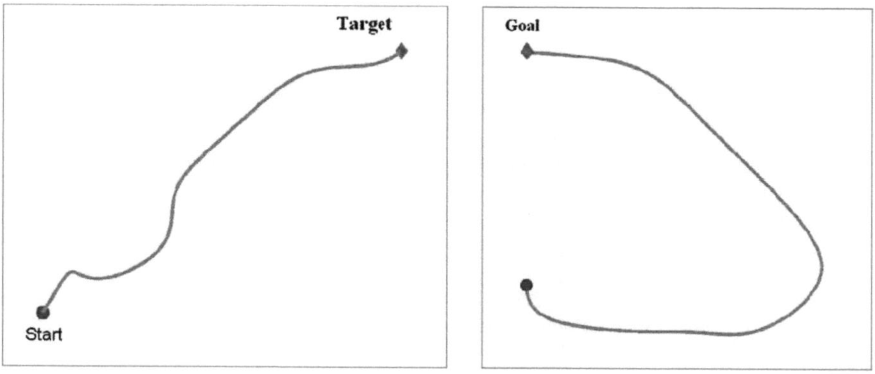

Fig. 5 Path traced by the look-ahead point L

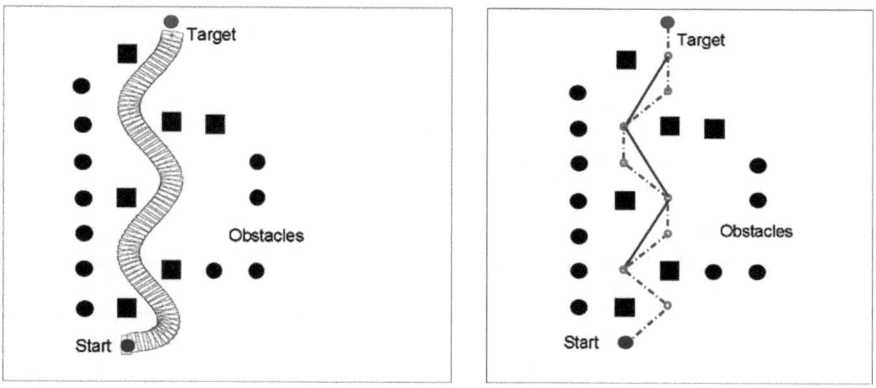

Fig. 6 Senario III

heading changes of their paths are not constrained to specific angles [16]. Then the optimised path is smoothened to create a continuous path using cubic spline interpolation.

5 Conclusion and Future Work

In this chapter, the dynamic model of the mobile robot has been simulated and the behaviour is compared with the Pioneer 3DX robot. Then the proposed model has been used for the path planning module. The heuristic based search algorithm, D^* lite algorithm is accomplished by using a cubic spline interpolation to create a smooth path using the kinematic constraints. Our future work will focus on the real time dynamic path planning algorithm in cluttered environment.

References

1. D.P. Miller, in *Robot Navigation*. Proceedings of the 11th International Joint Conferences on Artificial Intelligence, AAAI, IJCAI, (Detroit, MI 1989), pp. 1672–1674
2. E. Ivanjko, T. Petrinic, I. Petrovic, in *Modelling of Mobile Robot Dynamics*. Abstracts of the 7th EUROSIM Congress on Modelling and Simulation, Prague, (Czech Republic, 2010)
3. L.E. Dubins, On curves of minimal length with a constraint on average curvature, and with prescribed initial and terminal positions and tangents. Am. J. Math. **79**(3), 497–516 (1957)
4. F. Lamiraux, J. Lammond, Smooth motion planning for car-like vehicles. IEEE Trans. Rob. Autom. **17**(4), 498–501 (2001)
5. A.D. Luca, G. Oriolo, L. Paone, P.R. Giordano, M Vendittelli, in *Visual-Based Planning and Control for Nonholonomic Mobile Robots*. Abstract of the 10th Mediterranean Conference on Control and Automation, (Lisbon, Portugal, 2002)
6. J. Wung Choi, R. Curry, G. Elkaim, in *Curvature-Continuous Trajectory Generation with Corridor Constraint for Autonomous Ground Vehicles*. Abstract of the 49th IEEE Conference on Decision and Control (CDC), pp. 7166–7171, 15–17 Dec 2010. doi:10.1109/CDC.2010.5718154
7. G. Campion, G. Bastin, B. Dandrea-Novel, Structural properties and classification of kinematic and dynamic models of wheeled mobile robots. IEEE J. RA **12**(1), 47–62 (1996). doi:10.1109/70.481750
8. C. de La Cruz, R. Carelli, in *Dynamic Modeling and Centralized Formation Control of Mobile Robots*. IECON 2006: 32nd Annual Conference on IEEE Industrial Electronics, (Paris, 2006) pp. 3880–3885. doi:10.1109/IECON.2006.347299
9. B. Siciliano, O. Khatib (eds.), *Springer Handbook of Robotics* (Springer, Berlin, Heidelberg, 2008)
10. D. Ferguson, M. Likhachev, A. Stentz, in *A Guide to Heuristic Based Path Planning*. Abstract of the Workshop on Planning under Uncertainty for Autonomous Systems at the International Conference on Automated Planning and Scheduling (ICAPS), (2005)
11. I. Waheed, R. Fotouhi, Trajectory and temporal planning of a wheeled mobile robot on an uneven surface. J. Robot. **27**(4), 481–498 (2009). doi:10.1017/S0263574708004876
12. A. Bara, S. Dale, in *Dynamic Modelling and Stabilization of Wheeled Mobile Robot*, eds. by P. Leonid et al CONTROL'09: 5th WSEAS International Conference on Dynamical Systems and Control, Spain, 2009 (World Scientific and Engineering Academy and Society (WSEAS), Stevens Point, Wisconsin, USA, 2009) pp. 87–92
13. N. Sidek, N. Sarkar, in *Dynamic Modelling and Control of Nonholonomic Mobile Robot with Lateral Slip*. Abstract of the Third International Conference on Systems, (Washington, DC, USA, 2008) pp. 35–40
14. F. Aghili, *Modeling and Control of Mechanical Systems in Terms of Quasi-Velocities*, ed. by H. Abdellatif Robotics 2010 Current and Future Challenges, (InTech, 2010) pp. 45–62
15. Pioneer 3DX Operational Manual: Mobile Robot Inc. Tech Rep. (2007), http://www.mobile-robots.com/Mobile_Robots.aspx
16. A. Nash, K. Daniel, S. Koenig, A. Felner, in *Theta*: Any-Angle Path Planning on Grids*, ed. by A. Cohn AAAI'07: 22nd National Conference on Artificial Intelligence, vol. 2 (AAAI Press, Canada, 2007) pp. 1177–1183

Autonomous Navigation and Mapping with CoreSLAM

Oussama El Hamzaoui, Jorge Corsino Espino and Bruno Steux

Abstract In this chapter we describe a complete solution for autonomous navigation and exploration in indoor environment. We introduce some new concepts to achieve several tasks giving a robotic platform the ability to explore autonomously and safely its environment. We use a Simultaneous Localization and Mapping (SLAM) algorithm based on IML concept with low drift. We also introduce a new navigation method based on potential fields, with proven convergence. These methods and algorithms have been tested using a real mobile robot. This system proved its capabilities during the CAROTTE Robotics Contest. Our team Core-Bots won the first and second edition of this contest using these methods.

1 Introduction

In order to improve the capabilities of autonomous robots and to be able to achieve localization, mapping and terrain analysis, the CoreBots Team developed a robot to participate in the CAROTTE contest. This contest was initiated by the French National Research Agency (ANR) and the French Armament Procurement Agency (DGA) [1]. It lasted 3 years (2012 was the last year), increasing in difficulty along the time. It took place in an arena of 120 m^2. The environment contains several rooms, with variable grounds, walls (wood, fitted carpet, tiling, grid, sand, stones, etc.) and others items.

O. El Hamzaoui (✉) · J. C. Espino · B. Steux
Centre de Robotique CAOR, Mines-ParisTech, Paris, France
e-mail: oussama.elhamzaoui@gmail.com

J. C. Espino
e-mail: jorge.corsino@gmail.com

B. Steux
e-mail: bruno.steux@mines-paristech.fr

G. Sen Gupta et al. (eds.), *Recent Advances in Robotics and Automation*, 91
Studies in Computational Intelligence 480, DOI: 10.1007/978-3-642-37387-9_7,

Each team had to build an autonomous robot, which should be able to navigate autonomously in an indoor environment and recognize objects in the rooms, in order to build a map with semantic information.

This chapter presents the methods and algorithms used in our platform CoreBot 1 to give it the ability to navigate autonomously in the environment. First, we give a brief description of the mechatronic conception of the robot, followed by the Simultaneous Localization and Mapping (SLAM) algorithm. Section 4 will focus on the methods used to control the robot. Finally, we present some results obtained during the contest.

2 Platform

The platform used for our experiments is the WiFiBot (see Fig. 1), a low-cost robot characterized by its great flexibility, allowing it to be used in multiple environments and situations [2]. Its small size and light weight make it easily transportable and suited to explore narrow places. Its mechanical design and four-wheel drive allow this robot to navigate over irregular surfaces and even small obstacles.

For this project, the robot includes an Ethernet camera, two infrared lasers, one located in the middle with a range of 30 m, and the other on top with 4 m range.

The software architecture of the robot is build upon several components designed around a middleware called *Cables*, which manages communication and data exchange between components.

3 Localization and Mapping

3.1 Incremental Maximum Likelihood for SLAM

To solve the SLAM problem, CoreSLAM is based on the Incremental Maximum Likelihood [3] technique. The idea behind IML is based on finding the best match between the current observation (data from the laser) and the current map (combining the knowledge of the environment) at each step, and update the map using the result obtained. This is illustrated in Fig. 2.

The IML process is obviously divergent, since the localization step is used to update the map, and the updated map used to find the location of the robot. That is why the IML is rarely used in SLAM, because of its low *simultaneousness*. Researchers prefer estimation techniques mixing the localization of the robot in the system state, such as Kalman Filter [4, 5] or Rao-Blackwellised Particle Filter SLAM [6, 7].

Fig. 1 Platform used: the WiFiBot

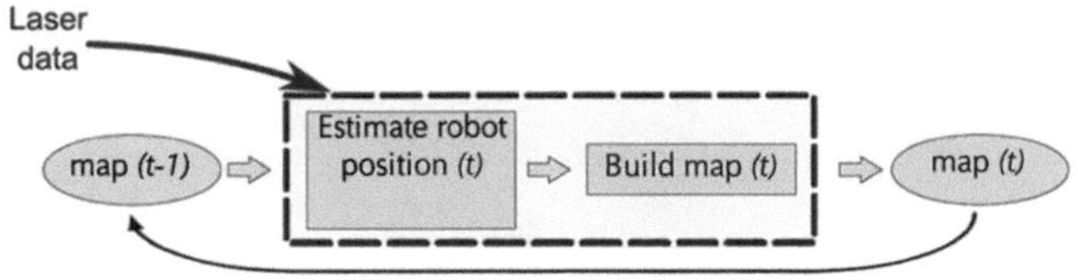

Fig. 2 General idea of IML in CoreSLAM

Our approach for solving the SLAM problem is mainly based on limiting the drift (or divergence) by taking advantage of the simplicity of the original concept of IML. The matching must be based on an optimization algorithm looking for the best position of the robot such that the observation (the last laser scan) matches with the current map. To be fast and accurate, this optimization algorithm must be based on a representation of the SLAM map that eases the search for a minimum distance between the laser scan data and the map. Our map is a hexagonal ternary map (obstacle—no obstacle—not explored) from which a map of distances to obstacles is calculated incrementally, overflowing around the obstacles on unexplored areas. More details on the map will be presented in Sect. 4.

3.2 Optimization Algorithm

During the scan matching [8], the optimization algorithm tries to find the best position for the robot sensor data. For this, it generates a set of hypotheses and assigns a score characterizing their quality. The higher the score, the lower the hypothesis is close to the actual position of the robot. The optimization algorithm used in CoreSLAM is based on a multi-scale search. It starts with a cube of $6 \times 6 \times 6$ hypotheses (in x, y, θ values of the robot position) around the assumed position of the robot. If the best hypothesis is on the edges of the cube, the cube is translated and scale unchanged. If the best hypothesis is found inside the cube, we move to the lower scale by generating a new $6 \times 6 \times 6$ cube centered around the best guess. Cubes explored are illustrated in Fig. 3, where there is a case of simple search.

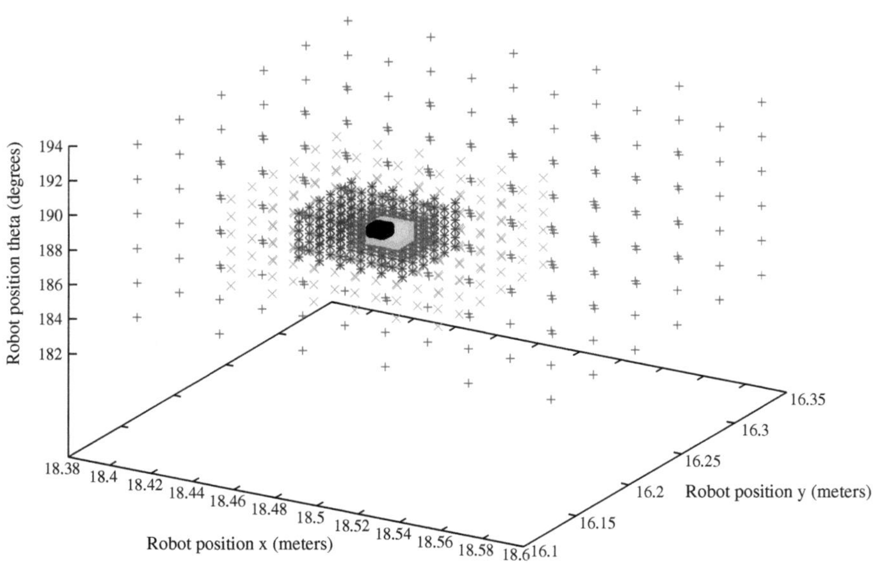

Fig. 3 Multi-scale search algorithm

We explore up to 7 cubes of $6 \times 6 \times 6$, which gives a localization resolution of the millimeter in position and of the 10th degree for orientation. The lower resolution of the cube (x, y) is $\frac{200}{6-1} = 40$ mm. This is divided by two at each iteration, if the local minimum is located inside the cube, resulting in a resolution of $\frac{40}{2^6} = 0.625$ mm. On the angular resolution, the algorithm starts with a resolution of $\frac{10}{6-1} = 2°$ to reach a possible resolution of $\frac{2}{2^6} = 0.03°$.

4 Driving

4.1 Distance Transform

The Distance Transform, also known as distance map or distance field, is a representation of a distance function to an object [9, 10], as shown in Fig. 4. The Distance Transform is an operator applied generally to binary images. The result of the transform is a grey level image where the intensities of points inside the foreground regions show the distance to the closest boundary from each point.

In our algorithm, we use the morphological distance proposed in [11]. This approach is based on the 6-connexity with the particularity that the shape changes between odd and even pixel columns as shown in Fig. 5. The distance maps are obtained by applying the filter mask to the whole image.

4.2 Path Planning

The most important issue in the design of a robot is the navigation process. Therefore, the representation of the space plays an important role in the navigation of an intelligent system.

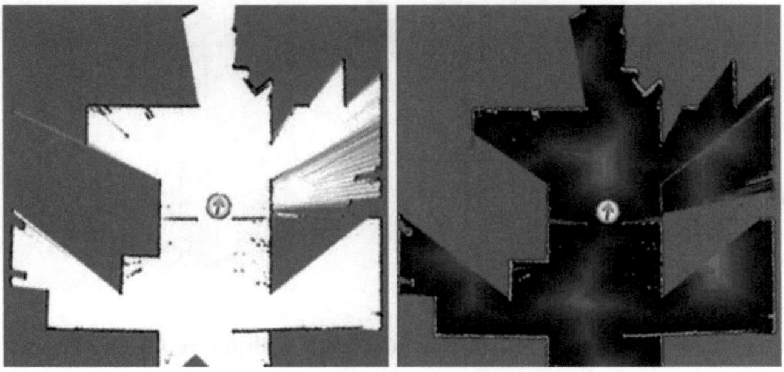

Fig. 4 The SLAM map (*left*) and the correspondent distance map (*right*). *The grey level changes* indicate the distance from obstacles. *The closer* is the pixel from the obstacle; *the darker* is its color

Fig. 5 Shape of 6-connexity for even and odd pixels

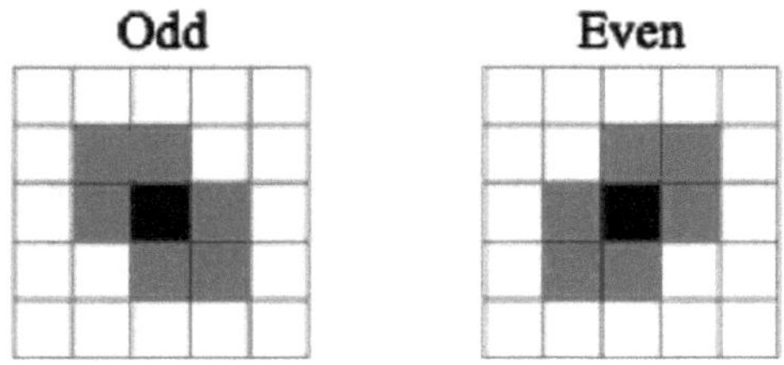

To plan the trajectory, we developed a slope that links the target and the robot position. The slope forces the robot to move on and avoid obstacles until it reaches the target. The user defines the safety distance D to obstacles. With this, a new slope is created around the obstacles. This repulsion slope is added to the borders of the path slope. Distance D is chosen so that the robot never hits an obstacle, this mainly depends on the robot model. D also limits the maw clearance distance between obstacles $(2xD)$. For instance, if the robot should be able to go through 80 cm wide doors, D must be set lower than 40 cm (otherwise, there is no path slope remaining).

This path planning of the mobile robot is similar to the traditional potential force field [12]. However, the slope force is different from the potential force in some aspects. In this approach, we do not have local minima and the repulsion slope from the obstacles and the width are always constant. In Fig. 6, we can see the shape of the slope in a 3D view and in the trajectory map.

To summarize, a continuous slope is created between the goal and the current position, using a propagation system until it reaches the robot, verifying that the slope is wide enough to let the platform safely navigate across it. To ensure a safe navigation, it is necessary to separate the slope from the obstacles. Therefore, a new slope is added at a distance D, determined empirically.

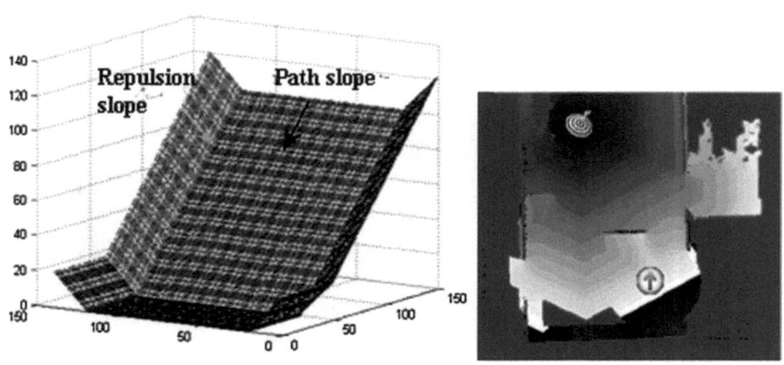

Fig. 6 Trajectory map. (*Left*) 3D view of the slope path with the repulsion slopes. (*Right*) View of the trajectory map

5 Control

We consider that the WiFiBot is a mobile robot made up of a rigid body and non-deformable wheels. It is assumed that the vehicle moves without sliding on a plane, i.e., there is a pure rolling contact between the wheels and the ground. The kinetic model is illustrated in Fig. 7.

The discrete-time model for the robot motion is defined as:

$$\begin{aligned} x(k+1) &= x(k) + v(k) \cdot \cos\theta(k)T \\ y(k+1) &= y(k) + v(k) \cdot \sin\theta(k)T \\ \theta(k+1) &= \theta(k) + w(k)T \end{aligned} \tag{1}$$

Given these characteristics, we control the robot using its orientation angle θ. This angle is always given by the SLAM algorithm. The potential field angle γ acts as the control signal to the robot as we can see in Eq. (2).

$$\dot{\theta}(t) = k\left(\gamma - \tilde{\theta}(t)\right) \tag{2}$$

A robot is a "system" that responds to sensor input by running a program without human intervention. This program has a delay. Therefore, the control system is considered as a corrector followed by a low pass filter or a delay.

In a closed loop, the impulse response of this control system is:

$$H_{cl}(s) = \frac{k}{\tau s^2 + s + k} \tag{3}$$

In the article [13], it has been demonstrated that the system control is stable. To prove that the system is a free obstacle path it is necessary to know the evolution of $\theta(t)$ to trace the trajectory. Using the inverse Laplace Transform we obtain:

$$\begin{aligned} \theta(t) &= \frac{\gamma}{2}\left(\frac{2 - e^{\frac{-t(1+\sqrt{\Delta})}{2\tau}}\left(-1 + \sqrt{\Delta} + e^{\frac{t\sqrt{\Delta}}{\tau}}\left(1 + \sqrt{\Delta}\right)\right)}{\sqrt{\Delta}}\right) \\ \Delta &= 1 - 4k\tau \end{aligned} \tag{4}$$

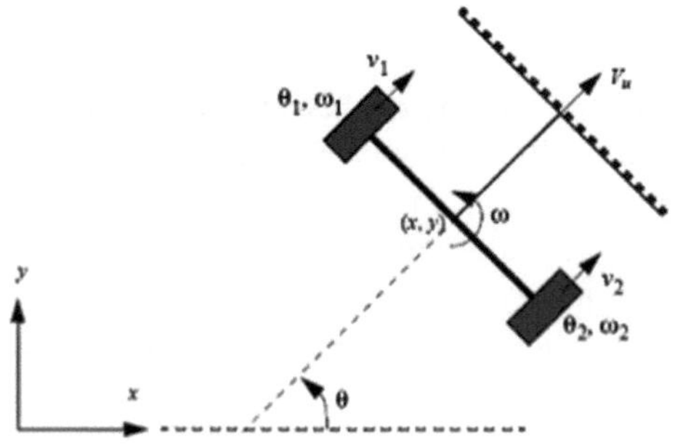

Fig. 7 Kinetic model of the platform

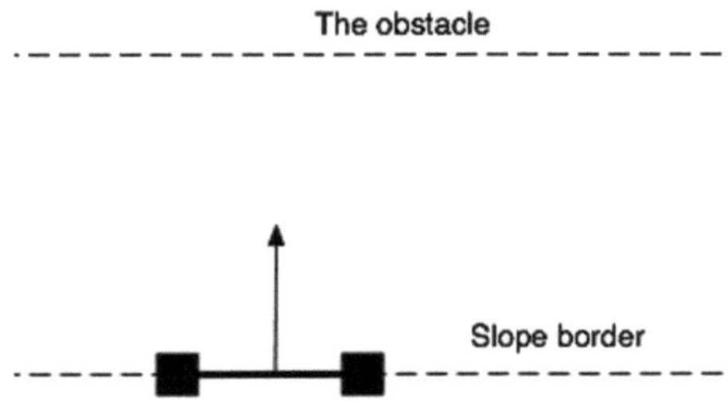

Fig. 8 Robot facing an obstacle

As we can see in Eq. (4), $\theta(t)$ depends only on k, τ and γ. Note that k is the correction factor of the command, τ is the delay of reaction of the robot and γ is the command angle.

In our approach, the worst situation during the navigation is given when the robot faces an obstacle. Then the gradient force is opposite to its orientation. The platform needs to turn around and avoid the obstacle. The gradient is opposite due to the repulsion slope added around the obstacles at a distance D. This situation is explained in Fig. 8.

For tests, we have fixed the potential field γ to π. This means that the robot needs to make a complete turn of 180°. The processing delay is assumed to be 10 ms. Therefore, the only parameter that we can modify is k.

We calculate the trajectories of the robot for different values of k: a high value, which represents a reactive control; a medium value and a low value for a softer control. For simplicity, we use a constant speed of 1 m/s and an obstacle is situated at a distance of 50 cm. Substituting these values in Eqs. (1) and (4), the trajectory is obtained.

We present here the result obtained for $k \approx \frac{1}{4\tau}$. The trajectory of the robot is explained in Fig. 9, where x and y are in meters. As illustrated, when the robot is facing the obstacle, it starts its trajectory from the origin turning right until it reaches the final orientation.

In this case, considering the origin as the initial robot position, the maximum distance covered in the y-axis is 23.04 cm. The resulting margin to the obstacle is 27 cm. It not possible to hit the obstacle.

6 Results

To evaluate the quality of mapping in CoreSLAM, we used data recorded during the CAROTTE contest. Arenas used to evaluate robotic systems are built using walls of 1 m. Figure 10 shows an example of arena.

The fixed size of walls used gives us a ground truth that helped us to evaluate the quality of mapping with CoreSLAM. Figure 11 shows the basic idea of this evaluation. In this map, each pixel represents 1 cm in reality. For both distances

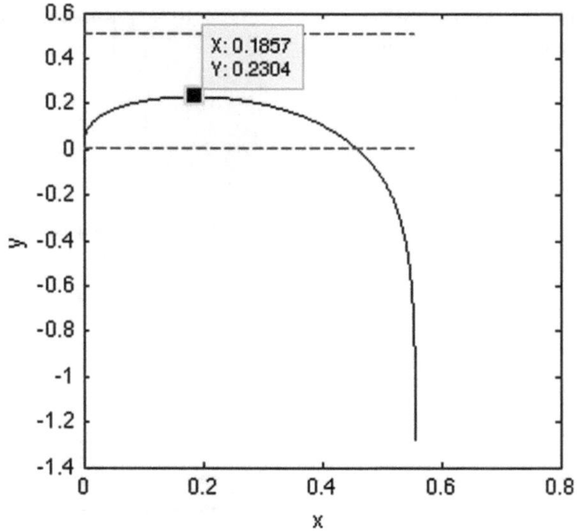

Fig. 9 Trajectory avoiding the obstacle for $k \approx \frac{1}{4\tau}$

Fig. 10 Example of an arena used in the CAROTTE Contest. Two measures are shown

marked, the mapping error is abs $abs(3 * 100 - 303) + abs(5 * 100 - 498) =$ 5 cm for 800 cm.

Applying this method to a set of 20 data recorded during CAROTTE, we obtained an average error of 1.15 %.

You can see results of our control method on YouTube videos indicated in the conclusion.

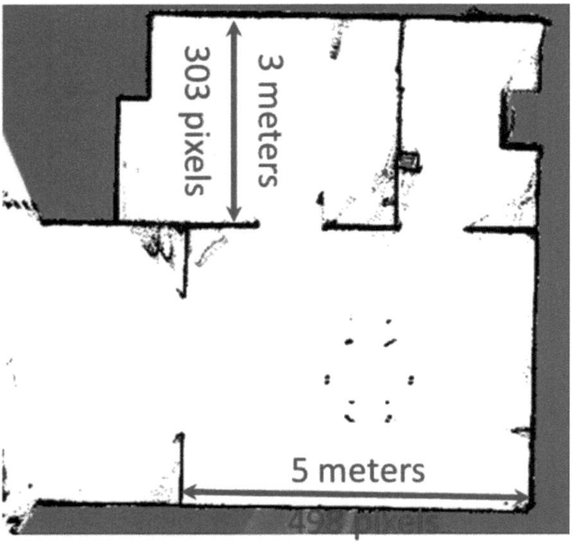

Fig. 11 Measuring the quality of mapping in CoreSLAM using the map of an arena of the CAROTTE Contest

7 Conclusions

In this chapter, we presented a complete system providing mapping and automated exploration of an unknown environment. We presented the two essential blocks that perform these tasks:

- The SLAM: to achieve mapping and automatic localization simultaneously.
- Control: to give the robot ability to navigate safely in an environment with several obstacles.

With our system, we won the first two editions of the CAROTTE Contest (2010 and 2011), thanks to the quality of its results, its speed and its performance.

Some examples of our navigation system are visible in:

- http://www.youtube.com/watch?v=4CWh4NmT2Mw
- http://www.youtube.com/watch?v=rpL47syUT3g
- http://www.youtube.com/watch?v=6CFv44TYTxQ

References

1. CAROTTE Contest Website. http://www.defi-carotte.fr
2. WifiBot Website. http://www.wifibot.com
3. G. Dissanayake, H. Durrant-Whyte, T. Bailey, in *A Computationally Efficient Solution to the Simultaneous Localization and Map Building (SLAM) Problem*. Working notes ICRA'2000 workshop W4: Mobile robot navigation and mapping, (2000)

4. R. Smith, P. Cheeseman, On the representation and estimation of spatial uncertainty. Int. J. Robot. Res. 56–68 (1986)
5. R. Smith, M. Self, P. Cheeseman, Estimating uncertain spatial relationships in robotics. Auton. Robot Veh. 167–193 (1990)
6. D. Hahnel, D. Fox, W. Burgard, S. Thrun, in *A Highly Efficient FastSLAM Algorithm for Generating Cyclic Maps of Large-Scale Environments from Raw Laser Range Measurements*, Proceedings of the international conference on intelligent robots and systems (2003)
7. A. Eliazar, R. Parr, in *DP-SLAM: Fast, Robust Simultaneous Localization and Mapping without Predetermined Landmarks*, proceedings of the international joint conference on artificial intelligence (2003)
8. D. Hahnel, D. Schulz, W. Burgard, in *Map Building with Mobile Robots in Populated Environments*, Proceedings of the international conference on intelligent robots and systems (IROS) (2002)
9. G. Borgefors, Distance transformations in digital images. Comput. Vis. Graph. Image Process. **34**(3), 344–371 (1986)
10. A. Rosenfeld, J. Pfaltz, Distance functions on digital pictures. Pattern Recogn. **1**(1), 33–61 (1968)
11. G. Matheron, Filters and lattices. Image Anal. Math. Morphol. **2**(3), 115–140 (1988)
12. J. Cheng, S. Foo, Dynamic directional gradient vector flow for snakes. IEEE Trans. Image Process. **15**(6), 1563–1571 (2006)
13. J.C. Espino, B. Steux,O. El Hamzaoui, in *Safe Navigating System for Indoor Environments*, International conference on automation, robotics and applications, pp. 419–423, (2011)

A 3D Path Planning Approach Extended by Bifurcation Theory for Formation Flights

Christoph Rasche, Claudius Stern, Lisa Kleinjohann and Bernd Kleinjohann

Abstract After big disasters, search and rescue missions take place. During these missions it is essential to obtain an overview of the overall situation to arrange efficient rescue tasks as soon as possible. Afterwards, this information has to be continuously updated during the complete mission in order to quickly respond to changing conditions. The use of unmanned aerial vehicles (UAVs) is a viable choice to obtain such an overview in a fast and efficient way. Using multiple UAVs, the problem of coordination has to be solved to decrease the time needed to explore an area. We present an approach for the coordination of UAVs by setting up formation patterns using bifurcation theory. We combine this theory with a potential field approach for exploration, based on harmonic functions.

Based on A Distributed Multi-UAV Path Planning Approach for 3D Environments [1], by Christoph Rasche, Claudius Stern, Lisa Kleinjohann and Bernd Kleinjohann which appeared in the Proceedings of the 5th International Conference on Automation, Robotics and Applications (ICARA 2011), © 2011 IEEE, we present an extension for formation establishment.

C. Rasche (✉) · C. Stern · L. Kleinjohann · B. Kleinjohann
Electrical Engineering and Mathematics, Department of Computer Science, C-LAB, University of Paderborn, Paderborn, Germany
e-mail: crasche@c-lab.de

C. Stern
e-mail: claudis@c-lab.de

L. Kleinjohann
e-mail: lisa@c-lab.de

B. Kleinjohann
e-mail: bernd@c-lab.de

G. Sen Gupta et al. (eds.), *Recent Advances in Robotics and Automation*, Studies in Computational Intelligence 480, DOI: 10.1007/978-3-642-37387-9_8,

1 Introduction

In recent years, the use of UAVs to solve exploration tasks has become more and more a viable choice. This led to the problem of how to coordinate multiple UAVs. One possible solution is to enable the UAVs to establish formation patterns which are maintained during their autonomous movements. Enabling UAVs to establish and fly in formations with precisely defined geometries has many advantages, like energy saving drag is taken into account. Another point is the fast exploration of given subareas where important information might be available. Using multiple UAVs in a well defined formation pattern will solve this task much faster than using single UAVs.

The challenge is to design a formation approach which is robust, computationally simple and able to react to external influences, e.g., to avoid obstacles. The approach should also be able to explore dynamic and complex areas, including buildings, tunnels, etc. Formations should be established if they are assumed to be beneficial. UAVs should be able to join and leave formations at any time and if no further benefit is achievable the formations should be disintegrated.

In this work we will introduce a method to set up different formation patterns which can change dynamically. The UAVs will not be just following some given leader but also setting up their own paths within the formation to avoid obstacles and other UAVs. The approach was combined with previous work [1] where we presented a 3D path planning system using potential fields for path planning of multiple UAVs.

In Sect. 2 we give an overview of related work. Afterwards, Sect. 3 introduces some issues relevant for establishing formations and also gives a brief introduction into bifurcation theory. Section 4 then describes the derivation of the resulting potential field. Section 5 shows some early results while Sect. 6 gives a conclusion and an outlook on future work.

2 Related Work

In recent years Multi-UAV cooperative control was a topic of extensive research. Several different methods to establish and guide a UAV cooperation were designed.

One widely used approach to solve this task is the model predictive control (MPC) or receding horizon control (RHC) method. It is an approach mainly used in industrial control processes that cannot be handled by classical PID controllers. RHC is an optimization-based approach which consists mainly of a feedback control scheme. The optimization of a trajectory is solved by iterating the optimization process in each subsequent step. Linear control as well as non-linear control is supported. When it is applied to UAV formation control the online computational costs are one big issue.

Duan and Liu [2] used RHC in combination with particle swarm optimization (PSO) to minimize the costs of the optimization problem. Chao et al. [3] use the RHC approach with a virtual reference point for formation configuration and a cost function to guarantee obstacle avoidance.

No et al. [4] presented a follower-leader system using a guidance law to control formations of fixed-wing UAVs. They use a branch of global leader and local leader/follower to represent the whole formation. This means that one UAV takes the role of the global leader. For each branch a local leader exists that follows the global leader. Additionally, each branch has further UAVs, the local followers, following the local leader.

Suzuki and Uchiyama [5] presented a guidance law similar to our approach. They use bifurcation theory to create different formation patterns and to guarantee obstacle and vehicle collision avoidance. They used fixed-wing UAVs and in addition to the guidance law a control law to calculate a motion vector to move the UAVs.

3 Formation Basics

To explore terrains using formations several requirements must be fulfilled. The UAVs must be able to actually establish a formation pattern. They can achieve this, e.g., by reaching a consensus about where and under which constraints a formation will be established. For that purpose an exchange of information is necessary. We use inter-UAV-communication based on UDP/IP to exchange all necessary information.

To establish a formation it is not necessary that each UAV can communicate with each other. We can describe the currently possible inter-UAV-communication as a graph $G = (V, E)$, where $V = \{1, 2, \ldots, N\}$ is the set of vertices representing N UAVs and $E \subset V \times V$ is the set of edges of G denoted by (i, j), representing that UAV i can exchange information with UAV j. One can proof now that consensus between all UAVs represented by the graph can be obtained if a minimum spanning tree in the graph exists [6].

The system to be designed is a dynamic, non-linear system. To ensure that the UAVs will be able to explore the terrain using a formation pattern it must be guaranteed that the formation provides a stable equilibrium. Without this guarantee the formation may break apart during its movement. Using bifurcation theory, Lyapunovs second method can be used to show that a global stability of the system converging at a desired final state can be guaranteed [7].

After establishing a formation, the UAVs have to move to a desired position. Three basic approaches for solving this task exist:

1. The follower leader approach where one UAV takes the part of a leader and the other UAVs follow the designated leader.

2. The virtual leader approach where the entire formation is treated as a single virtual UAV following a moving point.
3. The behavioral approach where each UAV has several desired behaviors, including formation keeping, goal seeking, collision/obstacle avoidance, etc.

Each of the approaches has its advantages and disadvantages. The most frequently used approach is the follower leader approach. The advantage of this approach is that it is possible to convert the problem of maintaining a formation to a standard tracking problem, if information about the leader is available. The big disadvantage is that it is not robust with respect to leader's failure. Using the virtual leader approach the formation handling is simpler than in the other approaches as the complete group is treated as a single object. The disadvantage is that the formation is only able to perform synchronized maneuvers which makes obstacle avoidance more challenging.

We use the behavioral approach to establish formation pattern. The UAVs have the capabilities to reach consensus about where and how a formation can be established and the UAVs are also able to identify areas, the formation has to explore next, by themselves. After establishing a formation, each UAV calculates its own path regarding to the formation pattern and the terrain properties, like obstacles and also with respect to the positions of other UAVs inside and outside the formation.

During an exploration of unknown terrain, it is not always beneficial to use all UAVs forming only one single formation. Due to this, the formation approach is used as a possible extension of our 3D potential field introduced in [1]. The 3D potential field is used for path calculations of single UAVs. For path calculations of UAVs participating at a formation, the 3D field is extended to a 4D potential field.

This extension is possible by the use of bifurcation theory. It allows us to calculate a formation potential which can simply be added to the previous calculated potential values. Path planning is then possible using the descent gradient.

3.1 Bifurcation Theory

Considering dynamical systems, a bifurcation takes place when a small smooth change made to the bifurcation parameters of a system causes a sudden qualitative or topological change in its behavior. An example is an upright standing stick which is fixed at the bottom and has a weight μ at its head. The angular deviation is denoted by x.

As long as the weight is small enough, $x = 0$ is a stable and balanced condition, i.e., for small deviations the stick adjusts itself back to its upright position $x = 0$. If μ is now continuously increased, the balanced condition becomes unstable at a given weight μ. Simultaneously, two new balanced conditions evolve as the stick bends down to the left or to the right. The change from one stable to two stable and one unstable balanced condition is called bifurcation, which is in this case a pitchfork bifurcation.

In general, at a bifurcation the local stability properties of equilibria or other invariant sets change. Figure 1 shows an example using the one dimensional bifurcation function $\mu|x| + \alpha|x^3|$, x: $-1 \leq x \leq 1$. Due to the change of the bifurcation parameter μ the number of equilibria changes from one to two if $\mu > 0$ changes to $\mu < 0$.

The bifurcation function used in our work is based on the pitchfork bifurcation which is a definite type of a bifurcation of a non-linear system with normal form

$$\frac{dx}{dt} = \mu x + \alpha x^3 \tag{1}$$

whereby μ depicts the bifurcation parameter and α can be used to change the bifurcation from a subcritical bifurcation ($\alpha = 1$) to a supercritical pitchfork bifurcation ($\alpha = -1$).

4 Potential Field

As mentioned before, we want to establish a system capable to explore terrains, including 3D structures, using UAVs. The UAVs must be able to explore the terrain on their own as well as they should have the capabilities to move in formation pattern.

Each UAV uses a configuration space (cf. [1]) representing the current knowledge of the UAV. The terrain is subdivided and represented by the leaves of an octree. We use a tree-based data structure to be able to change the subdivision. The advantage is that, e.g., nodes far away from the UAV can be combined, which decreases the computational complexity of calculating a potential field.

Using this kind of terrain representation, we can establish a potential field by calculating potential values ϕ for each leaf. We converted the problem of calculating these values into an optimization problem. The configuration space was

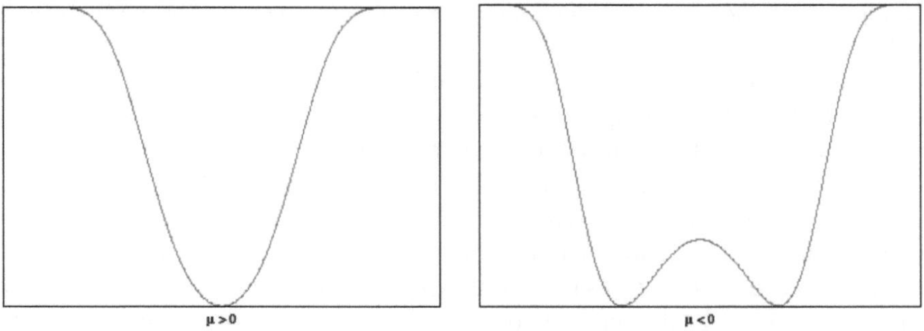

Fig. 1 Change of the equilibrium due to the change of the bifurcation parameter μ

modeled as a Dirichlet problem, which can be solved using a harmonic function [8]. Fixed potential values for occupied leaves ($\phi = 1$) and goal leaves ($\phi = 0$) are set a priori. We call these leaves bound leaves. The potential values for the remaining leaves, which we call unbound leaves, are then computed in a two step procedure. A first approximation computes potential values $0 < \phi < 1$ which are strictly increasing by increasing distance. Afterwards, an update function iteratively converges the potentials to a harmonic function until the resulting potential field is "good enough" to make path planning possible using the descent gradient [1], which pulls the UAV to the lowest potential value.

Harmonic functions are functions that satisfy Laplace's Equation in n dimensions:

$$\nabla^2 \phi = \sum_{i=1}^{n} \frac{\partial^2 \phi}{\partial x_i^2} = 0. \tag{2}$$

Here, ϕ denotes the potential value. An important requirement of our path planning approach is that the second partial derivations of ϕ must be zero and ϕ must be strictly increasing with the distance to the goal.

The use of harmonic functions [8] has several advantages, e.g., absence of local minima, compared to a usual potential field calculation. For a more detailed explanation of the advantages, disadvantages and mathematics we refer to [9].

4.1 Path Planning Potential

We start calculating a first approximation of the potential field by computing potential values $\phi^h_{x.y.z}$ for every unbound leaf using Eq. 3.

$$\phi^h_{x,y,z} = \frac{\log(\tau(x,y,z))}{\log(d)} \tag{3}$$

$\tau(x,y,z)$ represents the Euclidean distance from the center point (x,y,z) of the leaf to the target point. As one can easily see, the function is strictly increasing. Dividing by the logarithm of the diagonal d of the complete terrain normalizes ϕ such that $0 < \phi < 1$.

The first approximation of potential values will be iteratively optimized using a relaxation method. One property of a harmonic function is that the value at each point is the arithmetic mean of its surrounding points. Due to this fact Gauss–Seidel can be used to iteratively calculate the new potential values of the nodes. Using a dynamic octree the distances from the center point of a leaf to the center points of the neighbor leaves can be different. Hence, these distances have to be taken into account which leads to the following equation. To reduce the calculation complexity we only take the neighbors in six of the possible twenty six neighboring directions into account.

$$div = \tau_u\tau_d\tau_f\tau_b(\tau_l + \tau_r) + \tau_u\tau_d\tau_l\tau_r(\tau_t + \tau_b) + \tau_t\tau_b\tau_l\tau_r(\tau_u + \tau_d) \tag{4}$$

$$\phi^h_{x,y,z} = \frac{\tau_u\tau_d\tau_f\tau_b[\tau_r\phi_r + \tau_l\phi_l] + \tau_u\tau_d\tau_l\tau_r[\tau_f\phi_f + \tau_b\phi_b]}{div} + \frac{\tau_f\tau_b\tau_l\tau_r[\tau_u\phi_u + \tau_d\phi_d]}{div} \tag{5}$$

Here, τ_i represents the average distance from the current leaf (x,y,z) to the according neighbor leaves in that direction. As mentioned above we consider six directions in three dimensions: l and r are the neighbors in x dimension, f and b are the neighbors in y dimension and u and d are the neighbors in z dimension. Finally, ϕ_i represents the average potential value of the leaves in direction i. When using an octree, a leaf may has several neighbors in each direction. To reduce the calculation complexity, we approximate such that we only use the average distance and potential value of the neighbor nodes in each direction.

The above formula is used to iteratively optimize the potential values of the unbound nodes. With each iteration, the calculated potential field becomes more and more a harmonic function. The iteration stops, if the potential field is "good enough" for path planning, which is based on break conditions (cf. [1]).

4.2 Formation Potential

Several tests were presented [1, 9–11], to show that the designed system works as expected. The tests included the exploration of terrains, using different numbers of UAVs.

It was shown that the time needed for exploration decreases by increasing the number of UAVs. Nevertheless, it should be possible to decrease the exploration time even more by using a higher degree of coordination. To reach this goal, we want to enable the UAVs to establish formations. Therefore, we extend the potential field by formation potentials.

As mentioned before we use the pitchfork bifurcation as basis to establish formation patterns where the UAVs are pulled towards a global minimum. We extended the normal form (Eq. 1 from Sect. 3.1) to compute potentials suitable to establish 3D formation patterns, as shown in Eq. 8.

We first extended the normal form to take 3 dimensions into account which was simply done by adding a y and a z dimension. We can then use the formula

$$\phi^b_{x,y,z} = \mu(x + y + z) + \alpha(x^3 + y^3 + z^3) \tag{6}$$

to create such a potential. The value $\alpha|\alpha > 0$ is responsible for the amplitude of the potential. By changing the prefix of μ we are able to simply change, e.g., from a circle formation to a double circle formation pattern. To create formation patterns like lines or vortexes the influence of single dimensions must be changeable. By extending the formula to

$$\phi^{b}_{x,y,z} = \mu(a \cdot x + b \cdot y + c \cdot z) + \alpha(a \cdot x^3 + b \cdot y^3 + c \cdot z^3) \tag{7}$$

it is possible to specify weights for the single dimensions using the variables a, b, c, which leads, e.g., to ring or line formation pattern.

To combine the formation potentials with the potentials used for path planning, this formula has to be computed for each unbound leaf of the octree. Due to this fact, the positive distances normalized to values between 0 and 1 of the current leaf to the goal leaf represented by τ_x, τ_y and τ_z were taken into account.

$$\phi^{b}_{x,y,z} = \mu(a \cdot \tau_x + b \cdot \tau_y + c \cdot \tau_z) + \alpha(a \cdot \tau_x^3 + b \cdot \tau_y^3 + c \cdot \tau_z^3). \tag{8}$$

As shown in Eq. 9 we finally add the bifurcation potentials $\phi^{b}_{x,y,z}$ to the harmonic potentials $\phi^{h}_{x,y,z}$ calculated before. Dividing the resulting values by $\mu(a + b + c) + \alpha(a + b + c) + 1$ normalizes them so that $0 \leq \phi^{b}_{x,y,z} < 1$. Due to this division the values of the parameters a, b, c, μ and α have direct influence on the resulting potentials. As the resulting values are less than 1, we ensure that the so calculated results will never become higher than the potential values for occupied leaves, which are fixed set to 1 (cf. Sect. 4). Due to this and the use of the negative gradient for path planning, we are still able to guarantee obstacle avoidance.

$$\phi_{x,y,z} = \frac{\phi^{h}_{x,y,z} + \phi^{b}_{x,y,z}}{\mu(a + b + c) + \alpha(a + b + c) + 1} \tag{9}$$

5 Simulation Results

Based on the potential field calculated using Eq. 8 we are able to establish different formation patterns. Several results based on the parameter values shown in Table 1 are shown in Fig. 2.

We used a fictive terrain based on a height map. 8 UAVs were used to perform the tests. They started at some initial positions as shown in Fig. 2(a). The first task was to establish a line formation pattern in y direction. Therefore, the influence of the bifurcation potential was increased by increasing μ and the values for the

Table 1 Single values used to obtain the results shown in Fig. 2

Figure	μ	α	a	b	c
2b	65	1	1	0	1
2c	−1	500	1	0	10
2d	5	1	1	1	10

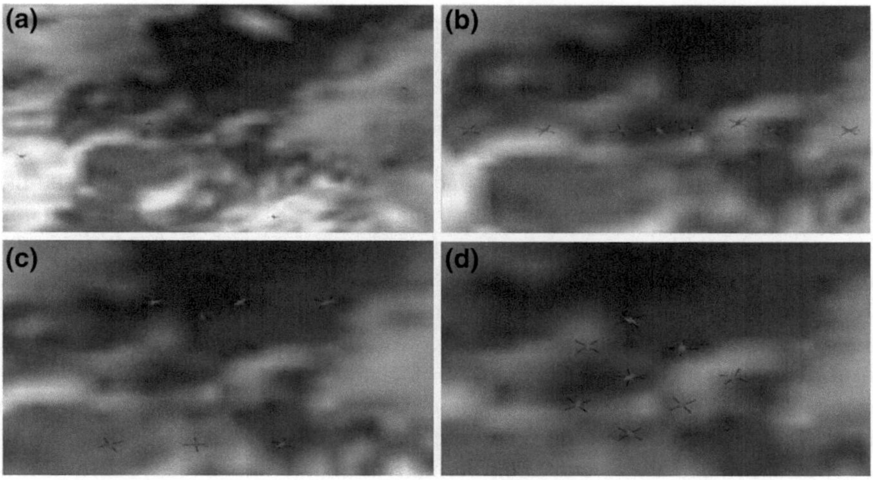

Fig. 2 The UAVs started at some given position of the terrain (**a**). In Figure (**b**) they established a line formation. By just changing m they moved to a *double line* formation (**c**) and finally by another parameter change they established *a ring* formation pattern

y direction were set to 0 as shown in the first row in Table 1. As one can see in Fig. 2(b) the UAVs reached a consensus and established a line. Due to the negation of the μ value the formation changed to a double line as shown in Fig. 2(c). Additionally, the amplitude of the bifurcation potential was increased to reach an adequate distance between the lines. Based on this formation pattern, the UAVs shall then move into multiple ring line formations. To reach this pattern, μ was set back to a positive value and each direction of the terrain was taken into account.

As depicted in Fig. 2(d) these value changes lead to a global minimum at the consensus point and consistent increasing values around that point. Therefore, one UAV moved to the minimum and the other UAVs arranged themselves around that one. 6 UAVs were needed to perform the first ring and therefore, the last UAV which is the rightmost in Fig. 2(d) started to set up another ring around the one the 6 UAVs performed.

As shown the presented approach is able to establish different formation patterns and one can simply change them by changing single parameter values.

6 Conclusion

We presented an approach to establish formations using potential field theory. The resulting potential field combines a harmonic function and bifurcation theory to allow formation flights. Therefore, we extended the 3Dpotential field used for path planning (cf. [1]) to a 4D potential field. Using this approach, different formation patterns, like lines, double lines, vortexes, circles, spheres, etc. can be established easily. Some possible formation patterns were presented in Sect. 5.

The next step will be the detailed evaluation of the designed system. The UAVs forming a formation have to compute smooth paths and must avoid all obstacles of the environment, including other UAVs of the formation as well as UAVs not participating in the current formation. The most interesting point will then be the comparison of the formation approach to the previous designed approach. Using formations should lead to a more efficient exploration of the terrain to justify the higher complexity.

To explore a complex terrain using only one static formation containing all UAVs does not make much sense. There will always be some small areas, e.g., inside buildings where a high number of UAVs exploring this areas simultaneously leads to a decrease of exploration efficiency since the UAVs disturb each other. Therefore, another point for future work is to enable the UAVs to dynamically establish and dissolve different formations and to join and leave them if it is beneficial.

Furthermore, efficient methods for obstacle avoidance have to be developed. UAV formations should, for instance, be able to split up to go around the obstacle and rejoin afterwards. They must also be able to automatically change the formation if necessary, e.g., forming a line to move through narrow corridors. In the next steps we plan to evaluate our system with regard to the requirements mentioned above and if necessary we will extend it even more to fulfill these tasks.

References

1. C. Rasche, C. Stern, L. Kleinjohann, B. Kleinjohann, in *A distributed multi-uav path planning approach for 3d environments*. The 5th International Conference on Automation, Robotics and Applications (ICARA). (Wellington, New Zealand, 2011)
2. H. Duan, S. Liu, Non-linear dual-mode receding horizon control for multiple unmanned air vehicles formation flight based on chaotic particle swarm optimisation. Control Theor. Appl. IET **4**(11), 2565–2578 (2010)
3. Z. Chao, L. Ming, Z. Shaolei, Z. Wenguang, Collision-free UAV formation flight control based on nonlinear mpc, ICECC (2011), pp. 1951–1956
4. T.S. No, Y. Kim, M.J. Tahk, G.E. Jeon, Cascade-type guidance law design for multiple-uav formation keeping. Aerosp. Sci. Technol. **15**(6), 431–439 (2011)
5. M. Suzuki, K. Uchiyama, Autonomous formation flight using bifurcating potential fields. 27th Congress of the International Council of the Aeronautical Sciences (2010)
6. W. Ni, D. Cheng, Leader-following consensus of multi-agent systems under fixed and switching topologies. Syst. Control Lett. **59**(3–4), 209–217 (2010)
7. D.J. Bennet, C.R. McInnes, Distributed control of multi-robot systems using bifurcating potential fields. Robot. Auton. Syst. **58**, 256–264 (2010)
8. C. Connolly, J. Burns, R. Weiss, Path planning using laplace's equation. IEEE ICRA **3**, 2102–2106 (1990)
9. C. Rasche, C. Stern, L. Kleinjohann, B. Kleinjohann, Coordinated exploration and goal-oriented path planning using multiple uavs. Int. J. Adv. Softw. IARA **3**, 351–370 (2010)

10. C. Rasche, C. Stern, L. Kleinjohann, B. Kleinjohann, in *Role-based path planning and task allocation with exploration tradeoff for uavs*. 11th International Conference on Control Automation Robotics Vision (ICARCV) (2010), pp. 417–422
11. C. Rasche, C. Stern, W. Richert, L. Kleinjohann, B. Kleinjohann, Combining autonomous exploration, goal-oriented coordination and task allocation in multi-uav scenarios. International Conference on Autonomic and Autonomous Systems (2010), pp. 52–57

Self-enhancing Featureless Robot Localization with Fixed Memory Tags

Felix Hackbarth

Abstract This chapter presents experimental results on localization of indoor mobile robots with limited sensor capabilities. On the basis of a bayesian approach with probabilistic global occupancy knowledge of the environment the robot is able to locate itself near obstacles. Far from obstacles the estimate gets inaccurate due to slippage. To locally increase the accuracy of the estimation in these regions we suggest integrating memory tags without initial knowledge on their position. The memory tags could also be RFID tags or fixed sensor nodes of a stationary sensor network. These local storages are used to save a processed actual estimate of the robot when it is within communication range. Hence the tags passively get an estimate of their own position which in turn is used by passing robots. Experiments show that even without initial position information for robot and local memory tags and despite vague and partially erroneous information of the robot, the position information of the tags converges towards a good stationary value. Even wrong initializations of the tag positions are tolerated and corrected. Thus positioning of any robot in communication range can be enhanced.

1 Problem Statement

Position estimation is an essential part in mobile robotics and often also needed for nodes of a sensor network. Information of different sensors can be combined to get

Based on Self-Enhancing Robot Localization via Local Memory Tags, by Felix Hackbarth which appeared in the Proceedings of the 5th International Conference on Automation, Robotics and Applications (ICARA 2011).

F. Hackbarth (✉)
Institute of Automation, Hamburg University of Technology,
Hamburg, Germany
e-mail: felix.hackbarth@tuhh.de

G. Sen Gupta et al. (eds.), *Recent Advances in Robotics and Automation*,
Studies in Computational Intelligence 480, DOI: 10.1007/978-3-642-37387-9_9,

a more accurate position estimate than from a single sensor. The data of relative sensors has to be put into relation with a global reference. Binary signals like the Eiffel Tower being in line of sight or not, can be used for positioning. This signal would belong to a relatively measuring sensor and the tower would be a feature with a global position reference. Without global information the estimation error increases due to slippage during robot movement.

An experimental setup is used with this approach. It is situated indoors in a spatially confined area defined as mission space $M_{Mission}$ with a dimension of 4 m by 3.2 m. There are three rectangular obstacles, one in an edge, one at one side and one inside the room. Figure 1 shows a picture of the upper left section of the experimental setup. In the mission space a mobile robot is acting and trying to estimate its position based on data of two different sensor types. The first sensor type is a rotary encoder. Three of these each mounted on an omni-directional wheel of the robot obtain the distance and direction relative to the previous position. The other sensor is an infrared distance sensor with a narrow measuring beam and a range of 40 cm. Nine of these measure the distance to obstacles in radial direction. The mission space is known to the extent that there exists a probability map of contained obstacles. The available sensors provide a vague estimate of the position, especially in the center of the mission space where there are no detectable obstacles to correct the position. Due to slippage the positional accuracy decreases with traveled distance.

The idea is to locally store position information to enhance positioning on later traverses. Because the used sensors are not capable of extracting and mapping landmark information for positioning, a featureless Bayesian approach is used for localization. Though designed to locally increase the position estimation of robots in the mission space the presented approach can be applied to dynamically determine the position of stationary sensor nodes or tags that previously were deployed without position information. Figure 1 shows the experimental setup that is used for validation.

2 State of the Art

Localization can be divided into three problem types: position tracking, global localization and kidnapped robot problem [1]. The global localization problem can be solved with either featureless localization or localization based on landmarks also called topological approach [2]. The methods of the latter are based on principles first presented by Smith et al. [3] and were largely extended over time. A good literature overview can be found in [4].

The featureless localization is based on a probability distribution of the robots position in a discrete grid. It was first presented by [5] who extended the certainty grids proposed in [6]. Occupancy grids were used by [7] to locally plan robot paths and avoid obstacles that are sensed and then noted in the grid. Likelihood fields methods are used for sensor integration as in [8]. In [9] probability grids

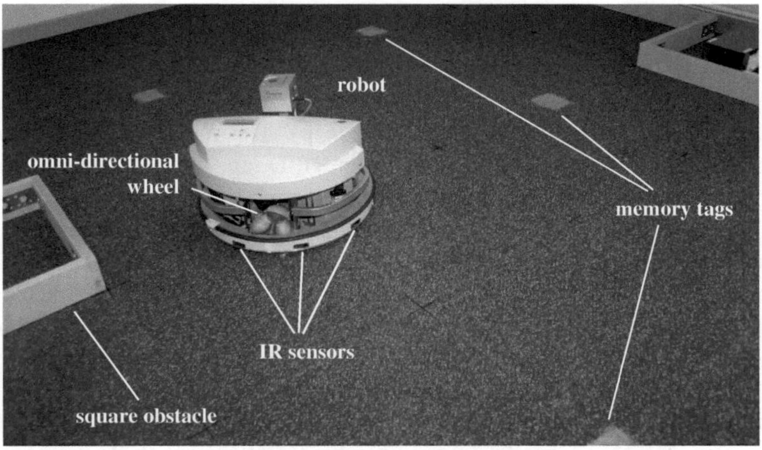

Fig. 1 Experimental setup with robot and memory tags. This perspective is showing the *upper left* part of the mission space. The robot is hiding another memory tag

were applied to the experimental setup used here. Topological and the probabilistic grid methods can also be combined [10]. Therein probability grids are used for local small scale positioning and mapped topological for large scale navigation. Fundamentals for both are described in [11].

The use of RFID tags as environmental marks has been demonstrated in [12]. There the localization of tags and robot is a two step process, where at first the position of the tags is determined by a robot and then the position information about the tags could be used to determine the position of the robot. In contrast, the approach presented here uses memory tags, which also could be RFID tags, to actively store an estimate of the position of the tag that could be used by other robots later on. Also the stored information does not have to fulfill specific requirements and can directly be used for the position estimate of the robot in a single step manner. In [12] an overview on publications is presented where systems of RFID positioning are proposed. Reference [13] proposes RFID tags for increasing GPS accuracy in downtown areas and tunnels. And also indoors, using a finer grid of RFID tags leads to position estimates in the 1 m range [14]. Also in [15] RFID tags were used to determine the position of mobile robots, but in contrast to the proposed approach the alignment of all tags and their position has to be known. Here no prior knowledge in form of position, number or alignment of tags is necessary.

3 Probabilistic Localization by Communication

The infrared distance sensor measurements and the odometry with its inherent kinematic model are considered in position probability grids with a discrete position vector

$$\mathbf{p}(k) = \begin{pmatrix} x \\ y \end{pmatrix} \quad \begin{matrix} x, y \in \mathbb{R} \\ k \in \mathbb{N} \end{matrix} \tag{1}$$

In this grid $P(\mathbf{p})$ describes the probability that the center of the actually considered object is located within this cell (at position $\mathbf{p}$). The object can be a robot, an obstacle, a sensor node, a memory tag or an RFID tag.

Figure 2 shows the block diagram for the information flow inside the model of each robot. The main objects and the knowledge about these are memory tag, robot and environment represented as boxes. The connecting elements, the boxes with the convolution symbol, are the transfer behavior or transmission between the main objects. In case of the communication this transmission is not always available; and in case of the odometry or IR sensor it is additionally dependent on the sensor value.

The only origin of new information in the system is the value of the infrared sensors. This value represented in K_{IR} in conjunction with probabilistic knowledge of the environment M_{env} leads to a position estimate of the robot. To transform the preceding robot position estimate to the following the kinematic model of the robot is represented as $K_{odo}(k)$ which is dependent on the distance to the previous robot position $\mathbf{p}(k-1)$. Because of slippage information is lost during this transformation.

To locally conserve the actual position information of the robot P_{robot} memory tags are introduced. P_{tag} is a probability grid describing the position of the center of one tag. The transfer behavior K_{comm} between robot position P_{robot} and memory tag position P_{tag} is equal in both directions.

3.1 Transforming Local into Global Sensor Data

Every sensor is locally measuring a specific feature of the environment. Since the feature is detectable but not identifiable like landmarks, this approach is called

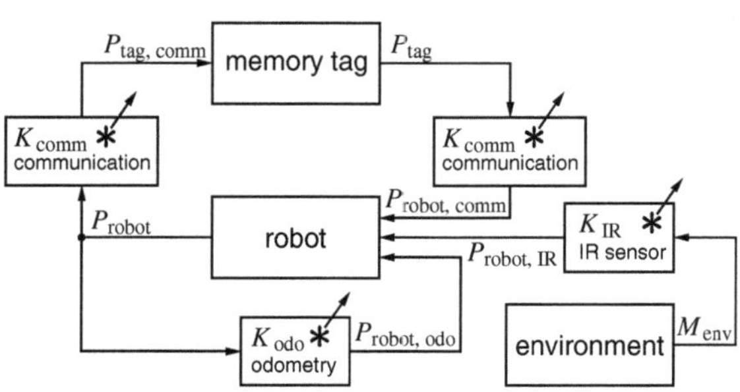

Fig. 2 Block diagram for the information flow in the system model of memory tag and robot in the environment. The *arrows* from the convolution symbol indicate that the kernel for the convolution could be variable or this link is available only temporarily. These are dependent on the observable variables

featureless localization. Thus the possibility of communication can also be seen as a binary sensor, like the Eiffel Tower being in line of sight mentioned in the first section. The global spacial feature distribution in our case is the global position of the communication partner.

The global spacial feature distribution is noted as probability distribution in a grid as $M_{feat.}(\mathbf{p})$ similar to the position probability grid $P(\mathbf{p})$. The relationship between this feature distribution and the sensor position with its actually measured value is established by the kernel $K_{feat.}(\mathbf{d}|s)$. For every sensor value s the kernel $K_{feat.}$ is a conditional probability distribution that relates the occurrence of the feature to the relative position of the sensor dependent on its relative displacement

$$\mathbf{d} = \begin{pmatrix} d_x \\ d_y \end{pmatrix} \qquad d_x, d_y \in \mathbb{R}. \tag{2}$$

If sensor kernel $K_{sensor}(\mathbf{d}|s)$ and feature distribution $M_{feat.}(\mathbf{p})$ are known, the position probability grid $P(\mathbf{p}|s)$ of the sensor with sensor value s can be calculated by convolution:

$$P_{global}(\mathbf{p}|s) = K_{feat.} * M_{feat.} = \sum_{d_x} \sum_{d_y} K_{feat.}(\mathbf{d}|s) \cdot M_{feat.}(\mathbf{p} - \mathbf{d}). \tag{3}$$

As the feature grid is dependent on x and y also the kernel can be dependent on these variables. Dependent on the type of the sensor, Eq. (3) differs in its variables and dimension. In addition there can be a dependency on φ like for infrared sensors, where the direction in which they are pointing has to be considered, but for the case of the memory tags presented in the following this is sufficient.

3.2 Fusing Global Sensor Data

To combine different global position probability grids, the grids have to be equal in size and have to contain comparable information, e.g. the position of the robot. Like probabilities the combination is done by a cell-by-cell multiplication but the probabilities in the resulting probability grid $P_{res.}$ do not sum up to 1 because of a scaling factor α. α is calculated by

$$\alpha = 1 / \sum_{\mathbf{p}} P_{res.}(\mathbf{p}). \tag{4}$$

The resulting scaled grid is the result of taking all sensor information into account. For each cell it contains the position probability of the currently considered object.

3.3 Extracting Position Information from Communication

For an additional source of position information memory tags are deployed in the environment. These have no initial knowledge about their position so in the beginning the probability grid of each tag is constant on the whole mission space:

$$\sum_i P_{tag}(p_i) = 1, \quad P_{tag}(p_j) \quad \forall i,j. \tag{5}$$

The communication process is the key to position enhancement. Position probability grids of robot and tag are exchanged and each has to perform its own calculation on the received grid to relate it to its own position. One communication act takes place when the robot enters the communication range of the tag. The next one is only initiated when the robot has left the communication range to gain additional environmental knowledge. Otherwise the Markov assumption of the independence of succeeding measurements would be violated.

The communication process itself equals a binary sensor reading s, so both communication partners sense if either the partner is within range $s = 1$ or not $s = 0$. Because there is no additional knowledge where the partner is located, each position within range has an equally weighted influence on the transformation from partner to own position probability grid and vice versa. When there is communication ($s = 1$), the kernel $K_{comm}(\mathbf{d})$ has the shape and dimension of the communication range with equal probabilities inside, and zero outside. Specific, also asymmetric communication properties can be handled with the values inside K_{comm} with the spatial dependence on $\mathbf{d}$. Here the communication range of the memory tags—when a robot is able to access data stored in it—is idealized to a rectangle with a side length of 25 cm with the memory tag placed in its center

$$K_{comm}(s = 1) = \frac{1}{25}\begin{pmatrix} 1 & 1 & 1 & 1 & 1 \\ 1 & 1 & 1 & 1 & 1 \\ 1 & 1 & \mathbf{1} & 1 & 1 \\ 1 & 1 & 1 & 1 & 1 \\ 1 & 1 & 1 & 1 & 1 \end{pmatrix} \tag{6}$$

with a discretization of 20 pixel per meter, as chosen for the whole implementation. The range has to be at least as large as the physical communication range, otherwise the convergence to the correct position could be inhibited. No communication takes place for $s = 0$, hence there is no kernel.

When a communication process takes place this kernel K_{comm} for both and the probability grid of the respective communication partner ($P_{robot}(\mathbf{p})$ or $P_{tag}(\mathbf{p})$) is taken and transformed into their own position probability according to Eq. (3). Combined with the previous position information the resulting position probability grids for robot $P_{robot,res}$ and memory tag are $P_{tag,res}$

$$P_{robot,res} = P_{robot} \cdot K_{comm} * P_{tag} \quad \text{and} \tag{7}$$

$$P_{tag,res} = P_{tag} \cdot K_{comm} * P_{robot}. \tag{8}$$

After the communication process the estimated position of the robot deteriorates by driving and is enhanced again by other sensor readings. The fixed memory tag has to wait for the next communication process to get additional data; in exchange the stored data does not deteriorate.

4 Experimental Results and Evaluation

Several experiments have been conducted. One of these with thirteen deployed memory tags without initial position information is presented here. Also the robot has to solve the global localization problem. The distribution of the tags is shown in the left column of Fig. 3 each of the thirteen gray/green marks represent one memory tag. The tag id is noted in the lowermost map.

The trajectory of the robot was generated by an algorithm that drives with constant velocity and continuously randomized angular velocity; in addition it is repulsed by walls. The experiment was run for 270 communication acts in total. At the end one tag ended with 6 and another one with 50 communication acts but an average of roughly 20 was reached by the rest (see Table 1).

4.1 Communication Benefit for Robot and Tag

Exemplarily Fig. 3 shows three communication acts (1, 5, 10) of the robot with memory tag with id 13. Generally at first there is mostly only a benefit for the tag. Then both robot and tag can enhance their position by communication. But sometimes also the robot position gets distorted by inaccurate information that was stored on the memory tag. When the position convergence of the tag is nearly done, the robot benefits by the communication. Prior to the communication act the accuracy of the robot position depends on the traveled trajectory, which is then fused with the probability distribution of the tag. Any error in the robot position is propagated to the tag and has to be corrected later on on the expense of the robot position.

In Table 1 the results of all changes in the robot position are shown. For each tag the sum of communication acts and the sum split into their positive and negative influences on the robot position are noted in the table. As noted the ones with a negative influence mostly are at the beginning of experiment. For each memory tag the average correction of the estimated robot position is shown in the last column. This is the distance that the post position of the robot is nearer to its actual position than the prior position. The actual position is measured by a camera reference system.

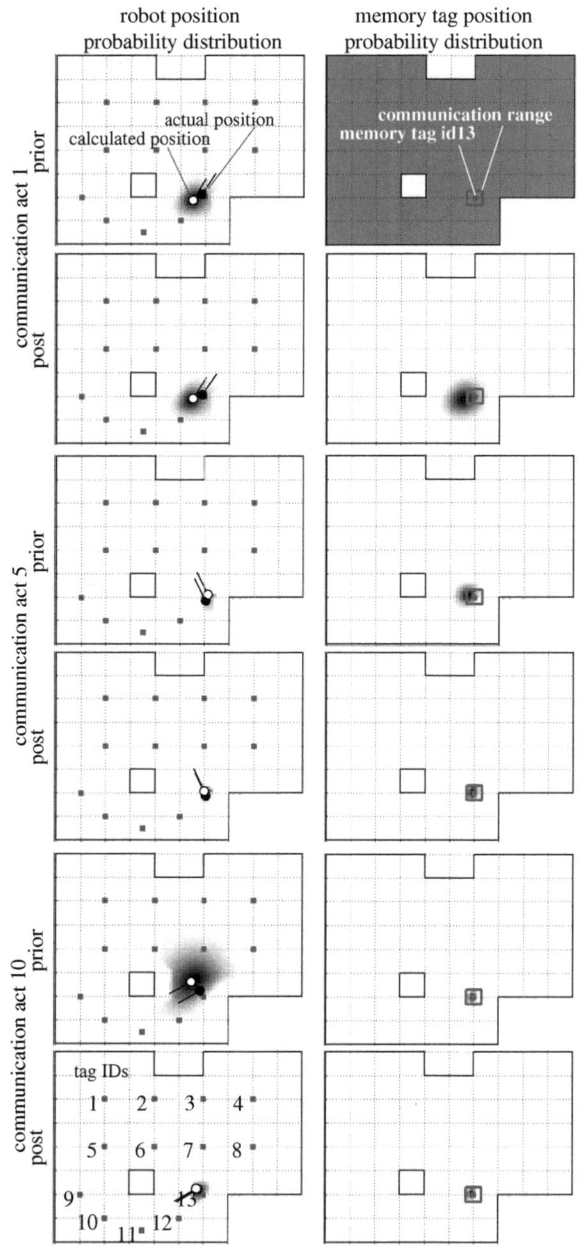

Fig. 3 Selected communication acts between robot and memory tag with id 13. Probability grids of the robot and the tag are shown in the *left* and in the *right column* respectively. For each communication act the prior probability distribution is shown in the *first row* and the posterior probabilities are shown in the *second row*. On the left the actual and the calculated robot position is indicated by the *filled and hollow needle*. The robot position cannot be corrected by the information in the first communication act, but in the fifth and tenth it is. For the tag there is a benefit in all shown communication acts

Significant is the negative influence of memory tag 6 that was caused by a slow convergence due to repetitive bad robot position information at the beginning. The same holds for memory tag 3. This is placed left to the tip of the lower right corner obstacle and the position probability distributions prior to the communication are shown in Fig. 4. There is a downwards displacement of the center of the position

Table 1 Average position benefit for the robot by each communication act during the experiment

Memory tag id	Communication acts		Average benefit (cm)
	Sum	(Positive/Negative)	
1	27	(18/7)	+5.8
2	13	(7/4)	+7.5
3	17	(6/8)	+2.0
4	16	(8/3)	+8.5
5	27	(17/8)	+2.7
6	12	(3/7)	−1.8
7	17	(10/4)	+2.4
8	16	(9/6)	+4.3
9	30	(17/6)	+6.1
10	50	(24/11)	+3.4
11	6	(2/0)	+2.2
12	21	(12/3)	+2.6
13	18	(11/4)	+2.6
Experiment	270	(144/71)	+3.9

probability cloud for the tag during the first communication acts. This error is corrected over time, with fast convergence at the beginning and slow convergence to the end. For the robot the first 5 communication acts perturbed its position estimate but from the 9th on the robot could benefit by the information and corrects its position on average by 2 cm which includes the negative influence at the beginning. Additional results can be found in [16].

4.2 Range and Quantity Considerations

The higher the frequency of communication, the faster is the convergence of the tags. Increasing the communication range increases the hitting chance while driving randomly and thus the frequency. On the other side a larger range has a

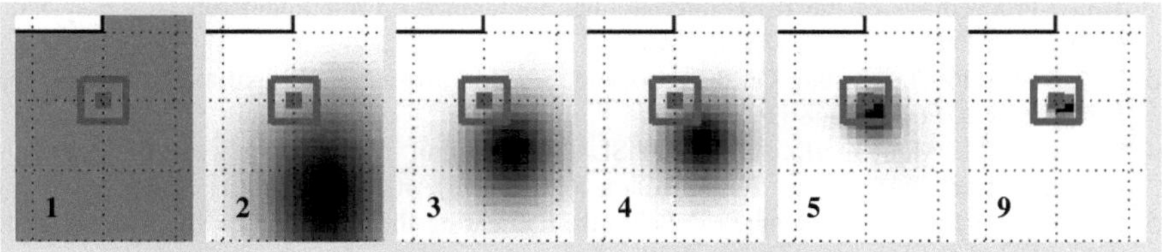

Fig. 4 Sequence of locally stored prior position probability grids, cropped to the region around the memory tag with id 3. The *black bar* on the *upper left side* is part of the mission space border. The *gray* values are scaled to fit the range from *black* to *white*

larger kernel with an increased uncertainty during the communication act thus leading to slower convergence. To enhance convergence, the tags should be placed in frequently traveled regions. To maximize position enhancement of the robot the tags should be placed in open areas, like tag 1,4 and 8. In an additional experiment with erroneous initialization of one tag with a probability of $P = 99.9\%$ at a single cell, it recovered within 4 steps if the position is far away from the actual position. If the faulty position is nearer to the actual position of the tag the convergence is slowed down. Similar to tag 3 and 6 the position then not is rejected but slowly corrected.

5 Conclusion

This approach is based on local memory in form of memory tags like RFID tags or sensor nodes. They store position data to enhance the positioning of mobile robots acting in that area. The robots determine and repetitively enhance the position accuracy of these local memory units. An experimental setup with memory tags builds the hardware basis. The conducted experiments show that despite vague sensor information and the lack of position information in the beginning of the experiment the locally stored position information converges rapidly into the region of the actual position and is steadily enhanced towards a stationary value. Even the partly corrupted and poor position data of the robot and a communication range of 25 cm only, lead to a good final position for the memory tags. This approach represents a self enhancing positioning environment that is applicable for a sensor network with stationary nodes or where memory tags can be or are deployed.

References

1. D. Fox, S. Thrun, W. Burgard, F. Dellaert, *Particle Filters for Mobile Robot Localization. In Sequential Monte Carlo Methods in Practice* (Springer Verlag, New York, 2001)
2. H.J. Chang, C.S.G. Lee, Y.C. Hu, Y.-H. Lu, in *Multi-Robot Slam with Topological/Metric Maps*. Proceedings of the IEEE international conference on intelligent robots and systems (RSJ), (San Diego, CA, 2007), pp. 1467–1472
3. R.C. Smith, P. Cheeseman, On the representation and estimation of spatial uncertainty. Int. J. Robot. Res. **5**(4), 56–68 (1986)
4. S. Thrun, M. Montemerlo, The GraphSLAM algorithm with applications to large-scale mapping of urban structures. Int. J. Robot. Res. **25**(5/6), 403–430 (2006)
5. W. Burgard, D. Fox, D. Hennig, T. Schmidt, Estimating the Absolute Position of a Mobile Robot Using Position Probability Grids. *Proceedings of the 14th National Conference on Artificial Intelligence AAAI/IAAI*, vol. 2, (1996), pp. 896–901
6. H.P. Moravec, Sensor fusion in certainty grids for mobile robots. AI Magazine **9**(2), 61–74 (1988)

7. J. Borenstein, Y. Koren, The vector field histogram—fast obstacle avoidance for mobile robots. IEEE Trans. Robot. Autom. **7**(3), 278–288 (1991)
8. A. Burguera, Y. González, G. Oliver, On the use of likelihood fields to perform sonar scan matching localization. Auton. Robots **26**(4), 203–222 (2009)
9. F. Hackbarth, Position probability grids for mobile robots obtained by convolution. *The 4th International Conference on Automation, Robotics and Applications* (ICARA 2009), (2009), pp. 578–583
10. J.-L. Blanco, J.-A. Fernández-Madrigal, J. González, Toward a unified bayesian approach to hybrid metric-topological slam. IEEE Trans. Robot. **24**(2), 259–270 (2008)
11. S. Thrun, W. Burgard, D. Fox, *Probabilistic Robotics* (The MIT Press, Cambridge, 2005)
12. D.Hähnel, W. Burgard, D. Fox, K.P. Fishkin, M. Philipose, Mapping and Localization with RFid Technology. *Proceedings of the IEEE International Conference on Robotics and Automation* (ICRA 2004), pp. 1015–1020
13. H.D. Chon, S. Jun, H. Jung, S.W. An, Using RFID for accurate positioning. J. Global Positioning Systems **3**(1–2), 32–39 (2004)
14. K. Zhang, M. Zhu, G. Retscher, F. Wu, W. Cartwright, *Three-dimension indoor positioning algorithms using an integrated rfid/ins system in multi-storey buildings. In Location Based Services and TeleCartography II* (Springer Berlin, Heidelberg, 2009)
15. G. Yang, G. Anderson, E. Tunstel, A RFID landmark navigation auxiliary system. World Automation Congress (WAC 2006), pp. 1–8
16. F. Hackbarth, Self-Enhancing Robot Localization via Local Memory Tags. *The 5th International Conference on Automation, Robotics and Applications* (ICARA 2011), pp. 395–400

Vision Based Control for Micro Aerial Vehicles: Application to Sense and Avoid

Luis Mejias, Iván F. Mondragón Bernal and Pascual Campoy

Abstract This work presents a collision avoidance approach based on omnidirectional cameras that does not require the estimation of range between two platforms to resolve a collision encounter. Our approach achieves minimum separation between the two vehicles by maximising the view-angle given by the omnidirectional sensor. Only visual information is used to achieve avoidance under a bearing-only visual servoing approach. We provide theoretical problem formulation, as well as results from real flight using small quadrotors.

Based on "Omnidirectional bearing-only see-and-avoid for small aerial robots", by Luis Mejias, Ivan Mondragón, Pascual Campoy which appeared in the Proceedings of the 5th International Conference on Automation, Robotics and Applications (ICARA 2011). © 2011 IEEE.

L. Mejias (✉)
Australian Research Centre for Aerospace Automation (ARCAA), Science and Engineering Faculty, EECS, Queensland University of Technology,
GPO Box 2434Brisbane 4000, Australia
e-mail: luis.mejias@qut.edu.au
URL: http://www.arcaa.aero/

I. F. Mondragón Bernal
Centro Tecnológico de Automatización Industrial CTAI,
Pontificia Universidad Javeriana, Carrera 7 No 40 -69,
CP 110311 Bogotá, D.C, Colombia
e-mail: imondragon@javeriana.edu.co

P. Campoy
Computer Vision Group U.P.M., Centro de Automática y Robótica -CSIC-UPM,
C/ José Gutiérrez Abascal 2 28006 Madrid, Spain
e-mail: pascual.campoy@upm.es
URL: http://www.vision4uav.com/

G. Sen Gupta et al. (eds.), *Recent Advances in Robotics and Automation*, Studies in Computational Intelligence 480, DOI: 10.1007/978-3-642-37387-9_10,

1 Introduction

A fast growing industry on Unmanned Aerial Systems (UAS) and their application in indoor and outdoor environments is motivating the research community [1–3], to put forward a new set of approaches to enable the use of these platforms in civilian context. With many research labs adopting small UASs as preferred testbed, it is envisaged that a new set of regulations will appear aiming to define minimum capabilities and compliance terms for small aerial robotic systems. In fact, UASs will soon be allowed to share civil airspace from 2015 onwards [4]. This will accelerate the refinement of regulatory frameworks that serve as guidelines for developers and researchers working on UASs. Previous efforts have included standards for see and avoid [5].

With some regulations already in place and the acknowledgement that UAS collision avoidance "see-and-avoid" is one of the most significant challenges for the integration of UASs into the civil airspace [6], this work contributes to the field of see-and-avoid proposing a omnidirectional see and avoid using visual sensors. Small UASs have inherent limitations in size weight and power (SWaP) which makes vision and other lightweight sensors a very attractive alternative. Vision as a sensor not only offers a rich source of information for navigational purposes, but it also offer the best chances for regulator approval [7]. The conventional means of resolving a collision encounter between two of more aerial vehicles is through the use of cooperative approaches that in most cases involve range estimation. Shakernia et al. developed a technique that estimates range by manoeuvring the owncraft finding that range converge in 7–15 s [8] (passive ranging).

An approach using a monocular camera was proposed by Driessen [9], in which bearing-rate combined with target growth rate in the image is used to estimate distance. In [10] cameras are combined with pulsed radar (to estimated range) proposing an approach for non-cooperative scenarios. Saunders and Beard [11] developed an approach for fixed-wing vehicles to avoid obstacles using image measurements similar to the approach presented here but their approach uses range to objects. Once more, Watanabe et al. [12] proposed a PN-based guidance law defining a collision cone using vision sensors, but their assumptions about knowledge of obstacle location may be difficult to validate in real flights, in the context of monocular vision. Similarly, Angelov et al. [13] have proposed a passive approach to collision avoidance assuming EO sensors. However, their bearing-rate approach presents many ambiguities when more than two encounter scenarios are considered. Mejias et al. [14] and Lai et al. [15] have done some pioneering work in the detection and conflict resolution using real flight test with manned and unmanned platforms.

Primarily so far, the tendency has been to use range or minimum separation as principal or complementary information to resolve collision conflicts. This work present a important contribution based on omnidirectional sensors that do not require explicit knowledge of obstacle range to avoid them. Also, we contribute with a visual servoing approach using omnidirectional sensors for aerial robots in

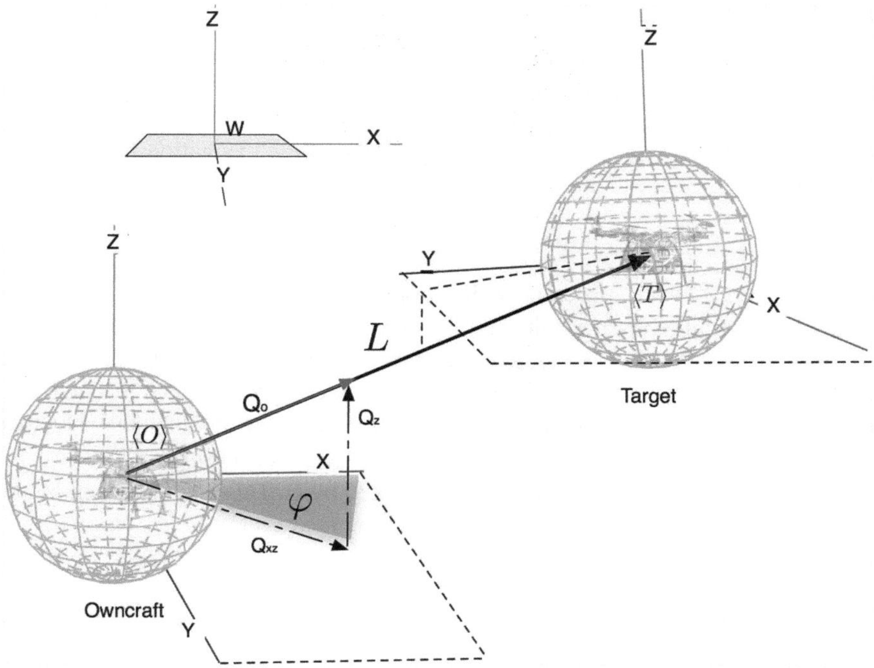

Fig. 1 Owncraft-Target setup in 3D. Each vehicle is enclosed inside a sphere of radius R that correspond to the minimum volume occupied by the vehicle. By maximising φ, we can guarantee that L will be higher than twice the radius of the sphere that contains each vehicle

real flights. We propose that the minimum separation between a vehicle and a target is achieved by maximising the view-angle (φ) given by the omnidirectional sensor (Fig. 1).

This chapter is structured as follows. Section 2 presents the problem statement. Section 3 introduces the camera projection model and tracking approach. Section 4 presents the control approach. Section 5 outlines the experimental setup and algorithm evaluation in simulations and real flight tests on a small quadcopter. Finally, conclusions and future work are presented.

2 Problem Description

This section presents the vehicle kinematic model and the formulation of our avoidance problem. This chapter considers a vehicle moving with constant altitude, therefore the kinematic of each vehicle is abstracted as an unicycle model. Next, we describe the details of these two approaches.

2.1 Vehicle Model

Let a Cartesian navigation frame (x_n, y_n, z_n) have origin at the location of a vehicle position when a target is detected at $t = 0$ (Fig. 2). Assume the vehicle has constant ground speed v then its translational motion in the navigation frame can be described by

$$\dot{x}_n = v\ cos\psi, \quad \dot{y}_n = v\ sin\psi, \quad \dot{\psi} = \omega \tag{1}$$

where $(x_n, y_n) \in \Re^2$ is the vehicle location and $\psi \in (-\pi, \pi)$ its heading. This model is constrained by the vehicle's linear velocity and maximum turning rate, i.e., $v_{min} < v < v_{max}$ and $|\omega| < \omega_{max}$, respectively. The two vehicles involved in the scenario are *owncraft* (O) (vehicle with camera) and *target* (T) (vehicle to avoid or static target). The control inputs of (O) and (T) are linear and angular velocities $[v_O\ \omega_O]^T$ and $[v_T\ \omega_T]^T$, where $v = [v_x\ v_y\ v_z]^T$ and $\omega = [\omega_x\ \omega_y\ \omega_z]^T$.

The view angle (or relative bearing) is denoted by φ. This angle is given by the onboard omnidirectional sensor (described in the next section).

2.2 Problem Formulation

With regards to Fig. 2, the whole system kinematics can be written as follows:

$$\dot{s} = J(s, L)u \tag{2}$$

where $s = [\phi\ \varphi]^T$ is the feature vector with ϕ and φ being the elevation and bearing of the target wrt owncraft,[1] $u = [v_O\ \omega_O]$ as defined above and $J(s, L) = J(s, L)_{trans} + J(s)_{rot}$ the jacobian which can be decomposed in a translational and rotational part (see [16, 17] for more details). Therefore, (2) can be expressed in a translational and rotational component (see Sect. 4). Let us define the control task such as

$$u = J(s, L)^{+}\dot{s} \tag{3}$$

where $\dot{s} = -\lambda(s - s^*)$ is the error signal, with $\lambda > 0$, s^* the desired feature value and $J(s, L)^+$ the pseudo-inverse [18, 19]. The aim is to minimise the error signal by maximising the desired value of our features ($\varphi \rightarrow \pm\frac{\pi}{2}$). This is achieved using visual servoing approach [18]. The details are presented in Sect. 4.

Our hypotheses are: (1) that collision avoidance is achieved when L (see Fig. 1) is higher than twice the radius of the sphere that contains each vehicle ($L > 2R$). Furthermore, L is maximum when $|\varphi|$ is maximum. We achieve minimum separation between a vehicle and a target by maximising the view angle given by the omnidirectional sensor. (2) that collision avoidance is achieved by following a

[1] Note: we assume that camera and vehicle cg are aligned, approximately in the same place. Also, the sensor provides ϕ and φ, but only the use of φ is used in this chapter.

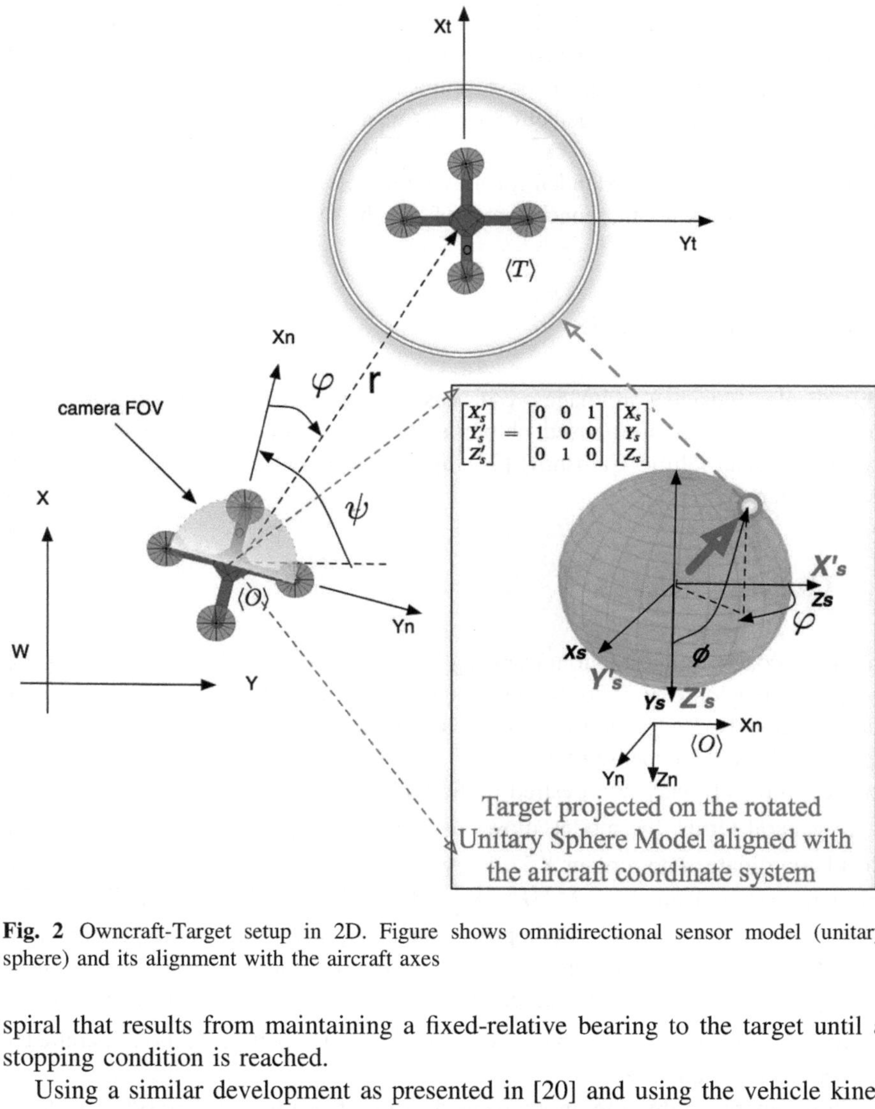

Fig. 2 Owncraft-Target setup in 2D. Figure shows omnidirectional sensor model (unitary sphere) and its alignment with the aircraft axes

spiral that results from maintaining a fixed-relative bearing to the target until a stopping condition is reached.

Using a similar development as presented in [20] and using the vehicle kinematic described by (1) expressed in polar coordinates ($x_n = f(r,\theta)$ and $y_n = f(r,\theta)$) we can demonstrate that the trajectory followed by our vehicle will be described by the following equation

$$r(u) = ae^{bu} \tag{4}$$

with $a = r_0$, $b = \cot(\varphi)$ and $u = \theta_0 - \theta$ which is the equation for an equiangular spiral in polar coordinates [20]. Here, r is used instead of L for polar representation.

When a stopping condition in the vehicle heading is defined (such as $f(\psi^*) \in [-\pi, \pi]$), the vehicle will avoid the collision keeping a minimum distance

r^*. A stopping condition can be define as $\theta^* = f(\psi^*) \in [-\pi, \pi]$ and r^* is the closest distance to target at ψ^*. Because r is always decreasing as $t \to \infty$, r^* is the closest range of approach to the object for stopping condition $\theta^* = f(\psi^*)$ and with $b \neq 0$ proves that as long as the detection occurs at a range $L > 2R$, it will avoid the target with $r^* > 2R$.

This theorem can be extended for moving targets. In this case it will result in a large set of spirals for the general case of which (4) is a special case.

3 Detection Approach

This chapter considers an image pre-processing approach that transforms the input image into a RGB normalised colour space [21] followed by a Continuously Adaptive Mean Shift (Camshift) [22] approach to detect and track persistent features (targets).

3.1 Camera Model

In our approach, omnidirectional sensing is achieved using fisheye lenses (dioptric system).The geometry follows the model proposed in [23–25]. These theories (unified model) can generally be used to approximate fisheye lenses by catadioptric model at the expense of accuracy [26]. However, they are typically accurate enough to reject many non desired outliers [27].

In general terms, the model allows us to project a point $X_w = (x_w, y_w, z_w)^T$ in world coordinates into a point $X_s = (x_s, y_s, z_s)^T$ on the unitary sphere by using the following relationship $X_s = \hbar^{-1}[H_c^{-1} X_i]$, where $X_i = H_c \cdot \hbar(X_w)$ is the projection of a world point on the catadioptric image, $H_c = R_c \cdot M_c$ is the collineation between R_c (camera rotation) and M_c (calibration matrix), $\hbar$ is a non-linear correspondence function between points in two different oriented projective planes defined as $\hbar(X_w) = \left(x_w, y_w, z_w + \xi\sqrt{x_w^2 + y_w^2 + z_w^2}\right)^T$, where ξ is a parameter defined by type lens used.

The camera calibration is performed using the method described in [28], obtaining the parameter ξ and the matrix M_c. Figure 3 shows the dioptric system employed and the projection of a captured image on the unitary sphere using the calibration data.

3.2 Colour Space and Target Tracking

Our focus is primarily to implement the colour space that best maintain object's apparent colour in outdoor settings as seen by a machine vision camera. We tested

Fig. 3 Omnidirectional system used. *Bottom Left*: The dioptric system employed. *Top*: Omnidirectional image taken with this system. *Bottom Right*: Projection on the unitary sphere of the omnidirectional image

several colour spaces and found consistencies with previous research [21, 29], finding that normalised RGB preserves better object's colour in outdoor setting.

We approach the problem of tracking by exploiting the colour characteristic of the target. We define a basic colour (orange) to the target by attaching a simple coloured mark to it and tracking this mark. We then use an algorithm that has proven to deal with outdoor lighting changes by dynamically adapting to changes in probability distributions, the *Continuously Adaptive Mean Shift* [22] (CamShift). This algorithm is based in the *mean shift* originally introduced by Fukunaga and Hostetler [30] (see Fig. 4).

3.3 Target Attitude Estimation

After detection and tracking is performed the target location in images is estimated obtaining at each iteration the image coordinates of the target point $X_{T_c} = (x_{t_c}, y_{t_c}, 1)^T$. The coordinates of the target point in the image plane are back projected on the unitary sphere model using the relationship defined in Sect. 3.1,

Fig. 4 *Top Left*: Input normalised RGB image used for segmentation. *Top Right*: Result of image segmentation. *Bottom*: Result of detection and angle estimation on the owncraft. Angle φ (*tb* in the image) is the relative bearing to the target. In this Figure, *tb* is target bearing, *db* is desired target bearing, *yC* is yaw command and *pC* is pitch command

obtaining the target projection point in the unitary sphere $X_{t_s} = (x_{t_s}, y_{t_s}, z_{t_s})^T$. Finally, Eq. 5 is used to obtain the target attitude wrt vehicle. This equation already includes the rotation between the camera unitary sphere and the aircraft coordinate systems.

$$\phi_t = \arctan\left(\frac{y_{t_s}}{z_{t_s}}\right); \varphi_t = \arctan\left(\frac{x_{t_s}}{z_{t_s}}\right) \tag{5}$$

where ϕ_t and φ_t represent the target elevation and bearing as seen by our omnidirectional camera. φ_t is measured from the camera optical axis which coincides with vehicle longitudinal x axis. Clockwise $\varphi_t > 0$, $\varphi_t \in [0, \frac{\pi}{2})$ and counterclockwise $\varphi_t < 0$, $\varphi_t \in [0, -\frac{\pi}{2})$.

4 Control Approach

In this work, control is achieved using a two stage control architecture. A low level controller is responsible for attitude stabilization, whilst a high-level vision-based controller sends desired attitude values based on image features, therefore a visual servoing approach is used and defined as follows.

Let's assume the two angles in (5) are obtained from the detection approach. We can then follow a similar approach to that used in [16, 18, 19, 31] to construct the image jacobian (previously introduced in Sect. 2.2) using our point feature vector $s = s(\phi, \varphi)$ and the desired $s^* = s^*(\phi^*, \varphi^*)$. However, we anticipate that our image Jacobian will have a similar form to those presented in [16, 32]. Recalling (2) and re-writing in a two component form

$$\dot{s} = J(s, L)u = \{J(s, L)_{trans} + J(s)_{rot}\}u \tag{6}$$

where each component of $J(s, L)$ is a 2×3 matrix. The translational part is a function of depth and location of the image features, whereas the rotational only depends on location. Thus, (6) can be re-written as

$$\begin{bmatrix} \dot{\phi} \\ \dot{\varphi} \end{bmatrix} = \frac{1}{L} J(s)_{trans} [v_O]^T + J(s)_{rot} [\omega_O]^T \tag{7}$$

This equation fully relates velocity of features in the image plane with camera (vehicle) linear and angular velocities. The dependency in L and invertibility of J imposes some constrains in the minimum number of features, type of servoing [33] and knowledge of feature depth. To deal with these constrains the following assumptions are made:

1. Vehicle altitude is kept constant during the experiments. This lead us to approximate $v_{z_O} = 0$
2. Velocity in x is constant ($v_x = 0.5, \forall t \neq 0$)
3. Velocity in y is set to zero initially, i.e, $v_y = 0,\ t = 0$ however $v_y \neq 0, \forall\ t > 0$
4. Vehicle roll and pitch are approximately constant, therefore angular velocities in x and y are negligible. $\omega_{x_O} \approx \omega_{y_O} \approx 0$
5. Target distance is large enough to consider features at infinity. This lead us to a direct relationship between optical flow and angular velocity. This is possible because of the constant elements 0 and 1 in the jacobian, indicating that elevation is invariant to rotation about the optical axis and relative bearing angle is invariant to optical axis translation.

Taking the vector containing $\dot{\varphi}$ in (7), and rewriting

$$\dot{\varphi} = \frac{1}{L} [J(s)^{21} J(s)^{22} 0] \begin{bmatrix} v_{xo} \\ v_{yo} \\ v_{zo} \end{bmatrix} + [J(s)^{21} J(s)^{22} 1] \begin{bmatrix} \omega_{xo} \\ \omega_{yo} \\ \omega_{zo} \end{bmatrix} \tag{8}$$

finally, using our assumptions we can express (8) as

$$\omega_{zo} = -k_z(\varphi_t - \varphi_t^*) \tag{9}$$

This control law is used to keep the target at a given relative bearing angle $\varphi_t > 0$ in the image plane while the vehicle moves with constant velocity.

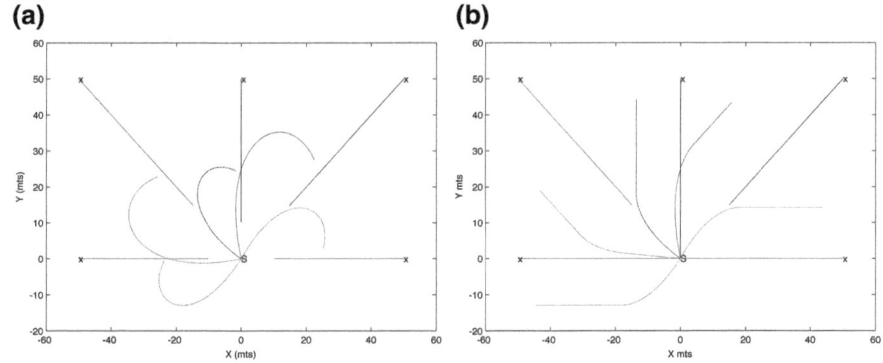

Fig. 5 Two simulated collision scenarios. *X* target starting position. *O* owncraft starting position. **a** No constrain has been imposed in owncraft heading. **b** Owncraft heading has been constrained

5 Approach Implementation and Results

5.1 Simulation

Prior to experimenting with real platforms we performed a series of simulations aiming at gaining more insights about the behaviour of our system. We simulated two vehicles using the dynamics depicted in (1). In particular, we were interested to validate our hypothesis of minimum distance (Sect. 2.1) and study the influence of desired view angle φ and constraints in vehicle heading (introduced in Sect. 2.1 as stopping condition) in the minimum avoidance distance. The two scenarios shown in Fig. 5 are representative of the simulations conducted. Figure 5a depict 5 trajectories generated with the owncraft starting at the same position with no constrains in owncraft heading. Figure 5b show the same scenarios but constraining owncraft heading. In these scenarios both aircraft initially moves towards each other in a collision path. In both cases the control law (9) was used to separate the vehicles with a desired view angle $\varphi_t^* = 50°$ (Fig. 6).

We were interested to study the influence of constraining owncraft heading in the minimum avoidance distance. It can be seen that in Fig. 6a which correspond to scenario (a) in Figure 5, the distance decreases 16.8 % when compared with scenario (b). This result indicates that constraining the heading angle guarantee greater minimum separation distance. Another factor to study is the view angle φ_t depicted in Fig. 6. Desired view angle and minimum separation distance are proportional which ratify our initial hypothesis, maximizing avoidance angle will maximize minimum avoidance distance.

It is important to note that although these results seem obvious, when using monocular sensors and no range to target is available, the issue of *when to break the avoidance maneuver* is still an open problem. Therefore, proposals such as constraining owncraft heading (implemented here), dynamic change of desired

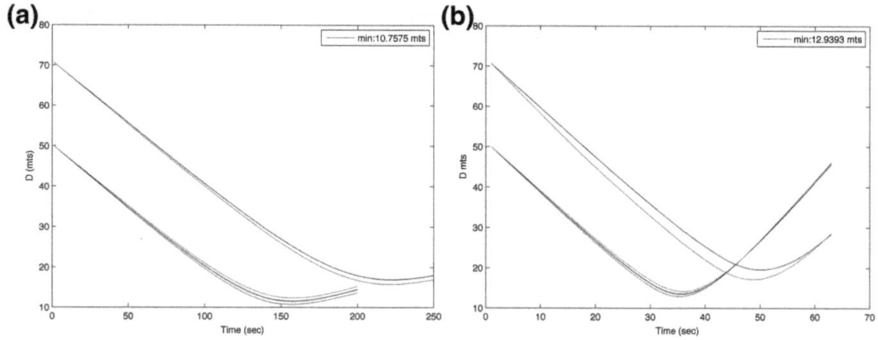

Fig. 6 Euclidean distance (r) owncraft-target for encounter scenarios in Fig. 5. **a** How distance decrease upto a minimum of 10.7 m for scenario in Fig. 5a. **b** Minimum distance of 12.9 m when owncraft heading is constrained, scenario in Fig. 5b

view angle (φ_t^*), thresholding line of sight rate, passive ranging are yet still to be tested in this context and benefits assessed.

5.2 *Flight Test*

We performed experimental tests using a Pelican quadrotor [34] and a static target (flight trial complexity is greatly reduced if a static target is used initially). The testbed is shown in Fig. 5. Onboard vision processing runs on a dual core Atom 1.6 GHz processor with 1 GB RAM, wireless interface and USB camera. Our vision-based controller communicates via serial interface with the low-level

Fig. 7 CVG-UPM [35] Pelican QuadCopter testbed used for sense and avoid experiments

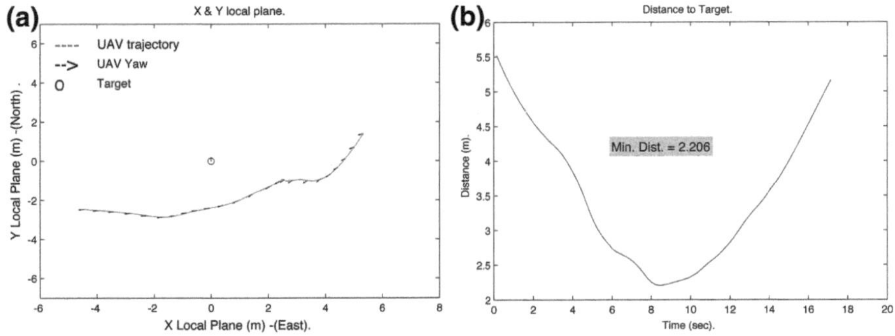

Fig. 8 Experiment 1, run 6. **a** The vehicle trajectory and location of the target (local tangent plane). **b** The evolution of L with a minimum value of 2.206 m

autopilot. This controller sends velocity references to the autopilot creating a *dynamic look-and-move* servoing architecture at 20 Hz.

Figures 8 and 9 present two of the most representative experiments conducted over several non-consecutive days. All plots show the vehicle trajectory, target location and distance L, where (O) is the starting vehicle position and (X) the target position. Vehicle evasive trajectory depends on initial target detection in the image. It will move left around the target if $\varphi_0 > 0$ setting $\varphi_t^* = 50°$, and right around the target if $\varphi_0 < 0$ setting $\varphi_t^* = -50°$.

During these experiments, the vehicle was set to keep constant longitudinal velocity of 0.5 m at fixed altitude. In all flights the vehicle starts heading toward the target with constant velocity, when the target is detected and consistently tracked, servoing commands are sent to the autopilot until the experiment stops.

In all cases, we can observe the evolution of the owncraft-target distance (L). We achieved a minimum separation of 2.2 m in the test-case depicted in Fig. 8 and 3.1 m in the test-case shown in Fig. 9. These distances are enough to achieve successful avoidance in this case given that our target had a radius of 0.2 m and our vehicle radius is 0.2 m approx (therefore, $L > 0.4$ m).

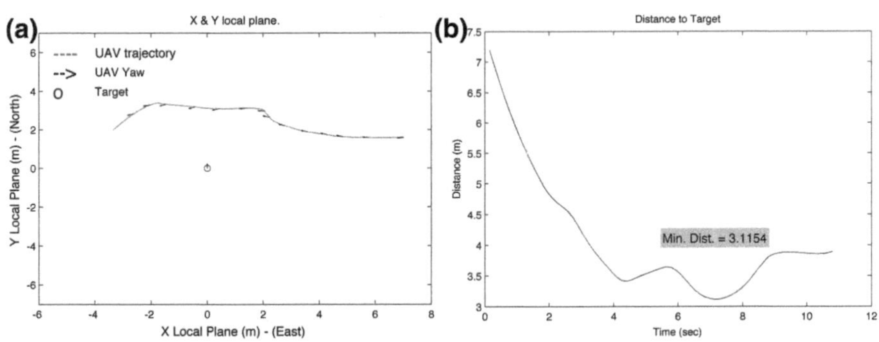

Fig. 9 Experiment 2, run 16. **a** The evasive trajectory of the vehicle and the location of the target (local tangent plane). **b** The evolution of L with a minimum value of 3.1154 m

In other three cases not shown here, the minimum separation achieved was 1.3 m, 2.34 m and 2.35 m, respectively. These magnitudes are well above 0.4 m which corroborates our hypothesis of $L > 2R$.

Additional experiments can be found in the project web page [35].

6 Conclusion

This chapter presented a real time method for UAS *see and avoid* based only in the visual information provided by an omnidirectional camera sensor. The system was tested on real UAS flights in collision scenarios, showing that the proposed method is an efficient technique for real time evasion. In addition, the proposed hypothesis that *maximizing avoidance angle will maximise minimum avoidance distance* has been demonstrated in real flights.

Experimental results have validated the proposed approach, showing a good response even if the evasion maneuver starts with a small range to the colliding target. Future work includes testing the approach in a dynamic scenario with two vehicles engaged in collision scenario.

Acknowledgments The work presented in this chapter is the result of an ongoing collaboration between the Computer Vision Group - Universidad Politécnica de Madrid and the Australian Research Centre for Aerospace Automation (ARCAA). This work has been supported by the European Commission and the Australian Academy of Science through a FP7-PEOPLE-IRSES-2008 grant (PIRSES-GA-2009-230797 - ICPUAS). The authors also would like to thank Universidad Politécnica de Madrid, Consejería de Educación de la Comunidad de Madrid and Fondo Social Europeo (FSE). The authors acknowledge the contribution and support from Troy Bruggemann, Miguel Olivares Mendez and Carol Martinez.

References

1. I. Dryanovski, W. Morris, J. Xiao, in *Multi-volume Occupancy Grids: An Efficient Probabilistic 3d Mapping Model for Micro Aerial Vehicles*. Intelligent robots and systems (IROS), 2010 IEEE/RSJ international conference on, pp. 1553–1559 (2010). http://dx.doi.org/10.1109/IROS.2010.5652494
2. S. Lupashin, A. Schö andllig, M. Sherback, R. D'Andrea, in *A Simple Learning Strategy for High-speed Quadrocopter Multi-flips*. Robotics and automation (ICRA), 2010 IEEE international conference on, pp. 1642–1648 (2010). http://dx.doi.org/10.1109/ROBOT.2010.5509452
3. N. Michael, D. Mellinger, Q. Lindsey, V. Kumar, The grasp multiple micro-uav testbed. Robot. Autom. Mag., IEEE **17**(3), 56–65 (2010). http://dx.doi.org/10.1109/MRA.2010.937855
4. B. Gielow, Congress sets 2015 deadline for unmanned aircraft systems to fly in the national airspace the u.s. house of representatives passed a compromise faa bill including uas provisions. AUVSI. Feb 2012. http://www.auvsi.org/news/#HouseFAAPass (2012)
5. ASTM, F2411–07 - standard specification for design and performance of an airborne sense-and-avoid system (2007)

6. US-OSoD, Unmanned systems roadmap 2009–2034, 4th edn. Office of the Secretary of Defense, Washington, District of Columbia (2009)
7. B. Karhoff, J. Limb, S. Oravsky, A. Shephard, in *Eyes in the Domestic Sky: An Assessment of Sense and Avoid Technology for the Army's Warrior Unmanned Aerial Vehicle*. IEEE systems and information engineering design symposium, pp. 36–42 (2006). http://dx.doi.org/10.1109/SIEDS.2006.278710
8. O. Shakernia, W.Z. Chen, V.M. Raska, in *Passive Ranging for uav Sense and Avoid Applications*. Proceedings of the AIAA infotech@aerospace conference, Arlington, Virginia, pp. 1–10 (2005)
9. J. Driessen, Object tracking in a computer vision based autonomous see-and-avoid system for unmanned aerial vehicles. Master's thesis, Department of Numerical Analysis and Computer Science. Royal Institute of Technology, Stockholm, Sweden (2004)
10. G. Fasano, Multisensor based fully autonomous non-cooperative collision avoidance system for uavs. Phd thesis, University of Naples (2008)
11. J. Saunders, R. Beard, in *Vision-based Reactive Multiple Obstacle Avoidance for Micro Air Vehicles*. 2009 American control conference, pp. 5253–5258 (2009)
12. Y. Watanabe, A.J. Calise, E.N. Johnson, in *Vision-based Obstacle Avoidance for uavs*. Proceedings of the AIAA guidance, navigation and control conference, Hilton Head, South Carolina, pp. 20–23 (2007)
13. P. Angelov, C.D. Bocaniala, C. Xideas, C. Patchett, D. Ansell, M. Everett, G. Leng, in *A Passive Approach to Autonomous Collision Detection and Avoidance*. Proceedings of the 10th international conference on computer modeling and simulation, Cambridge, UK, pp. 64–69 (2008)
14. L. Mejias, S. Mcnamara, J. Lai, J.J. Ford, in *Vision-based Detection and Tracking of Aerial Targets for uav Collision Avoidance*. Proceedings of the 2010 IEEE/RSJ international conference on intelligent robots and systems (IROS), pp. 87–92 (2010)
15. J. Lai, L. Mejias, J.J. Ford, Airborne vision-based collision-detection system. J. Field Robot. **28**(2), 137–157 (2011)
16. P. Corke, in *Spherical Image-based Visual Servo and Structure Estimation*. Robotics and automation (ICRA), 2010 IEEE international conference on, pp. 5550–5555 (2010)
17. P. Corke, S. Hutchinson, A new partitioned approach to image-based visual servo control. Robot. Autom., IEEE Trans. **17**(4), 507–515 (2001). http://dx.doi.org/10.1109/70.954764
18. F. Chaumette, S. Hutchinson, Visual servo control. i. basic approaches. Robot. Autom. Mag., IEEE **13**(4), 82–90 (2006). http://dx.doi.org/10.1109/MRA.2006.250573
19. F. Chaumette, S. Hutchinson, Visual servo control. ii. advanced approaches [tutorial]. Robot. Autom. Mag., IEEE **14**(1), 109–118 (2007). http://dx.doi.org/10.1109/MRA.2007.339609
20. C. Baumgarten, G. Farin, Approximation of logarithmic spirals. Comput. Aided Geom. Des. **14**(6), 515–532 (1997). http://www.sciencedirect.com/science/article/pii/S01678396960%0043X
21. A. Tews, J. Roberts, K. Usher, in *Is the sun too Bright in Queensland? An Approach to Robust Outdoor Colour Beacon Detection*. Australasian conference on robotics and automation (ACRA) (2005)
22. G.R. Bradski, Computer vision face tracking for use in a perceptual user interface. Intel. Technol. J. (No. Q2) (1998)
23. S. Baker, S.K. Nayar, A theory of single-viewpoint catadioptric image formation. Int. J. Comput. Vis. **35**(2), 1–22 (1999)
24. J.A. Barreto, H. Araujo, in *Issues on the Geometry of Central Catadioptric Image Formation*. Computer vision and pattern recognition, 2001. CVPR 2001. Proceedings of the 2001 IEEE computer society conference on 2, II-422-II-427 vol. 2 (2001). http://dx.doi.org/10.1109/CVPR.2001.990992
25. C. Geyer, K. Daniilidis, A unifying theory for central panoramic systems and practical applications. ECCV (2), 445–461 (2000)

26. X. Ying, Z. Hu, in *Can We Consider Central Catadioptric Cameras and Fisheye Cameras within a Unified Imaging Model*. Computer vision - ECCV 2004, Lecture notes in computer science, vol. 3021, pp. 442–455. (Springer, Berlin, Heidelberg, 2004)
27. D. Scaramuzza, Omnidirectional vision: from calibration to robot motion estimation. Ph.D. thesis, ETH Zurich (2008)
28. C. Mei, P. Rives, in *Single View Point Omnidirectional Camera Calibration from Planar Grids*. IEEE international conference on robotics and automation (2007)
29. D. Austin, N. Banes, in *Red is the New Black—Or Is It?*. Australasian conference on robotics and automation (ACRA) (2003)
30. K. Fukunaga, L. Hostetler, The estimation of the gradient of a density function, with applications in pattern recognition. Inf. Theory, IEEE Trans. **21**(1), 32–40 (1975). http://dx.doi.org/10.1109/TIT.1975.1055330
31. L. Mejias, J. Ford, J. Lai, in *Towards the Implementation of Vision-based uas Sense-and-avoid*. Proceedings of the 27th international congress of the aeronautical sciences (ICAS 2010 CD-Rom) (2010)
32. S. Benhimane, E. Malis, in *A New Approach to Vision-based Robot Control with Omni-directional Cameras*. Robotics and automation, 2006. ICRA 2006. Proceedings 2006 IEEE international conference on, pp. 526–531 (2006). http://dx.doi.org/10.1109/ROBOT.2006.1641764
33. S. Hutchinson, G. Hager, P. Corke, A tutorial on visual servo control. Robot. Autom., IEEE Trans. on **12**(5), 651–670 (1996). http://dx.doi.org/10.1109/70.538972
34. AscendingTechnologies: Ascending technologies. http://www.asctec.de (2010)
35. CVG-UPM: Universidad Politécnica de Madrid. Computer Vision Group. Vision for UAV Project. http://www.vision4uav.com (2010)

Design and Characterisation of a Full-Field Range Imaging Camera for Use in Mobile Applications

Ben Drayton, Dale A. Carnegie and Adrian A. Dorrington

Abstract This chapter will outline the design and testing of a compact, configurable, indirect time of flight range imager. Potential applications include areas such as mobile robotics, where the introduction of robots into more complex environments necessitates the development of more advanced sensor systems. This chapter evaluates the range imager, particularly with regard to motion, and demonstrates that an improvement in the linearity of the motion error is possible by changing the order of frame measurements.

1 Introduction

An ideal range sensor should provide a full field range measurement with high accuracy and precision in real time. In emerging fields such as mobile robotics, fast and accurate range sensors are necessary to improve the viability of real world applications such as urban search and rescue [1], healthcare [2], the military [3] and space exploration [4]. Without adequate sensor systems that allow mobile robots to quickly gather detailed range information, robotic applications in these fields are limited to human control and guidance.

Based on The Development of a Time of Flight Range Imager for Mobile Robotics, by Ben Drayton, Dale A. Carnegie and Adrian A. Dorrington which appeared in the Proceedings of the 5th International Conference on Automation, Robotics and Applications (ICARA 2011).

B. Drayton · D. A. Carnegie (✉)
School of Engineering and Computer Science,
Victoria University of Wellington, Wellington, New Zealand
e-mail: dale.carnegie@vuw.ac.nz

A. A. Dorrington
School of Engineering, University of Waikato, Hamilton, New Zealand

G. Sen Gupta et al. (eds.), *Recent Advances in Robotics and Automation*,
Studies in Computational Intelligence 480, DOI: 10.1007/978-3-642-37387-9_11,

A number of different types of sensors are widely available that provide range information. Common sensors used are position sensitive detectors (PSD), ultrasonic sensors and laser range finders. However as of yet there is no single sensor that provides a high accuracy full field range image in real time with a high frame rate.

A promising technology is using the indirect time of flight (ToF) method. This method uses the properties of modulated light and a specialized camera to take a full field of view range measurement in real time. This chapter will outline the development of an Amplitude Modulated Continuous Wave indirect ToF sensor system suitable for use in mobile robotic applications and the current challenges facing the use of these systems. It will then outline a method for improving the performance of such systems.

2 Background Theory

Homodyne indirect time of flight cameras operate by modulating both a light source and a detector at the same high frequency (generally 10–100 MHz). The time taken for the light to travel to the object and return introduces a phase shift, φ, between the reflected wave and the modulation of the shutter, which is related to the distance, d, by the equation

$$d = \frac{c\varphi}{4\pi f} \tag{1}$$

where c is the speed of light and f is the modulation frequency.

Typically an intensity based camera is used to measure the overlap between the returning light and the shutter; however intensity is not solely dependent on the relative phase of the returning light and the shutter but also on the reflectivity of the object, the inverse square relationship caused by waves spreading from a point source and background light levels. To get a phase measurement independent of these factors, multiple frames are recorded with a differing phase shift introduced between the emitted light and the shutter. A Discrete Fourier Transform is used to calculate the phase. It is common to use four frames per measurement as this simplifies the phase calculation to

$$\varphi = \tan^{-1}\left(\frac{I_1 - I_3}{I_0 - I_2}\right), \tag{2}$$

where I_n is the intensity of frame n. The intensity I_n can be modeled as

$$I_n = A\cos(\theta - \varphi) + B, \tag{3}$$

where A is a gain factor that includes the gain of the sensor, the power of the laser diodes and the reflectivity of the object, θ is the introduced phase shift between the

Table 1 Comparison of commercial indirect ToF cameras

Feature	CamCube 3.0	SR4000
Resolution	200 × 200 Pixels	176 (h) × 144 (v) pixels
Range	0.3–7 m	0.8–5 m
Speed	40 FPS	50 FPS
Field of view	40° × 40°	43.6° (h) × 34.6° (v)
Repeatability (1σ)	3 mm	4 mm
Cost	~$12,000 USD	~$10,000 USD

modulation of the light and the camera and B is an offset due to the DC offset of the ADC and background light.

Due to the high shutter speeds required, mechanical shutters are not suitable for indirect time of flight systems. Previously image intensifiers were used to provide high speed shuttering [5] however, these intensifiers are expensive, high voltage, bulky devices that are not suitable for mobile robotics. The recent development of solid state CMOS cameras which can be electronically switched at high frequencies [6] has allowed indirect time of flight measurement to become suitable for mobile robotics applications. Because of the additional circuitry required to make them electronically modulatable, these sensors do not have the same resolution traditional digital cameras can achieve. Resolutions of up to 200 × 200 pixels are currently available, and resolutions of 320 × 200 have been demonstrated [7], although it is not unreasonable to expect sensors approaching 1 MegaPixel in the near future.

Several commercial indirect time of flight sensor systems are currently available. Two of the most prominent commercial options are the PMD[vision]® CamCube 3.0 released by PMDTechnologies in June 2010[1] and the SR4000 released by Mesa Imaging in 2008.[2] Table 1 shows a comparison of the main features of these two systems.

Although these cameras can provide high quality range images, they have a high cost and lack the customizability often required for academic research. Because of this it was decided to develop a flexible custom range finding system suitable for mobile applications research.

3 Design of a Compact Range Imaging System

The Victoria Range Imaging System is designed specifically for mobile robotic research. It is based on a prototype system which was desktop sized and required several external power supplies to operate [8]. Because of this, size and

[1] http://www.pmdtec.com/

[2] http://www.mesa-imaging.ch/

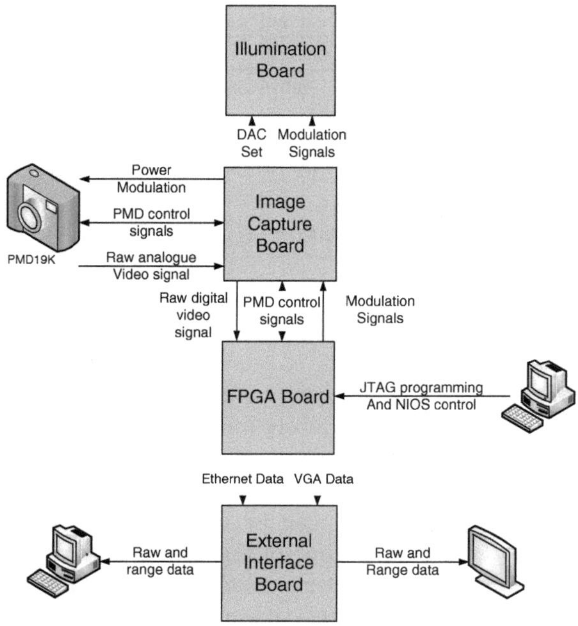

Fig. 1 Data flow between the boards in the compact range imaging system

customizability are the most critical design parameters. Special attention also had to be paid to power distribution throughout the system. A modular design utilizing four separate boards stacked on top of each other is used as modularity means upgrading or changing the design does not require an entirely new system to be made. It also provides greater physical flexibility as different optical lenses can be used while keeping the laser diodes in a concentric circle aligned with the end of the lens. Using a stacked design also creates a smaller physical footprint than a single large board would.

The four boards implemented in this design are the External Interface Board, the FPGA Board, the Image Capture Board and the Illumination Board. Each board is supplied with unregulated DC power in a daisy chain configuration. This distributed design provides the most flexibility as each board is then a self contained system independent of the other boards.

Figure 1 shows how data flows between the boards. The FPGA is programmed via a JTAG interface. This interface also allows a control computer to communicate with a NIOS II soft processor that is implemented on Altera FPGAs. This processor is used to modify the parameters of the system during run time without having to reprogram the FPGA. It also controls the capturing of data frames via the Ethernet interface to a computer for long term data storage.

A custom FPGA board implementing a Cyclone III FPGA is used to provide programmable logic for central control of the system. Table 2 shows a comparison of the features of a selection of relevant FPGAs. The Cyclone III family of FPGAs

Table 2 Comparison of FPGAs

FPGA	# of logic elements	# of RAM bits	# of PLLs	Cost (USD)
Stratix III EP3SL150 F1152C2N	142,500	6,543,360	12	$3,777
Cyclone III EP3C120 F780C7	119,088	3,981,312	4	$502
Cyclone III EP3C40 F780C6	39,600	1,161,216	4	$167

are approximately an order of magnitude cheaper than the Stratix III family and with the addition of external memory can provide the same functionality. The board designed for this system is compatible with both the EP3C120 and EP3C40 FPGAs.

Each pixel is stored as a 16 bit number. For our sensor's current resolution of 160×120 pixels this means each frame requires 300 Kb of RAM. Six frames are stored in buffers at any time requiring 1,800 Kb of RAM. There is also some RAM required for operation of the NIOS soft processor meaning the amount of memory required for range imaging using our current sensor is approximately 2 Mbits. Because of this, the EP3C120 was used for this build since for this sensor no external memory was required which greatly simplifies this initial design.

To allow for future improvements in sensor technology external memory chips are included on the FPGA board. DDR2 memory was chosen for its high transfer speeds and large memory sizes.

A Flash memory chip is included to provide non-volatile storage of the FPGA configuration. On power-up the FPGA is programmed by the Flash memory meaning an external computer using the JTAG interface is only necessary if the program is being changed. A photograph of the range imaging system is shown in Fig. 2.

The Illumination Board provides modulated light to illuminate the scene being measured. It has sixteen 658 nm laser diodes run in a constant current configuration providing 800 mW of optical power. The current is set using an AD5311 DAC set by the FPGA via a two wire interface (TWI). To avoid overdriving and damaging the laser diodes, the current is slowly increased to allow the diode's temperature to stabilise at a safe operating temperature before applying full power. Because the distance between the Illumination Board and the Image Capture Board must be able to accommodate different sized lenses, both the TWI signals and the modulation signals are sent via SATA interfaces.

Fig. 2 Photograph of the compact range imaging system [8]

The Image Capture Board contains the modulation drivers for the PMD19K2 sensor. The sensor is split into four 40 × 120 pixel modulation blocks. Each block has two modulation inputs (ModA and ModB, each of which has an associated analogue video output). Eight Ultra High Current pin drivers are used to drive the modulation inputs.

Analogue video signals from the sensor are converted into digital frames using a AD9826 16 bit imaging signal processor ADC from Analog Devices. A 172 pin high speed mezzanine connector connects the Image Capture Board to the FPGA board.

The External Interface Board provides three connections for interfacing with the range finding system. A VGA connection allows the display of data in real time on an external 640 × 480 monitor. This display is split into four zones which can be configured to display data at any stage of the processing steps completed by the FPGA; from raw data through to range images. An Ethernet connection is provided to transfer images to an external computer for long term storage. A USB connection provides a secondary means to store image data onto an external computer and can also be used to send commands to configure and control the imaging system. The External Interface Board is connected to the FPGA board by a 172 pin high speed mezzanine connector.

4 Problems with Motion

Because indirect time of flight measurement requires a number of successive frames to be acquired, object motion introduces errors. Obviously motion errors have a large impact on the viability of a system for mobile robotics applications. These errors can be classified as lateral motion errors from movement across the field of view and axial motion errors from movement along the viewing axis. All current indirect range finding cameras suffer from these motion errors.

If an object moves laterally within the ranger's field of view this causes problems for the pixels at the edges of the object. Existing methods of solving this include the use of a 2D camera for edge detection [9] and optical flow algorithms [10].

Axial motion has a more subtle effect on range imaging than lateral motion as the change in distance for successive frames is generally smaller. If we allow for an object to move with a constant velocity v in (2), and using (3) to model the intensity of the light, the phase measured by the system φ_{m} is described by the equation

$$\varphi_m = \tan^{-1}\left(\frac{\cos\left(\frac{\pi}{2} - \varphi_1\right) - \cos\left(\frac{3\pi}{2} - \varphi_3\right)}{\cos(-\varphi_0) - \cos(\pi - \varphi_2)}\right) \tag{4}$$

Equation (5) is used to relate the phase at any frame φ_{n}, where n is the frame number, to the phase at the start of the measurement φ_0, where t_i is the frame period and $\alpha = 4\pi f t_i / c$, which is constant for a given camera configuration.

$$\begin{aligned}\varphi_n &= \varphi_{n-1} + \frac{4\pi f v t_i}{c} \\ \varphi_n &= \varphi_0 + \frac{4\pi n f v t_i}{c} \\ \varphi_n &= \varphi_0 + n v \alpha\end{aligned} \tag{5}$$

By substituting (5) into (4) and simplifying the equation relating the measured phase φ_m with the actual phase at the first frame, φ_0 is shown to be

$$\varphi_0 = \tan^{-1}\left(\frac{\tan(\varphi_m)\cos(2v\alpha) + \tan(\varphi_m) - \sin(v\alpha) - \sin(3v\alpha)}{\cos(v\alpha) + \cos(3v\alpha) + \tan(\varphi_m)\sin(2v\alpha)}\right) \tag{6}$$

This theoretical error is shown against actual distance in Fig. 3 for three different velocities. The error has an average offset that is proportional to velocity and also has an oscillatory relationship with distance. This large non-linearity is highly undesirable.

5 Improving the Motion Response of Indirect ToF Cameras

A promising approach to improving the motion response of indirect time of flight cameras is changing the order in which the samples for the Discrete Fourier Transform are taken. If the order of the phase steps is changed from the traditional 0°, 90°, 180°, 270° then (6) can be generalised to

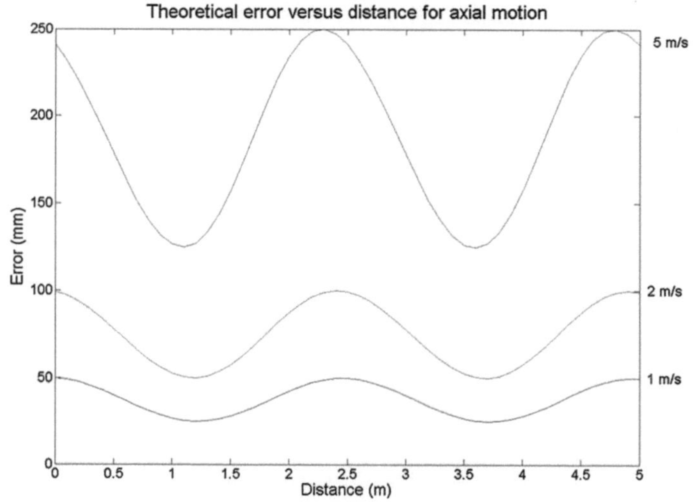

Fig. 3 Theoretical motion error versus actual distance for various velocities

$$\varphi_0 = \tan^{-1}\left(\frac{\tan(\varphi_m)[\cos(m_0 v\alpha) + \cos(m_2 v\alpha)] - \sin(m_1 v\alpha) - \sin(m_3 v\alpha)}{\cos(m_1 v\alpha) + \cos(m_3 v\alpha) + \tan(\varphi_m)[\sin(m_0 v\alpha) + \sin(m_2 v\alpha)]}\right) \quad (7)$$

where $m_{0,1,2,3}$ is some ordering of [0 1 2 3] indicating the order in which frames were measured. Figure 4 shows the result that changing the order of frames has on the error. Only four orderings are shown, all other permutations produce the same results to those shown with different phase shifts. While the average error remains unchanged, the change in error with distance can be greatly reduced by a careful selection of sampling order.

Experimental results were taken using a linear table to move a target axialy. The same motion path was repeated 100 times and the mean value was taken. The target was then moved to the same position it was at the start of each frame and 200 static frames were recorded and averaged. The error was then taken as the difference between the static and dynamic cases. Measurements were taken using both the standard order and a theoretically better order. These are compared with data taken from a SwissRanger 4,000 camera (See Foot note 2).

Figure 5 shows that the relationship between velocity and the observed motion error for both the Swiss Ranger camera and the Victoria Ranger camera (using the standard phase calculation method) is as predicted by theory both in terms of average offset and oscillatory amplitude. Further as also predicted, by changing the order of the measurements, the oscillatory amplitude of the error can be reduced significantly although the average error is still the same.

The measured distance versus actual range was tested by moving a target to set distances and recording 100 samples. The mean distance of these measurements is

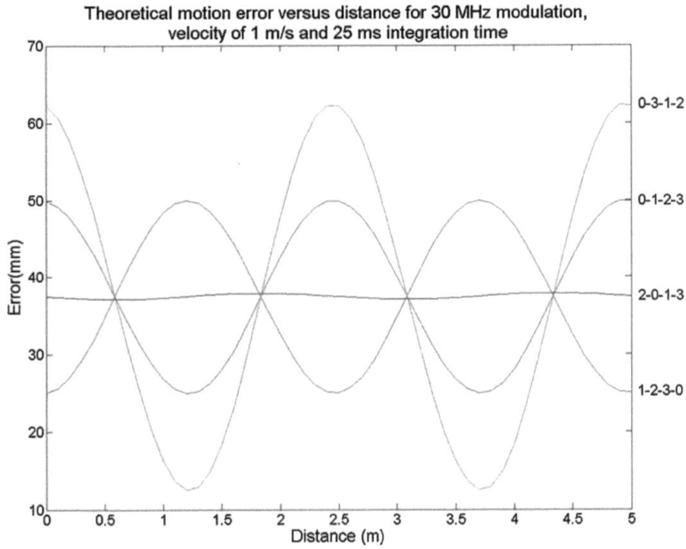

Fig. 4 Theoretical motion error versus distance for selected orderings of frame offset

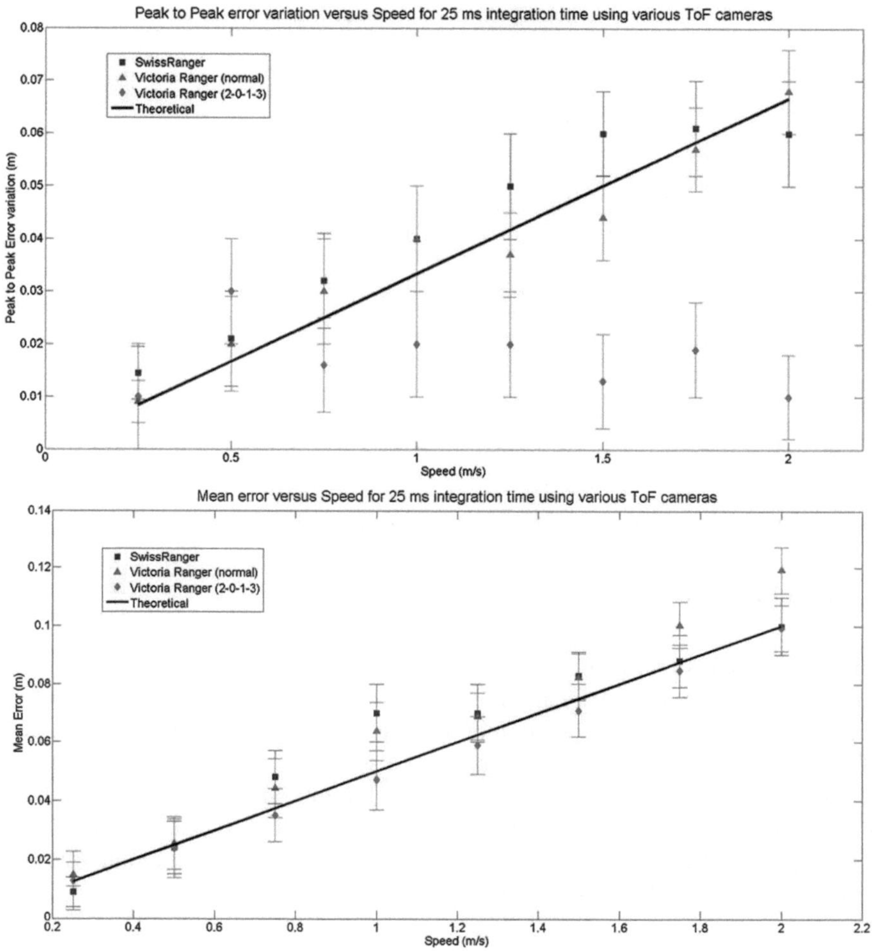

Fig. 5 Experimental motion error versus distance for selected orderings of frame offset

shown in Fig. 6. As the distance increases, the amplitude of the returning light decreases so the precision of the system decreases. Error bars indicate one standard deviation. This figure shows that using a different order of measurement has not degraded the linearity or precision of the system.

These data support the theoretical model outlined in this chapter and show that at least one commercial camera has not implemented an effective method to improve on the velocity errors described above. Using a different ordering of measurements provides significantly decreased ripple in the distance response to a moving object with no increased computational effort or effect on the static quality of the measurements.

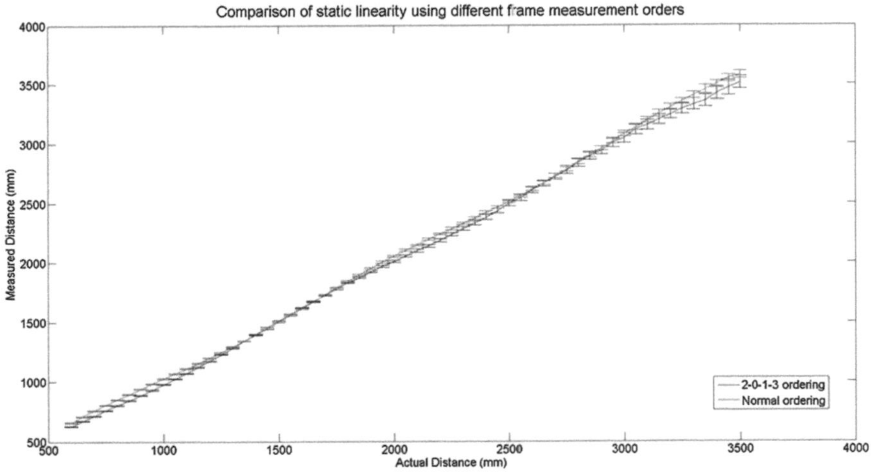

Fig. 6 Static measured distance versus actual distance for different orders

6 Conclusions

This chapter has described the development of a compact indirect time of flight range imager designed specifically for mobile robotics applications and research. It has outlined the limitations due to motion both of this system and of indirect time of flight cameras in general and has presented solutions or improvements that can be made to make the system more suitable for this application. Experimental results show that an improvement in the linearity of motion error is possible by changing the order of the frame measurements. Future work is still required to find a method to reduce the average offset error.

References

1. R. Murphy, Trial by fire [rescue robots]. IEEE Robot. Autom. Mag. **11**(3), 50–61 (2004)
2. E. Broadbent, R. Stafford, B. MacDonald, Acceptance of healthcare robots for the older population: review and future directions. Int. J. Soc. Robot. **1**(4), 319–330 (2009)
3. D. Voth, A new generation of military robots. Intell. Syst. IEEE **19**(4), 2–3 (2004)
4. K. Daniel, R. Some, NASA advances robotic space exploration. Computer **36**(1), 52–61 (2003)
5. M.J. Cree, A.A. Dorrington, R. Conroy, A.D. Payne, D.A. Carneigie, The waikato range imager. Image Vis. Comput. N.Z. 233–238 (2006)
6. T. Ringbeck, T. Möller, B. Hagebeuker, Multidimensional measurement by using 3-D PMD sensors. Adv. Radio Sci. **5**, 135–146 (2007)
7. Canesta3D: Point Cloud Demo, using Canesta's 320 × 200 3D Tof image sensor. http://www.youtube.com/watch?v=4xIXsJuH74c. Accessed 2 Mar 2012

8. J. McClymont, D. Carnegie, A. Jongenelen, B. Drayton, *The Development of a Full-Field Image Ranger System For Mobile Robotic Platforms*. Sixth IEEE International Symposium on Electronic Design, Test and Applications (DELTA), pp. 128–133, 17–19 Jan 2011
9. O. Lottner, A. Sluiter, K. Hartmann, W. Weihs, *Movement Artefacts in Range Images of Time-Of-Flight Cameras*. International Symposium on Signals, Circuits and Systems ISSCS, (2007), pp. 1–4
10. M. Lindner, A. Kolb, *Compensation of Motion Artifacts for Time-Of-Flight Cameras*. DAGM Workshop on Dynamical 3D Imaging, (Springer, Berlin, 2009), pp. 16–27

Improving Robustness of Vision Based Localization Under Dynamic Illumination

Jared Le Cras, Jonathan Paxman and Brad Saracik

Abstract A dynamic light source poses significant challenges to vision based localization algorithms. There are a number of real world scenarios where dynamic illumination may be a factor, yet robustness to dynamic lighting is not demonstrated for most existing algorithms. Localization in dynamically illuminated environments is complicated by static objects casting dynamic shadows. Features may be extracted on both the static objects and their shadows, exacerbating localization error. This work investigates the application of a colour model which separates brightness from chromaticity to eliminate features and matches that may be caused by dynamic illumination. The colour model is applied in two novel ways. Firstly, the chromaticity distortion of a single feature is used to determine if the feature is the result of illumination alone. These features are removed before the feature matching process. Secondly, the chromaticity distortion of features matched between images is examined to determine if the monochrome based algorithm has matched them correctly. The evaluation of the techniques in a Simultaneous Localization and Mapping (SLAM) task show substantial improvements in accuracy and robustness.

Based on Vision Based Localization under Dynamic Illumination, by Jared Le Cras, Jonathan Paxman and Brad Saracik which appeared in the Proceedings of the 5th International Conference on Automation, Robotics and Applications (ICARA 2011).

J. Le Cras · J. Paxman (✉) · B. Saracik
Curtin University, Bentley, Western Australia, Australia
e-mail: j.paxman@curtin.edu.au

G. Sen Gupta et al. (eds.), *Recent Advances in Robotics and Automation*, Studies in Computational Intelligence 480, DOI: 10.1007/978-3-642-37387-9_12,

1 Introduction

Dynamically illuminated environments pose a challenge to vision based localization and mapping that is often overlooked due to the prevalence of static light sources in environments used for the evaluation of such systems. Successful vision based localization algorithms [1–5] have all been evaluated in scenarios with a primary static light source. Scenarios exist, however, where a primary static light source cannot be guaranteed, yet there are no localization algorithms focused on this challenging application. Examples of scenarios with dynamic illumination include slow moving equipment affected by the movement of the sun, aerial photographs taken at different times of day and the primary motivation for this work: underground mining environments where dynamic vehicles carry their own light sources. We found that algorithms such as Scale Invariant Feature Transform (SIFT) [6] and Features from Accelerated Segment Test (FAST) [7] corner detection frequently extract shadow boundaries as potential features for tracking. This is not a problem within statically illuminated environments, where shadows are as reliable features as the objects that cast them. However, when the primary light source is dynamic, the inconsistency of stationary objects and moving shadows causes difficulties for localization algorithms, as demonstrated in the examples presented in Sect. 4.3.

This work is motivated by our efforts in the surveying and mapping of underground mine systems [8, 9]. We examine an approach to distinguish object based features from shadow based features in order to improve localization by tracking only those considered to be robust to variations in illumination. We also consider a technique to remove mismatches produced by monochrome matching algorithms in dynamically lit scenes by making use of colour information.

Our work is structured as follows: Firstly, a review of research related to the field is presented, focusing on work improving the robustness of feature matching and localization in difficult lighting conditions. The techniques applied and developed in this chapter are then described, examining the advantages of including colour information to improve the illumination robustness of feature matching. This is followed by an extensive evaluation on simulated datasets that allow the position of the camera and light source to be explicitly controlled to accurately assess the effects of dynamic illumination on localization. The evaluation is then extended to a real world dataset to demonstrate the robust nature of the developed algorithm.

2 Related Work

Vision based localization has been an area of active research for the last 20 years. Successful localization has been realized by many research groups [1–5]. However, all of the scenarios used to evaluate these localization algorithms have involved a primarily static light source and therefore, the results cannot necessarily be transferred to scenarios that involve a primarily dynamic light source. To improve the robustness of vision based localization to dynamic illumination, the properties of the features used for localization must be examined. Two feature extraction techniques that are prevalent in vision based localization algorithms are the Scale Invariant Feature Transform (SIFT) [6] and Features from Accelerated Segment Test (FAST) corner detection [7]. SIFT has been proven to be a robust technique for extracting and matching features across multiple images [10], whereas FAST corner detection is designed to work with real-time video frame rates. Filtering the features extracted by these two techniques can improve localization as well as partially improve illumination robustness.

SIFT is scale invariant and partially invariant to orientation and slight changes in illumination [6]. A significant problem experienced by those that use SIFT for feature extraction and localization is the large number of features extracted from each image. Improving the efficiency of feature matching between images is still an area of active research.

FAST corner detection can produce a smaller number of features than SIFT and is consequently faster and more suitable for real time implementation. A region is determined to be a feature if a centroid pixel is significantly darker or lighter than the surrounding circumference of 16 pixels. To improve accuracy when operating with low numbers of features, Chekhlov et al. [11] use an estimation of camera pose during the matching process.

A prediction of feature locations based on a movement model is used to reduce search regions for matching and improve matching robustness, slightly improving illumination invariance. This technique is applied by Civera et al. [1] in a (six degree of freedom) monocular EKF SLAM algorithm. The filtering techniques developed in our work are evaluated with a localization algorithm based on Civera's.

One limiting factor of these modest improvements to illumination invariance is the use of monochrome images. Colour information has been successfully exploited in other areas of computer vision research to improve illumination invariance. Swain and Ballard [12] were the first to use a three dimensional RGB histogram to describe pixel colour values for recognition, known as 'colour indexing'. Finlayson et al. [13] expanded upon this approach to produce basic, lighting invariant, object recognition using 'colour angles'. They demonstrated improved efficiency over colour indexing, at the cost of number of correct recognitions. An alternative approach trialed by Geusebroek et al. [14] estimated the original colour of an object mathematically given known lighting conditions. This technique was robust to viewing direction, surface orientation, highlights, illumination direction, illumination intensity, illumination colour and inter-reflection, but was limited to a range of known materials and strictly controlled lighting conditions. Successful shadow detection in a dynamic environment was achieved by Horprasert et al. [15] to detect moving objects against a static background. An RGB histogram compared successive pixel colours between frames to produce a colour model based on two components: brightness and chromaticity. These two components were used to examine a change in pixel colour over time such that a differentiation could be made between static background, shadow and dynamic foreground. The concept behind the technique is that a shadow cast on an object will result in a large change in brightness, whereas the chromaticity difference between pixels will remain small. Our approach to the filtering of SLAM features is based on the brightness and chromaticity model and will be examined in detail in Sect. 3.

The use of colour information to improve feature extraction algorithms is not a novel concept. Both SIFT and FAST corner detection use only the monochrome version of an image, so it has long been assumed that the wealth of information available in the colour image could be used to improve descriptors and matching. Abdel-Hakim and Farag [16] combine both colour and geometrical information in the feature description, resulting in an increased number of extracted and matched features. Burghouts and Geusebroek [17] integrate the Gaussian opponent colour model into SIFT; this model consists of intensity, red-green and yellow-blue channels. The separation of intensity enhances illumination invariance and results in improved matching when compared to standard SIFT over a range of lighting conditions. This work was aimed at a general object recognition problem where the constraints on computation time are less strict than those involved in work on real time localization, which is the problem addressed in our chapter.

The colour model employed by this chapter in Sect. 3 also separates intensity (brightness) from the colour channels. These colour characteristics are utilized to produce two novel techniques which aid in the accurate use of extracted features for localization in dynamically illuminated environments.

There are alternative approaches to improving illumination robustness in localization as demonstrated by the promising results achieved by Silveira and Malis [18], Silveira et al. [19] by modeling illumination change as a surface that can evolve over time. The model is combined with projective geometry to produce basic planar visual tracking. This approach requires no a priori knowledge about the light source(s) or the subject surface materials. A planar feature in an image is 'warped' to match the same feature seen from a different viewpoint in a subsequent image. The 'warp' required to match the features is then used to perform basic localization. The need for high quality planar features is the limiting factor in this technique and therefore it is less applicable, despite its illumination robustness.

Alternative approaches to improving robustness to dynamic illumination have been produced by other researchers [20–23]; however, these approaches all rely on additional sensors or a priori information, which our work aims to avoid.

The lack of a robust localization technique for dynamically illuminated environments led to the development of the technique detailed in the following section. This technique uses the colour information associated with extracted image features to improve robustness to lighting while requiring only a stream of colour images from a single camera.

3 Colour Model for Feature Filtering

In order to correctly distinguish between physical features and illumination artifacts, it is desirable to estimate the true colour of a feature (as opposed to a perceived colour which may be the result of lighting conditions). The difficulty in determining the true colour of an object is due to the sheer number of factors that affect perceived colour [14]. An approximation of true colour is provided by the brightness and chromaticity model proposed by Horprasert et al. [15]. The model compares two pixel colours, E and I, in RGB space. A line OE passing through the origin and the first pixel colour (E = [E_R, E_G, E_B]) is called the expected chromaticity line and is used to determine the 'distortion' in chromaticity and brightness of the second pixel. If the second pixel colour (I = [I_R, I_G, I_B]) is on this line, then there is a distortion in brightness only (α), otherwise there is also a distortion in chromaticity (CD), see Fig. 1.

The brightness distortion (α) is a scalar value which represents the point on the expected chromaticity line that is closest to the comparative colour (I). It is defined as:

$$\alpha(E, I) := \arg\min_{\alpha} \|I - \alpha E\| \tag{1}$$

Brightness distortion will be equal to 1 if the comparative colour has the same brightness, less than 1 if it is darker and greater than 1 if it is brighter.

Chromaticity distortion (CD) is the orthogonal distance between the expected chromaticity line and the comparative colour (I). The chromaticity distortion of the second pixel is given by:

$$CD := \min_{\alpha} \|I - \alpha E\| = \|I - \alpha E\|. \tag{2}$$

Our approach applies this colour model to features extracted from images rather than individual pixels. Both SIFT and FAST corner detection are used as feature extraction algorithms. SIFT is used for offline applications due to its popularity in machine vision and robustness [10], whereas FAST corner detection is used for the real time localization due to its ability to limit search time, improving speed. A

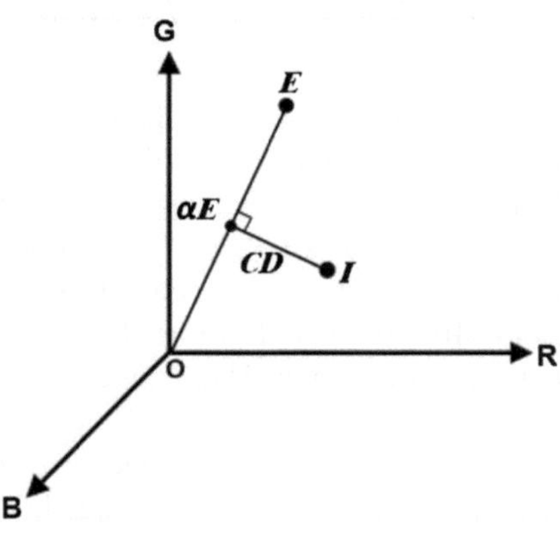

Fig. 1 Determining the brightness and chromaticity distortion between two colours (E, I) [15]

3×3 pixel region is used to represent the colour information of features extracted by either technique. SIFT has scale invariant features, so the 3×3 pixel region is scaled to the magnitude of the extracted feature. In order to maintain the 3×3 pixel colour information, the feature is divided into a 3×3 grid and the mean colour of the pixels in each grid square are stored as a single pixel representative. FAST corner detection uses a 3 pixel radius window having a 16 pixel circumference. A 3×3 pixel region is again used to characterize the colour of the center of the extracted feature. Since FAST features are not scaled, this simple approach works for all extracted features. The colour of a feature (E) is represented by the RGB values of the 3×3 pixel region i.e.

$$E = \begin{bmatrix} E_{R0} & E_{R1} & \dots & E_{Rn-1} \\ E_{G0} & E_{G1} & \dots & E_{Gn-1} \\ E_{B0} & E_{B1} & \dots & E_{Bn-1} \end{bmatrix}. \tag{3}$$

The mean colour of the feature is characterized by:

$$\mu(E) = \begin{bmatrix} \mu(E_R) \\ \mu(E_G) \\ \mu(E_B) \end{bmatrix} = \begin{bmatrix} \mu\{E_{R0} & \dots & E_{Rn-1}\} \\ \mu\{E_{G0} & \dots & E_{Gn-1}\} \\ \mu\{E_{B0} & \dots & E_{Bn-1}\} \end{bmatrix}. \tag{4}$$

The standard deviation of the RGB values for each feature is given by finding the standard deviation of each row of E.

$$\sigma(E) = \begin{bmatrix} \sigma(E_R) \\ \sigma(E_G) \\ \sigma(E_B) \end{bmatrix} = \begin{bmatrix} \sigma\{E_{R0} & \dots & E_{Rn-1}\} \\ \sigma\{E_{G0} & \cdots & E_{Gn-1}\} \\ \sigma\{E_{B0} & \dots & E_{Bn-1}\} \end{bmatrix} \tag{5}$$

Dividing each pixel value by the standard deviation of the entire feature produces an output scaling that emphasizes the difference between changes in brightness and changes in chromaticity. This emphasis simplifies the thresholding process.

3.1 Shadow Feature Removal

The first novel technique applied in this work is the use of the chromaticity distortion colour model to detect features extracted from the edges of shadows. Dynamically illuminated environments contain dynamic shadows. The tracking of dynamic shadows will lead to erroneous localization; therefore, for illumination invariance, features extracted from these shadows are identified and removed. Both SIFT and FAST corner detection tend to extract features from the 'edges' of objects in the image. These edges are therefore the focus of identifying a feature as shadow-based or object-based.

To determine the likelihood of a feature being extracted from a shadow, the 'chromaticity distortion' of the whole feature is calculated. This is a measure of the

RGB value of each pixel in the 3 × 3 pixel feature window compared to the mean colour of the feature. The resulting range of chromaticity distortions is then compared to a threshold to identify a feature as shadow-based or object-based. A feature extracted from the edge of a shadow is likely to have a low chromaticity distortion due to the 'edge' containing a variation in illumination only. A feature extracted from the edge of an object will usually have a higher chromaticity distortion due to the higher likelihood that the edge contains a variation in colour.

Brightness and chromaticity distortion are determined by comparing each of the 9 pixels in the 3 × 3 feature window E_i to the mean colour value of the overall feature $\mu(E)$.

$$\alpha_i = \frac{\frac{E_{Ri}\mu(E_R)}{\sigma^2(E_R)} + \frac{E_{Gi}\mu(E_G)}{\sigma^2(E_G)} + \frac{E_{Bi}\mu(E_B)}{\sigma^2(E_B)}}{\left[\frac{\mu(E_R)}{\sigma(E_R)}\right]^2 + \left[\frac{\mu(E_G)}{\sigma(E_G)}\right]^2 + \left[\frac{\mu(E_B)}{\sigma(E_B)}\right]^2} \tag{6}$$

$$CD_i = \sqrt{\left[\frac{E_{Ri} - \alpha\mu(E_R)}{\sigma(E_R)}\right]^2 + \left[\frac{E_{Gi} - \alpha\mu(E_G)}{\sigma(E_G)}\right]^2 + \left[\frac{E_{Bi} - \alpha\mu(E_B)}{\sigma(E_B)}\right]^2} \tag{7}$$

The standard deviation of the feature is used here to emphasize small changes in colour across the feature. Even a minor colour deviation in a feature with only a slight colour deviation will produce a large chromaticity distortion vector. Therefore, only features that are truly illumination artefacts will have a chromaticity distortion range below a selected threshold (τ_{SFR}). These features are removed as they are not considered likely to be robust to changes in illumination. The remaining features can then be used for matching, increasing the probability of illumination robustness during localization.

3.2 Colour Based Feature Matching

The second novel technique applied in this work is the use of the same colour model to increase the constraints on feature matching. The technique filters standard feature matching results and so can be applied to any existing feature matching algorithm.

A correctly matched feature is likely to exhibit a small difference in true colour, whereas an incorrectly matched feature is likely to show a significant difference in true colour. Both SIFT and FAST corner detection use monochrome images for feature matching and therefore, can be misled by lighting conditions that cause different objects to look similar in grey scale.

To identify a match as being of similar true colour, the chromaticity distortion between the matched features is calculated and compared to a threshold. The distortion value will be low, regardless of lighting conditions, if a correct match has been made. A standard feature matching algorithm is used to produce a set of matched features. The brightness and chromaticity distortion is determined by comparing the colour of one feature (E) to the colour of the matched feature (I).

$$\alpha = \frac{\frac{\mu(I_R)\mu(E_R)}{\sigma(I_R)\sigma(E_R)} + \frac{\mu(I_G)\mu(E_G)}{\sigma(I_G)\sigma(E_G)} + \frac{\mu(I_B)\mu(E_B)}{\sigma(I_B)\sigma(E_B)}}{\left[\frac{\mu(E_R)}{\sigma(E_R)}\right]^2 + \left[\frac{\mu(E_G)}{\sigma(E_G)}\right]^2 + \left[\frac{\mu(E_B)}{\sigma(E_B)}\right]^2} \tag{8}$$

$$CD = \sqrt{\frac{(\mu(I_R) - \alpha\mu(E_R))^2}{\sigma(I_R)\sigma(E_R)} + \frac{(\mu(I_G) - \alpha\mu(E_G))^2}{\sigma(I_G)\sigma(E_G)} + \frac{(\mu(I_B) - \alpha\mu(E_B))^2}{\sigma(I_B)\sigma(E_B)}} \tag{9}$$

The standard deviation of colour across each of the features is again used to emphasize minor differences in colour. This is particularly powerful in scenes with recurring similar features as even slight differences in colour result in large chromaticity distortion vectors. Concordantly, matched features that have a chromaticity distortion above a selected threshold (τ_{CBM}) are discarded as they are considered likely to be a false match due to lighting conditions.

4 Experimental Results

4.1 Comparison of Colour Models

The chromaticity distortion model has never before been applied to a localization task. To confirm that it was an appropriate choice of colour model, it was compared to two other models that separate colour from intensity—HSV and colour angles. Hue Saturation Value (HSV) is a cylindrical coordinate representation of RGB colour. It presents hue on a circular colour chart and then separately defines saturation and 'darkness' values. Colour angles describe the difference between two colours in RGB space as the angle between the two RGB vectors. The image of a handle casting a shadow on a door (Fig. 2a) was used to compare these three techniques by assessing their ability to differentiate between a true colour difference and a perceived colour difference. SIFT is used to identify about 3,000 features in the image, each colour model is then tuned to reject around 1,000 features based on a comparison of each pixel in the 3×3 grid to the mean feature colour, as described in Sect. 3.

The results in Fig. 2b show that the chromaticity distortion model correctly rejects the features on the edge of the shadow and on the highlight. HSV and colour angles were also evaluated but poorly identified the shadow and highlight. The chromaticity distortion technique was trialed on other images to test the repeatability of the results and similar outcomes were achieved. When standard deviation based scaling was added to HSV and colour angles, similar results to chromaticity distortion were achieved. However, chromaticity distortion demonstrated a significant advantage in computational efficiency. On average

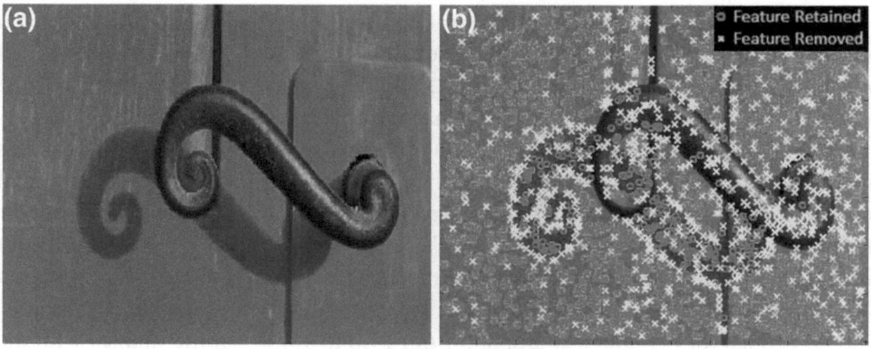

Fig. 2 **a** The original image. **b** Classification using chromaticity distortion

chromaticity distortion was processing three times faster than colour angles. To use HSV, the RGB colours in the features must be converted using a lookup table. This conversion significantly exceeded the chromaticity distortion processing time before any calculations using HSV values could be produced. Since the focus of this work was on real time applications for localization, chromaticity distortion was the obvious choice.

4.2 Feature Matching Under Dynamic Illumination (Simulated Scene)

To develop the concept further, a computer generated dynamic illumination scenario was developed in POV-Ray (the Persistence of Vision Raytracer http://www.povray.org). This allowed explicit control over the light source properties and position. By controlling camera and light source movement, scenes could be rapidly developed and used for the evaluation of the derived techniques. SIFT was again used as the image extraction algorithm. The scenario in Fig. 3 is a scene containing a chess board with a fixed viewpoint and varying lighting conditions. This series of images was used to test three approaches derived in this work: The removal of shadow based features before matching, SIFT match rejection based on colour and applying a combination of both techniques.

Fig. 3 Simulated scene with dynamic lighting

The use of a computer generated scene allowed only the lighting to be altered between images; this meant that features could be accurately classified as matched or mismatched between images due to the identical nature of the remainder of the scene. The results in Table 1 show a dual improvement: a reduction in the percentage of incorrect matches and a reduction in the overall number of matches made. Colour based matching was tested by first running the SIFT extraction and matching algorithms, then discarding matches based on discrepancies in colour determined by chromaticity distortion. The results in Table 1 show a significant reduction in the percentage of incorrect matches and again a reduction in the total number of matches. A combination of the two techniques provided the best results. The percentage of mismatches dropped from 17.6 to 1.4 % for the images with the smallest change in lighting (A–B) and from 29.9 to 3.0 % for the images with the largest change in lighting (A–D).

4.3 Localization Under Dynamic Illumination (Simulated Scene)

To prove the concept further, a dynamic illumination scenario was introduced using computer generated images (created using POV Ray). This allowed explicit control over the camera and lighting conditions. A series of images depicting movement through a scene with dynamic lighting allowed full assessment of the effect on localization caused by chromaticity distortion filtering. The camera moved with a constant velocity through a scene containing a chess board. The motion path was a straight line parallel to the edge of the chess board, while the single light source moved from slightly behind the chess board to directly overhead, causing dynamic shadows (see Fig. 4a).

Table 1 Feature matches resulting from chromaticity distortion

Images	Threshold	Total matches	Mismatches	% Error
A–B	None	954	168	17.6
	$\tau_{SFR} = 1.4$	253	38	15.0
	$\tau_{CBM} = 0.05$	250	6	2.4
	$\tau_{SFR} = 1.4$, $\tau_{CBM} = 0.05$	71	1	1.4
A–C	None	657	147	22.4
	$\tau_{SFR} = 1.4$	185	37	20.0
	$\tau_{CBM} = 0.05$	169	9	5.3
	$\tau_{SFR} = 1.4$, $\tau_{CBM} = 0.05$	58	0	0.0
A–D	None	484	145	29.9
	$\tau_{SFR} = 1.4$	145	31	21.4
	$\tau_{CBM} = 0.05$	87	4	9.4
	$\tau_{SFR} = 1.4$, $\tau_{CBM} = 0.05$	33	1	3.0

The SLAM implementation for this experiment was based on the six degree of freedom monocular SLAM algorithm written by Civera et al. [1]. The algorithm was modified by the addition of the chromaticity distortion filter developed in this chapter. Figure 4c & d are a top view representation of the 3D localization produced by the monocular SLAM algorithm. Each image shows the localization path as a line starting at the origin (0, 0) and finishing in the center of the triangle representing the final orientation of the camera. The ellipsoids in each image center on a tracked feature and represent the locational uncertainty of that feature. The remaining lines point to features that were observed in only a single image and hence have no depth estimate. The threshold used in each test was a constant and based on effective vales used in previous experiments. All compensation techniques were assessed, starting with uncompensated (Fig. 4c), shadow features removed, colour based matching and finally both techniques combined (Fig. 4d).

The ground truth for the test is shown in Fig. 4b. The distances are dimensionless as the scene is computer generated and were scaled by identifying key features in the image. The path is a straight line with no change in camera orientation. The first test introduced the dynamic light source and was run using the unfiltered SLAM algorithm. The dynamic shadows have a significant influence on the effectiveness of the localization, resulting in the heavily distorted path seen in Fig. 4c.

The second test applied the feature filtering approach outlined in Sect. 3.1. This approach produced a significant improvement in localization, yet still contained evidence of the distortion caused by the dynamic light source. The third test filtered feature matches based on their chromaticity distortion as detailed in Sect. 3.2. The localization path produced by this approach completely removed the distortion caused by the dynamic light source. The consequence of this

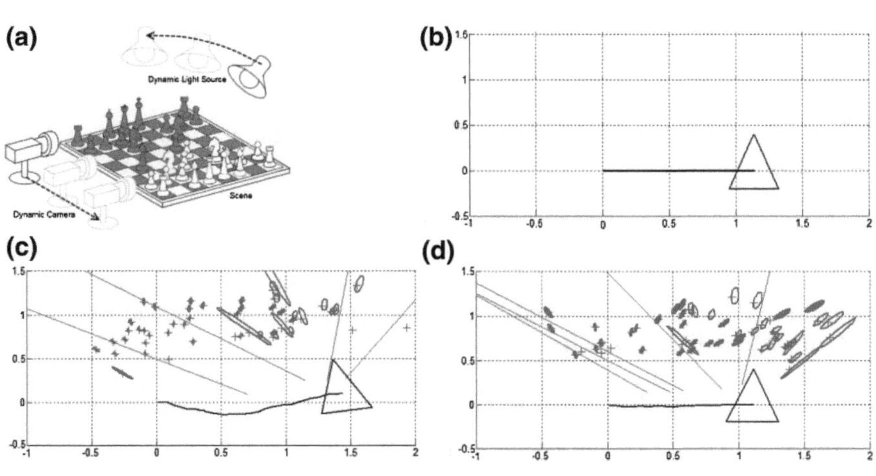

Fig. 4 **a** Depiction of scene. **b** The ground truth of the camera movement. **c** Uncompensated localization under dynamic lighting. **d** Localization using SFR and CBM

improvement was a large reduction in the number of maintained matches, resulting in inconsistent localization that had poor correlation to the ground truth.

The final test combined both filtering techniques. The resulting localization path (Fig. 4d) demonstrated the effectiveness of the approach. The distortion produced by the dynamic light source was completely removed and the high level of maintained features produced a highly accurate displacement. These results demonstrate the effectiveness of the two novel techniques derived in this work as approaches to overcoming the challenges of localization in dynamically illuminated environments.

4.4 Localization Under Dynamic Illumination with Camera Defects (Simulated Scene)

Computer generated images are a useful way to quickly assess the performance of filtering based on chromaticity distortion; however, to completely assess the competence of the algorithm, camera defects need to be included that would be found in real world datasets. Images taken by an actual camera contain noise such as lens distortion, unfocused features, lens flare, graininess and poor contrast. These five types of noise were all added to the simulated dataset used in Sect. 4.3 to test the algorithm's robustness.

Figure 5a and b compare an image from the dataset before and after the addition of camera defects. Figure 5c shows the uncompensated SLAM algorithm localization path and Fig. 5d shows the compensated path. Using simulated camera defects rather than a real world dataset allowed the effect of the defects to be predicted. The comparison of the noisy localization path (Fig. 5d) to the noise free localization path (Fig. 4f) shows that the noise does not affect the robustness of the algorithm.

4.5 Dynamic Scene with Dynamic Lighting Localization (Real World)

Section 4.4 demonstrated the robustness of chromaticity distortion based filtering to simulated camera defects. However, to be recognized as an applicable technique, the algorithm must work in the real world. To analyze the real world performance, a dataset was recorded of movement through a scene with dynamic lighting under laboratory conditions. The camera moves through the scene sideways, five meters to the left in a straight line, the light source moves parallel to the camera yet at a slower speed producing shadows that drift away from the direction of movement of the camera (see Fig. 6a).

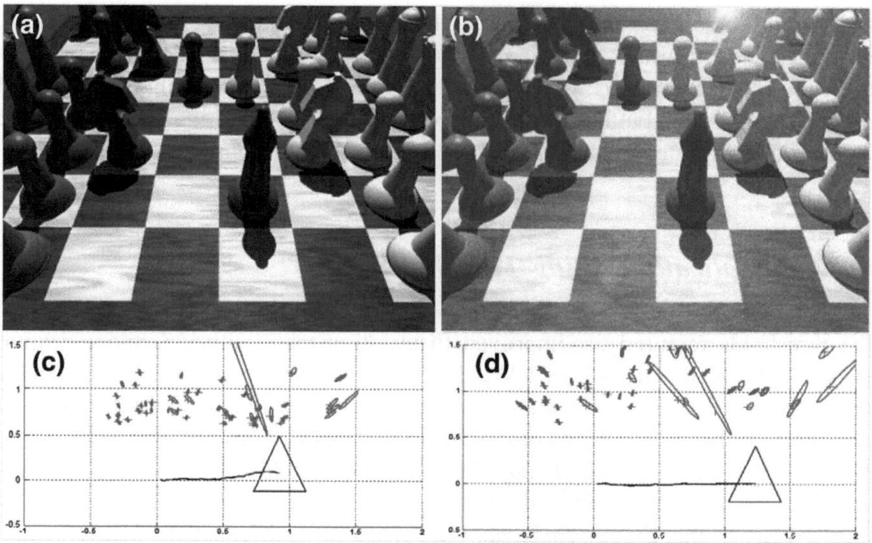

Fig. 5 **a** Image from scene with no noise. **b** Image from scene with simulated noise. **c** Uncompensated localization from noisy images. **d** Localization from noisy images using SFR and CBM

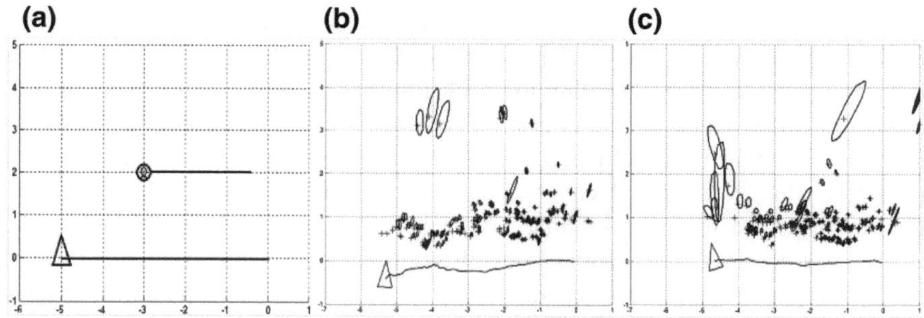

Fig. 6 **a** Camera and lighting movement ground truth. **b** Uncompensated localization in real world scene. **c** Localization in real world scene using SFR and CBM

This dataset was run through the uncompensated monocular SLAM algorithm (Fig. 6b) and then the compensated algorithm (Fig. 6c). The localization results are again presented in the same format as previous results. Running the dataset through the uncompensated algorithm produced accurate localization for a brief window of time, as expected by the slight illumination invariance characteristic of most feature extraction techniques. As the dataset continues, however, the tracking of dynamic shadow features leads to the incorrect localization path seen in Fig. 6b. Far superior results are produced by the algorithm incorporating chromaticity distortion based filtering (Fig. 6c) where the localization path is highly correlated to the ground truth. Future work will expand real world testing to examine the effect of using the technique outside of laboratory conditions.

4.6 Automatically Scaling Thresholds

The thresholds required for improved localization using either Shadow Feature Removal (SFR) or Colour Based Matching (CBM) are dependent upon the colour range of the image. Images that are considered to be low in colour contrast need to have the SFR threshold relaxed otherwise all features will be classified as shadows. However, the CBM threshold must be stringent otherwise the colour difference between mismatched features will always pass through the chromaticity distortion filter. To ensure that the system is valid in a range of environments without manual tuning, automatically scaling thresholds were implemented based on the colour content of an image. The chromaticity distortion colour model was again utilized to determine the colour range of an image. The chromaticity distortion of a sample of random pixels was compared to the mean colour of the entire image. The standard deviation of the sample of random pixels correlated to the colour range present in each image and was used to scale the threshold.

The standard deviation of 400 random pixels per image over a 100 image sequence can be seen in Fig. 7a. The five lines in the plot represent five executions of the random sampling algorithm to check for consistency. The localization path of the 100 image dataset without any compensation can be seen in Fig. 6b, this is the same dataset as used in Sect. 4.5. The localization path of the sequence with a fixed SFR threshold of 1.0 can be seen in Fig. 7b. This threshold value was selected based on the results Table 1. The auto scaling threshold applied to the dataset was based on the standard deviation range seen in Fig. 7a. The standard deviation of each frame was scaled such that the resultant threshold was always between 0.5 and 1.5. These vales represent the extreme operating conditions of the filter. Below this limit, the majority of features are passed, making the filter redundant. Above the limit, a substantial number of features are classified as shadow features and are rejected, significantly increasing the processing time or even arresting the localization altogether. The localization path resulting from auto scaling the threshold can be seen in Fig. 7c. The same approach can be applied to the CBM thresholds, also improving localization precision. The thresholds required for the chromaticity distortion based filter can therefore be determined in real-time without prior knowledge of the dataset or lighting conditions.

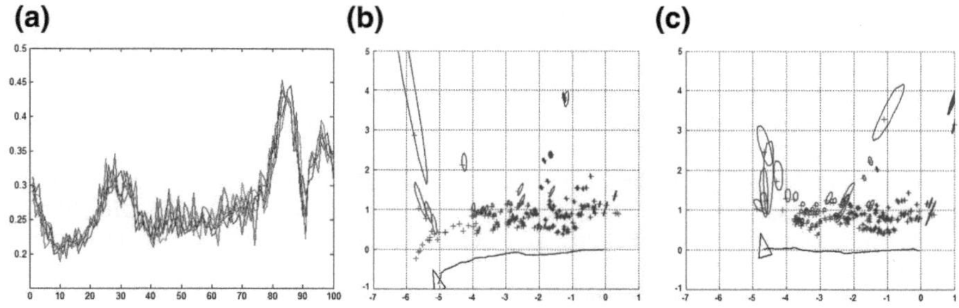

Fig. 7 **a** Standard deviation of image colour across scene. **b** Fixed threshold compensation localization. **c** Automatically scaling compensation localization

5 Conclusion

This work has demonstrated substantial improvements in the robustness of localization under dynamic illumination by applying a chromaticity distortion model.

These findings were the result of the successful integration of a modified version of the chromaticity distortion model with a fully implemented, six degree of freedom monocular SLAM algorithm. The two techniques resulting from this pairing were the removal of shadow based features and the colour based filtering of feature matching.

The application of these techniques has resulted in localization under dynamic illumination which is almost as accurate as localization under static illumination in simulation. Testing in a real world scenario shows significant reductions in localization distortion under dynamic illumination, resulting in a localization path which is close to ground truth. The use of automatically scaling thresholds demonstrated that the filter can be tuned without a priori information.

In future work, these techniques will be integrated into a hybrid SLAM system to improve localization and mapping for deployment in underground mining environments. As we have noted, there are a number of application areas involving dynamic illumination that could also benefit from this work.

References

1. J. Civera, O. Grasa, A. Davison, J. Montiel, 1-Point RANSAC for EKF filtering: application to real-time structure from motion and visual odometry. J. Field Robot. **27**(5), 609–631 (2010)
2. E. Mouragnon, M. Lhuillier, M. Dhome, F. Dekeyser, P. Sayd, Generic and real-time structure from motion using local bundle adjustment. Image Vis. Comput. **27**(8), 1178–1193 (2009)

3. S. Se, D. Lowe, J. Little, in *Vision-Based Mobile Robot Localization and Mapping Using Scale-Invariant Features*. Proceedings IEEE international conference on robotics and automation, vol. 2 (2001), pp. 2051–2058
4. J. Tardif, Y. Pavlidis, K. Daniilidis, in *Monocular Visual Odometry in Urban Environments Using an Omnidirectional Camera*. international conference on intelligent robots and systems (2008), pp. 2531–2538
5. J. Wolf, W. Burgard, H. Burkhardt, Robust vision-based localization by combining an image-retrieval system with Monte Carlo localization. IEEE Trans. Rob. **21**(2), 208–216 (2005)
6. D. Lowe, Distinctive Image Features from Scale-Invariant Keypoints. Int. J. Comput. Vision **60**(2), 91–110 (2004)
7. E. Rosten, T. Drummond, Machine learning for high-speed corner detection. Lect. Notes Comput. Sci. **3951**, 430–443 (2006)
8. A. Jarosz, in *Development of inspection system for evaluation of ore-passes at Grasberg mine, PT freeport, Indonesia*. Proceedings of the 21st world mining congress & expo (2008)
9. J. Le Cras, J. Paxman, B. Saracik, A. Jarosz, in *An inspection and surveying system for vertical shafts*. Proceedings Australasian conference on robotics and automation 2009 (2009)
10. K. Mikolajczyk, C. Schmid, A performance evaluation of local descriptors. IEEE Trans. Pattern Anal. Mach. Intell. **27**(10), 1615–1630 (2005)
11. D. Chekhlov, M. Pupilli, W. Mayol, A. Calway, in *Robust Real-Time Visual Slam Using Scale Prediction and Exemplar Based Feature Description*. Proceedings of the international conference on computer vision and pattern recognition (2007)
12. M. Swain, D. Ballard, Color indexing. Int. J. Comput. Vision **7**(1), 11–31 (1991)
13. G. Finlayson, S. Chatterjee, B. Funt, in *Color Angular Indexing*. Proceedings of the 4th European conference on computer vision, vol. 2 (1996), pp. 16–27
14. J. Geusebroek, R. van den Boomgaard, A. Smeulders, H. Geerts, Color invariance. IEEE Trans. Pattern Anal. Mach. Intell. **23**(12), 1338–1350 (2001)
15. T. Horprasert, D. Harwood, L. Davis, in *A Robust Background Subtraction and Shadow Detection*. Proceedings of the Asian conference on computer vision (2000)
16. A. Abdel-Hakim, A. Farag, in *CSIFT—A SIFT Descriptor with Color Invariant Characteristics*. Proceedings of IEEE conference on computer vision and pattern recognition, vol. 2 (2006), pp. 1978–1983
17. G. Burghouts, J. Geusebroek, Performance evaluation of local colour invariants. Comput. Vis. Image Underst. **113**, 48–62 (2009)
18. G. Silveira, E. Malis, in *Real Time Visual Tracking Under Arbitrary Illumination Changes*. Proceedings of the 2007 IEEE conference on computer vision and pattern recognition (2007) pp. 1–6
19. G. Silveira, E. Malis, P. Rives, An efficient direct approach to visual slam. IEEE Trans. Rob. **24**(5), 969–979 (2008)
20. H. Bischof, H. Wildenauer, A. Leonardis, Illumination insensitive recognition using eigenspaces. Comput. Vis. Image Underst. **95**, 86–104 (2004)
21. G. Steinbauer, H. Bischof, Illumination insensitive robot self-localization using panoramic eigenspaces. Lect. Notes Comput. Sci. **3276**, 84–96 (2005)
22. A. Sunghwan, C. Jinwoo, C. Minyong, C. Wan Kyun, in *Metric Slam in Home Environment with Visual Objects and Sonar Features*. Proceedings of the 2006 IEEE international conference on intelligent robots and systems (2006), pp. 4048–4053
23. A. SungHwan, C. Jinwoo, D. Nakju Lett, C. Wan Kyun, A practical approach for EKF-SLAM in an indoor environment: fusing ultrasonic sensors and stereo camera. Auton. Robots **24**(3), 315–335 (2008)

Optical-Inertial Tracking System with High Bandwidth and Low Latency

Göntje C. Claasen, Philippe Martin and Frédéric Picard

Abstract We propose an optical-inertial tracking system for a servo-controlled handheld tool in a computer-assisted surgery system. We present a mathematical system description and a data fusion algorithm which integrates data from optical and inertial sensors. The algorithm is a right-invariant Extended Kalman Filter (EKF) which takes into account system symmetries to improve the filter convergence. The tracking system has a high bandwidth thanks to the inertial sensors and a low latency thanks to a direct approach where sensor data is used directly in the data fusion algorithm without previous computations. Experimental data show that the optical-inertial system can indeed track a moving object.

1 Introduction

Computer-assisted surgery systems have become more and more common in operating rooms. These systems often rely on optical tracking systems for tracking the patient and surgical tool positions. An optical tracking system consists of

Based on High-Bandwidth Low-Latency Tracking Using Optical and Inertial Sensors, by Göntje C. Claasen, Philippe Martin and Frédéric Picard which appeared in the Proceedings of the 5th International Conference on Automation, Robotics and Applications (ICARA 2011).

G. C. Claasen (✉) · P. Martin
Centre Automatique et Systèmes, MINES ParisTech,
60 boulevard St-Michel 75006 Paris, France
e-mail: caroline.claasen@mines-paristech.fr

P. Martin
e-mail: philippe.martin@mines-paristech.fr

F. Picard
Department of Orthopaedics, Golden Jubilee National Hospital, Glasgow, UK
e-mail: frederic.picard@gjnh.scot.nhs.uk

G. Sen Gupta et al. (eds.), *Recent Advances in Robotics and Automation*,
Studies in Computational Intelligence 480, DOI: 10.1007/978-3-642-37387-9_13,

stationary cameras and optical markers attached to the object which is being tracked.

Over the last few years, handheld tools [1, 2] for computer-assisted surgery have been developed. They can potentially save operating time and reduce the amount of surgical material in the operating room compared to mechanical cutting or drilling guides. These tools have to be servo-controlled to keep them in a desired plane. Servo-control demands for a fast tracking system which has to be faster than human reaction in order to correct small errors the surgeon might make and to correct deviations caused by a change of bone structure. The tracking system has to have a bandwidth of at least 200 Hz and a low latency. Existing optical tracking systems used in computer-assisted surgery have a low bandwidth of 10–60 Hz and an important latency which make them unsuitable for servo-control of a handheld tool.

One possibility to overcome problems of optical tracking is to combine it with another sensor type. Inertial sensors are an appropriate choice because of their high bandwidth, small size and low cost. They cannot be used on their own because of their inherent drifts. Combining information from both sensor types can be achieved with a data fusion algorithm. It gives an estimation of the object pose at the higher of the two sensor bandwidths which would be the inertial bandwidth here.

For our tool-tracking problem, we are interested in the configuration of a stationary camera and an inertial measurement unit (IMU) and optical markers which are rigidly connected and fixed to the object being tracked. Several systems have been presented in the literature which combine optical and inertial data for object tracking. They all use a Kalman Filter to fuse the data from both sensor types.

In [3], the system tracks a pen-like tool to which is attached an IMU and whose tip is detected by four cameras placed on a curved line. The sample rates are 100 Hz for the IMU and 20 Hz for the cameras. The authors of [3] find that with this setup the position error is reduced compared to optical stereo tracking. Roetenberg [4] presents a setup consisting of a Vicon optical tracking system with 6 cameras and an optical marker attached to an IMU which is fixed to the object being tracked. With an optical sample rate of 10 Hz and the IMU at 100 Hz, the system accuracy is found to be as good as the Vicon's at 100 Hz. The authors of [5] combine an optical tracking system with two cameras and three markers with a sample rate of about 55 Hz with an IMU being sampled at 500 Hz which is attached to the optical markers. They find that pose estimation is possible even during short marker occlusions, as long as at least one marker is visible to the cameras. In [6], only one camera is used which tracks the position (and the yaw angle when the object is moving) of a fiducial marker pattern which is attached to an IMU being sampled at 400 Hz. The camera sample rate is 5–10 Hz. The authors of [6] report that this system can track a moving object and compensate short-time marker occlusions.

Among the presented systems, [3] and [4] combine optical and inertial data with the goal of reducing the optical sample rate, thus saving processing time compared to optical tracking. They use inertial sensors with a low sample rate of 100 Hz and do not use all of their potential. Tobergte et al. [5] and Hartmann et al. [6] use inertial sensors with high bandwidths of 400 and 500 Hz, respectively, which would be suitable for servo-control. However, they use marker positions as given by an optical tracking system as measurements (indirect Kalman Filter) which does not solve the problem of their latencies.

To make our tracking system suitable for servo-controlling a handheld tool, it has to have a high bandwidth and a low latency. This defines our goals: (a) augment the bandwidth of existing optical tracking systems to 200–300 Hz and (b) make the pose estimation delays as small as possible. Our optical-inertial tracking system consists of a stationary stereo camera pair and a sensor unit. The sensor unit is made up of an IMU and optical markers and is attached to a handheld tool.

Section 2 presents the mathematical model for this system. Section 3 addresses the data fusion algorithm for the presented system. Estimating the sensor unit pose is achieved by fusing the data from the camera and the IMU with a modified Extended Kalman Filter. This filter uses a direct approach which reduces latencies. Our experimental setup for the optical-inertial system is described in Sect. 4. Using experimental data, we show that our system can track motion at a high bandwidth and compare it to purely optical tracking.

2 System Model

The motion of the sensor unit will be expressed in camera coordinates which are denoted by C and are fixed to the right camera center. Their unit vectors are $E_1 = [1,0,0]^T$, $E_2 = [0,1,0]^T$ and $E_3 = [0,0,1]^T$. The camera's optical axis runs along E_1. Image coordinates are expressed in the image sensor coordinate system S which is attached to one of the corners of the camera's image sensor. The left camera coordinate system is denoted by CL and the image sensor coordinate system by SL. The left camera unit vectors are $\tilde{E}_1$, $\tilde{E}_2$ and $\tilde{E}_3$. The body coordinates, denoted by B, are fixed to the origin of the IMU frame and are moving relative to the camera frames. ${}^{B}a$ is an acceleration in body coordinates, for example. Finally, we also use an Earth-fixed world coordinate system, denoted by W.

2.1 System Dynamics and Output

The following equations describe the sensor unit dynamics:

$$ {}^{C}\dot{p} = {}^{C}v \tag{1} $$

$$ {}^{C}\dot{v} = {}^{C}G + {}^{BC}q * {}^{B}a * {}^{BC}q^{-1} \tag{2} $$

$$^{BC}\dot{q} = \frac{1}{2}{}^{BC}q * {}^{B}\omega \tag{3}$$

where ${}^{C}G = {}^{WC}q * {}^{W}G * {}^{WC}q^{-1}$ is the gravity vector expressed in C coordinates. ${}^{W}G = [0, 0, g]^T$ is the gravity vector in the world frame with $g = 9.81\,\frac{m}{s^2}$ and ${}^{WC}q$ describes the (constant) rotation from world to camera coordinates. ${}^{C}p$ and ${}^{C}v$ are the sensor unit position and velocity, respectively. The orientation of the sensor unit with respect to camera coordinates is represented by the quaternion ${}^{BC}q$. ${}^{B}a$ and ${}^{B}\omega$ are the sensor unit accelerations and angular velocities.

The output is equal to the optical marker images. To project the markers to the camera we use a standard pinhole model [7]:

$${}^{C}y_i = \frac{f}{\langle {}^{C}m_i, {}^{C}E_1\rangle}\begin{bmatrix}\langle {}^{C}m_i, {}^{C}E_2\rangle \\ \langle {}^{C}m_i, {}^{C}E_3\rangle\end{bmatrix} \tag{4}$$

with ${}^{C}m_i = {}^{C}p + {}^{BC}q * {}^{B}m_i * {}^{BC}q^{-1}$ where y_i is the 2D image of marker i with $i = 1, \ldots, l$ (l is the number of markers). f is the camera's focal distance. ${}^{C}m_i$ and ${}^{B}m_i$ are the position of marker i in camera and body coordinates, respectively. $\langle a, b\rangle$ denotes the scalar product of vectors a and b.

The 2D image can be transformed from camera to sensor coordinates by a translation ${}^{S}y_i = {}^{C}y_i + {}^{S}u$ where ${}^{S}u$ is the camera's principal point.

The camera coordinate system in which the sensor unit pose is expressed is attached to the right camera of the stereo camera pair. The transformation between the left and right camera coordinates is expressed by R_{St} and t_{St}:

$${}^{CL}p = R_{St}{}^{C}p + t_{St}. \tag{5}$$

2.2 Model with Noises and Biases

The accelerometer and gyroscope measurements a_m and ω_m of ${}^{B}a$ and ${}^{B}\omega$ resp. contain noise which is assumed white and biases:

$$a_m = {}^{B}a + v_a + {}^{B}a_b, \quad \omega_m = {}^{B}\omega + v_\omega + {}^{B}\omega_b.$$

The accelerometer biases ${}^{B}a_b$ and gyroscope biases ${}^{B}\omega_b$ can be represented by a sum of two components: a constant one and a varying one which is modeled as a rate random walk [8]:

$${}^{B}\dot{a}_b = v_{ab}, \quad {}^{B}\dot{\omega}_b = v_{\omega b}. \tag{6}$$

Since the IMU consists of a triad of identical accelerometers and a triad of identical gyroscopes and since we consider the inertial sensor noises to be independent white noises with zero mean, we can write for the auto-covariance

$$E(v_j(t)v_j^T(t+\tau)) = \xi_j^2 I_3 \delta(\tau) \tag{7}$$

for $j \in \{a, \omega, ab, \omega b\}$ where δ is the Dirac function.

The marker images are measured in the sensor frame and are corrupted by noise η_y which is also assumed white: $y_m = {}^S y + \eta_y$.

The complete model with noises then reads:

$${}^C\dot{p} = {}^C v \tag{8}$$

$${}^C\dot{v} = {}^C G + {}^{BC}q * (a_m - v_a - {}^B a_b) * {}^{BC}q^{-1} \tag{9}$$

$${}^{BC}\dot{q} = \frac{1}{2}{}^{BC}q * (\omega_m - v_\omega - {}^B\omega_b) \tag{10}$$

$${}^B\dot{a}_b = v_{ab} \tag{11}$$

$${}^B\dot{\omega}_b = v_{\omega b}. \tag{12}$$

The measured outputs for the right and the left camera:

$$y_{im} = {}^S y_i + \eta_{yi} = \frac{f_R}{\langle {}^C m_i, {}^C E_1 \rangle} \begin{bmatrix} \langle {}^C m_i, {}^C E_2 \rangle \\ \langle {}^C m_i, {}^C E_3 \rangle \end{bmatrix} + {}^S u_R + \eta_{yi} \tag{13}$$

$$y_{(i+l)m} = {}^{SL} y_i + \eta_{y(i+l)} = \frac{f_L}{\langle {}^{CL} m_i, {}^{CL}\tilde{E}_1 \rangle} \begin{bmatrix} \langle {}^{CL} m_i, {}^{CL}\tilde{E}_2 \rangle \\ \langle {}^{CL} m_i, {}^{CL}\tilde{E}_3 \rangle \end{bmatrix} + {}^{SL} u_L + \eta_{y(i+l)} \tag{14}$$

Indices R and L refer to right and left camera resp. (e.g. f_L is the focal distance of the left camera).

The six accelerometer and gyroscope measurements $u = [a_m, \omega_m]$ are considered as the inputs of our system and the marker images

$$y_m = y + \eta_y = [{}^S y_1, \ldots, {}^S y_l, {}^{SL} y_1, \ldots, {}^{SL} y_l] + [\eta_{yi}, \ldots, \eta_{y2l}] \tag{15}$$

as its outputs. y is a vector of length $2l + 2l = 4l$. The state vector which is to be estimated by the data fusion filter in Sect. 3 has the form $x = [{}^C p, {}^C v, {}^{BC} q, {}^B a_b, {}^B \omega_b]$ and is of dimension 16.

2.3 *Quaternions*

A quaternion q [9] consists of a scalar $q_0 \in \mathbb{R}$ and a vector $\tilde{q} \in \mathbb{R}^3$: $q = [q_0, \tilde{q}^T]^T$. The quaternion product of two quaternions s and q is defined as

$$s * q = \begin{bmatrix} s_0 q_0 - \tilde{s}\tilde{q} \\ s_0 \tilde{q} + q_0 \tilde{s} + \tilde{s} \times \tilde{q} \end{bmatrix}. \tag{16}$$

The cross product for quaternions reads $s \times q = \frac{1}{2}(s * q + q * s)$. A unit quaternion can be used to represent a rotation: $R(q)a = q * a * q^{-1}$ where $a \in \mathbb{R}^3$ and $R(q)$ is the rotation matrix associated with quaternion q. If q depends on time, we have $\dot{q}^{-1} = -q^{-1} * \dot{q} * q^{-1}$. If

$$\dot{q} = q * a + b * q \tag{17}$$

with $a,b \in \mathbb{R}^3$ holds true, then $\|q(t)\| = \|q(0)\|$ for all t.

3 Data Fusion Filter

Estimating the sensor unit pose relative to camera frame is achieved by a modified extended Kalman filter (EKF). After describing the general EKF we will present a modified filter for the optical-inertial tracking system, called right-invariant EKF.

3.1 General EKF and Modified Version for Optical-Inertial System

3.1.1 Extended Kalman Filter

We consider a system

$$\dot{x} = f(x, u) + Mv \tag{18}$$

$$y = h(x, u) + N\eta \tag{19}$$

where v and η are independent white noises. M is of dimension $n \times r$ and N of dimension $s \times s$ where n is the number of states, r is the number of input noises and s is the number of outputs. The input u and the output y are known. x is the state which is to be estimated. The EKF calculates an estimate $\hat{x}(t)$ of the state $x(t)$ according to

$$\hat{x} = f(\hat{x}, u) - K\big(y - h(\hat{x}, u)\big) \tag{20}$$

$$\dot{P} = AP + PA^T + Q - PC^T R^{-1} CP \tag{21}$$

where $K = PC^T R^{-1}$, $A = \partial_1 f(\hat{x}, u)$ and $C = \partial_1 h(\hat{x}, u)$. In this notation, ∂_i is the partial derivative with respect to the ith argument. The matrices Q (of dimension $n \times n$) and R (of dimension $s \times s$) depend on $\tilde{Q}$ and $\tilde{R}$ through

$$Q = M\tilde{Q}M^T \tag{22}$$

$$R = N\tilde{R}N^T \tag{23}$$

where $\tilde{Q}$ contains the covariance parameters of v and $\tilde{R}$ those of η.

The estimation error $\Delta x = \hat{x} - x$ satisfies up to higher-order terms the linear equation

$$\Delta\dot{x} = (A - KC)\Delta x - Mv + KN\eta. \tag{24}$$

Recall that when A and C are constant, e.g. around a steady state point, the EKF converges when (A, M) is stabilisable and (A, C) detectable [10, pp. 704].

3.2 Right-Invariant EKF for the Optical-Inertial System

We propose the following right-invariant EKF for the system presented in Sect. 2.2. This filter is invariant w.r.t. to a rotation by the right i.e. applying this transformation to the filter equations leaves them unchanged. This is a logical step since the system is also invariant w.r.t. this rotation. These modifications should improve the filter convergence and reduce the computational cost.

The filter equations read

$${}^{C}\dot{\hat{p}} = {}^{C}\hat{v} + K_p E \tag{25}$$

$${}^{C}\dot{\hat{v}} = {}^{C}G + {}^{BC}\hat{q} * (a_m - {}^{B}\hat{a}_b) * {}^{BC}\hat{q}^{-1} + K_v E \tag{26}$$

$${}^{BC}\dot{\hat{q}} = \frac{1}{2}{}^{BC}\hat{q} * (\omega_m - {}^{B}\hat{\omega}_b) + K_q E * {}^{BC}\hat{q} \tag{27}$$

$${}^{B}\dot{\hat{a}}_b = {}^{BC}\hat{q}^{-1} * K_a E * {}^{BC}\hat{q} \tag{28}$$

$${}^{B}\dot{\hat{\omega}}_b = {}^{BC}\hat{q}^{-1} * K_\omega E * {}^{BC}\hat{q} \tag{29}$$

with $E = y_m - \hat{y}$. The gain $K = -[K_p, K_v, K_q, K_a, K_\omega]^T$ is calculated according to $K = PC^T R^{-1}$ where P satisfies $\dot{P} = AP + PA^T + Q - PC^T R^{-1} CP$.

To calculate matrices A and C, we consider the state error $e = [e_p, e_v, e_q, e_a, e_\omega] = [\hat{p} - p, \hat{v} - v, {}^{BC}\hat{q} * {}^{BC}q^{-1}, {}^{BC}q * (\hat{a}_b - a_b) * {}^{BC}q^{-1}, {}^{BC}q * (\hat{\omega}_b - \omega_b) * {}^{BC}q^{-1}]$ for the proposed system.

We linearize the error system around $\overline{e} = (0, 0, 1, 0, 0)$ to bring it into the form (24) and obtain the following matrices:

$$A = \begin{bmatrix} 0 & I_3 & 0 & 0 & 0 \\ 0 & 0 & -2[R(^{BC}\hat{q})(a_m - \hat{a}_b)]_\times & -I_3 & 0 \\ 0 & 0 & 0 & 0 & -0.5I_3 \\ 0 & 0 & 0 & A_1 & 0 \\ 0 & 0 & 0 & 0 & A_1 \end{bmatrix} \tag{30}$$

$$A_1 = [R(^{BC}\hat{q})(\omega_m - \hat{\omega}_b)]_\times \tag{31}$$

$$C_i = f_R \frac{1}{\alpha^2} \begin{bmatrix} \beta & -\alpha & 0 \\ \gamma & 0 & -\alpha \end{bmatrix} \begin{bmatrix} I_3 & 0 & -2[R(^{BC}\hat{q})^B m_i]_\times & 0 & 0 \end{bmatrix} \tag{32}$$

$$C_{i+l} = f_L \frac{1}{\chi^2} \begin{bmatrix} \mu & -\chi & 0 \\ \kappa & 0 & -\chi \end{bmatrix} R_{St} \begin{bmatrix} I_3 & 0 & -2[R(^{BC}\hat{q})^B m_i]_\times & 0 & 0 \end{bmatrix} \tag{33}$$

$$M = \begin{bmatrix} 0 \\ diag([-I_3, -0.5I_3, I_3, I_3]) \end{bmatrix} \tag{34}$$

$$N = I_{4l} \tag{35}$$

where $^{C}\hat{m}_i = [\alpha, \beta, \gamma]^T$ and $^{CL}\hat{m}_i = [\chi, \mu, \kappa]^T$. C_i and C_{i+l} each correspond to two lines of the matrix C, associated with marker i. For the cross product, we use the notation $[a]_\times b = a \times b$.

The transformation group considered for this filter is

$$\begin{pmatrix} p^\circ \\ v^\circ \\ q^\circ \\ a_b^\circ \\ \omega_b^\circ \end{pmatrix} := \varphi_{q_0} \begin{pmatrix} p \\ v \\ q \\ a_b \\ \omega_b \end{pmatrix} = \begin{pmatrix} p \\ v \\ q * q_0 \\ q_0^{-1} * a_b * q_0 \\ q_0^{-1} * \omega_b * q_0 \end{pmatrix}.$$

φ is indeed a transformation group since we have $\varphi_{q_1}(\varphi_{q_0}(x)) = \varphi_{q_0 * q_1}(x)$. The system (8)–(12) is invariant w.r.t. φ; for example, we have for (10):

$$\begin{aligned} ^{BC}\dot{q}^\circ &= \dot{\overbrace{^{BC}q * q_0}} = {^{BC}}\dot{q} * q_0 = \frac{1}{2}(^{BC}q * q_0) * q_0^{-1} * (\omega_m - v_\omega - {^B}\omega_b) * q_0 \\ &= \frac{1}{2} {^{BC}}q^\circ * (\omega_m^\circ - v_\omega^\circ - {^B}\omega_b^\circ) \end{aligned}$$

The filter (25)–(29) is also invariant w.r.t. φ. The transformation group φ represents a quaternion multiplication by the right. This is why the system and filter is called "right-invariant", determining the name of this type of EKF.

4 Experimental Results

Figure 1 shows our sensor unit which is made of an ADIS16355 IMU and infrared LEDs. The IMU is fixed to the back of the sensor unit. The camera is a Wiimote image sensor which we unsoldered and put on a PCB [11] to use it on its own. The Wiimote is the remote control of Nintendo's game console. This image sensor sees up to four luminous points and outputs the point coordinates. We chose this low-cost sensor because it does image processing internally and can easily be interfaced and synchronized with a microcontroller. To build a stereo system, we fixed two image sensors relative to each other in a stereo rig as shown in Fig. 1. The cameras were calibrated using [12] to find their intrinsic parameters and the stereo transformation (5) using a target with four LEDs in a plane. Note that the size of this experimental sensor unit is much bigger than that of the final sensor unit which would be attached to a handheld tool. This size was chosen to be able to adapt the setup during the development stage.

Data from all sensors were acquired by an Atmega2560 microcontroller which communicated with the camera via the Inter-Integrated Circuit (I2C) protocol (we used an I2C switch to change between the two cameras) and with the IMU via the Serial Peripheral Interface (SPI). The sensor readings were synchronized with a camera sample rate of 50 Hz and an IMU sample rate of 250 Hz and the data were sent to a PC serial port. Our direct RIEKF estimated the sensor unit pose from these data in real-time with an xPC Target application with Matlab/Simulink. To compare our results to optical tracking, we used the same data offline to calculate the sensor unit pose from the marker images using [7] and [13].

To have a reference for the sensor unit motion, we mounted it on a linear rail (see Fig. 1) which measures its carriage position with a resolver. We used this measurement as a ground truth. In order to compare this 1D measurement to the

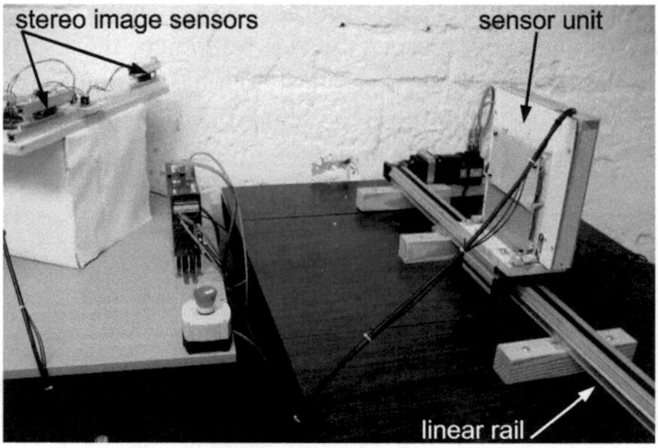

Fig. 1 Experimental setup with stereo rig, sensor unit and linear rail

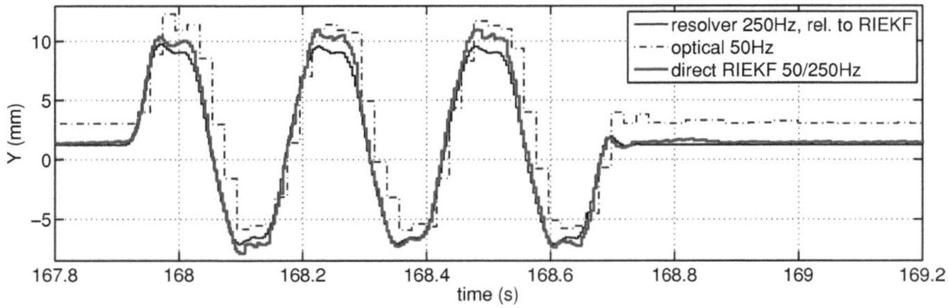

Fig. 2 oscillating motion: *Y* coordinate of reference, optical and optical-inertial position

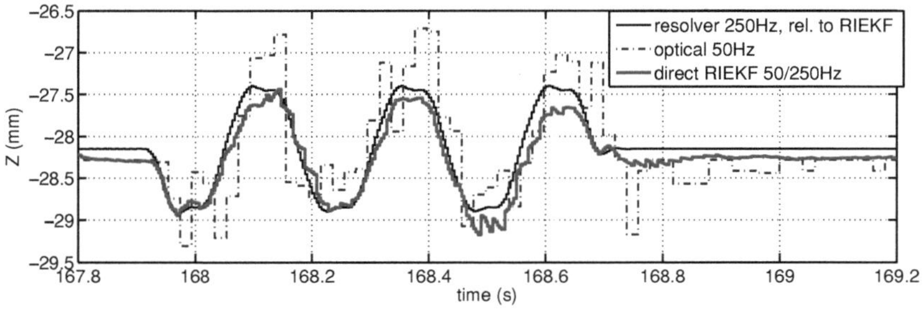

Fig. 3 oscillating motion: *Z* coordinate of reference, optical and optical-inertial position

RIEKF and optical pose estimations, we calculated the position and direction of the rail using RIEKF position estimates and then computed resolver positions relative to this line.

To show that our optical-inertial tracking system can follow the motion of the sensor unit with a high bandwidth, we chose an oscillating motion on the rail with an amplitude of 1 cm. Figures 2 and 3 show the horizontal and vertical position coordinates, comparing the rail reference, the estimation from the direct RIEKF and optical tracking. The purely optical system sees only a very approximate form of the oscillating motion while the optical-inertial system detects the motion correctly with only a few deviations.

5 Conclusion and Future Works

We presented an optical-inertial tracking system consisting of a stereo camera pair and a sensor unit with an inertial measurement unit and optical markers. This sensor unit can be attached to a handheld tool, i.e. in a computer-assisted surgery system.

We presented the system model and equations for a direct right-invariant Extended Kalman Filter (RIEKF) estimating position, velocity and orientation of the sensor unit as well as the accelerometer and gyroscope biases. The motion of the handheld tool is tracked at a high rate of 250 Hz and with small latencies which makes servo-control possible. With our experimental setup we conducted an experiment on a linear rail to show that the optical-inertial system with a direct RIEKF estimates the sensor unit motion correctly and with a high bandwidth.

In the future, the experimental setup could be improved, i.e. with a more performing camera system, and the sensor unit should be calibrated more precisely.

References

1. G. Brisson, T. Kanade, A.M.D. Gioia, B. Jaramaz, *Precision Freehand Sculpting of Bone*. Medical Image Computing and Computer-Assisted Intervention (Springer, LNCS, 2004), pp. 105–112
2. H. Haider, O.A. Barrera, K.L. Garvin, Minimally invasive total knee arthroplasty surgery through navigated freehand bone cutting: Winner of the 2005 HAP Paul Award. J. Arthroplasty **22**(4), 535–542 (2007)
3. N. Parnian, S.P. Won, F. Golnaraghi, *Position Sensing Using Integration of a Vision System and Inertial Sensors*. 34th Annual Conference of the IEEE Industrial Electronics Society (2008), pp 3011–3015
4. D. Roetenberg, Inertial and magnetic sensing of human motion. PhD thesis, Universiteit Twente, 2006
5. A. Tobergte, M. Pomarlan, G. Hirzinger, *Robust Multi Sensor Pose Estimation for Medical Applications*. IEEE/RSJ International Conference on Intelligent Robots and Systems (2009), pp. 492–497
6. B. Hartmann, N. Link, G.F. Trommer, *Indoor 3D Position Estimation Using Low-Cost Inertial Sensors and Marker-Based Video-Tracking*. IEEE/ION Position Location and Navigation Symposium (2010), pp. 319–326
7. R. Hartley, A. Zisserman, *Multiple View Geometry in Computer Vision*, 2nd edn. (Cambridge University Press, Cambridge, 2003)
8. IEEE, IEEE standard specification format guide and test procedure for single-axis interferometric fiber optic gyros (IEEE Std 95–1997).
9. B.L. Stevens, F.L. Lewis, *Aircraft Control and Simulation* (Wiley, Hoboken, New Jersey, 2003)
10. G.C. Goodwin, S.F. Graebe, M.E. Salgado, *Control System Design* (Prentice Hall, Englewood Cliffs NJ, 2001)
11. U. Jürss, W. Rudolph, Tracking Hot Spots. elektor 384 (2008)
12. J.Y. Bouguet, Camera calibration toolbox for Matlab, http://www.vision.caltech.edu/bouguetj/calib_doc
13. S. Umeyama, Least-squares estimation of transformation parameters between two point patterns. IEEE. T. Pattern. Anal. **13**(4), 376–380 (1991)
14. D. Simon, *Optimal State Estimation: Kalman, H_∞, and Nonlinear Approaches* (Wiley, Hoboken, New Jersey, 2006)

VUIR: A Vehicle Undercarriage Inspection Robot

Robert Ross, John Devlin and Anthony de Souza-Daw

Abstract This chapter describes a mobile robot system used to image vehicle undercarriages. The system uses a visual processing technique to identify and parameterise tyres as fiduciary points, from which pose estimation can be performed. Following this, a path planning algorithm is used to navigate the mobile robot underneath the target vehicle whilst imaging the undercarriage using a wide-angle catadioptric camera. The images of the undercarriage are re-projected and template matched with a odometry-controlled dynamic region of interest to generate a mosaic view of the vehicle undercarriage. The mosaic images are visually evaluated, with the inclusion of simulation improvised explosive devices (IEDs), and the proposed system is shown to be significantly more versatile and safer for operators compared to alternate methods of undercarriage inspection.

Based on "Catadioptric Vehicle Undercarriage Imaging with Visual Path Planning", by Robert Ross, John Devlin and Anthony de Souza-Daw, which appeared in the Proceedings of the 5th International Conference on Automation, Robotics And Applications (ICARA 2011) © IEEE.

R. Ross (✉) · J. Devlin
Department of Electronic Engineering, La Trobe University,
Victoria 3086, Australia
e-mail: R.Ross@latrobe.edu.au

J. Devlin
e-mail: J.Devlin@latrobe.edu.au

A. de Souza-Daw
Centre for Technology, RMIT International University,
Ho Chi Minh City, Vietnam
e-mail: Anthony.deSouza-Daw@rmit.edu.vn

G. Sen Gupta et al. (eds.), *Recent Advances in Robotics and Automation*,
Studies in Computational Intelligence 480, DOI: 10.1007/978-3-642-37387-9_14,

1 Introduction

Motor vehicles have become a tool of choice for terrorists worldwide to deliver explosives into restricted areas and for assassinations of specific targets [1]. The simplest, out-of-sight region on a vehicle where explosives are planted is on the undercarriage. An explosive package can be magnetically attached in seconds—avoiding detection due to car alarms when breaking into a vehicle.

Vehicle undercarriages are routinely checked by security at two significant instances. Firstly, at a checkpoint, when a vehicle is entering a secure facility like a military establishment or an airport, to ensure the safety of personnel and infrastructure. Secondly, before a designated VIP approaches the car to prevent against assassination attacks like those used to kill Zelimkhan Yandarbiyev [2].

The remainder of this chapter is organized as follows: Sect. 2 commences with an overview of existing techniques used to search vehicle undercarriages. Following this, Sect. 3 goes on to describe the proposed mobile robot imaging solution. In Sect. 4 results of undercarriage imaging are presented and discussed which is followed by the conclusion in Sect. 5.

2 Survey of Existing Techniques

Across the world, four methods have been prominently used to detect under-vehicular IEDs; mirror inspections, canine inspections, embedded road cameras [3, 4] and various mobile robots [1, 5, 6]. The different methods are described and evaluated here with a summary provided in Table 1.

Mirror inspection of vehicle undercarriages is the oldest and the most pervasive of all detection techniques [1]. In its basic form, an operator uses a vehicle undercarriage inspection mirror (VUIM)—essentially a mirror mounted to the end of a long stick. The operator, whilst standing next to the vehicle, uses the mirror to visually inspect the vehicle undercarriage. More advanced tools have been produced for this technique, including the inclusion of lights to illuminate the undercarriage, roller wheels to improve operator comfort and camera/screen based systems which are compact and could conceivably include a recording functionality. Although this approach is inexpensive to implement, it requires attentive

Table 1 Summary of vehicle undercarriage inspection methods

	Mirror	Canine	Road camera	Mobile robot	Proposed system
Portability	High	Moderate	Some models	Moderate	High
Cost	Low	Moderate	High	High	Moderate
Stationary vehicle	Yes	Yes	No	Yes	Yes
Operator is safe	No	No	No	Yes	Yes
Skill required	Moderate	High	Low	High	Low

security personnel who are at risk by inspecting areas suspected of having explosives [7].

Specialised canines, trained to search for the scent of explosive residue, are used to perform undercarriage inspections and are useful for rapidly searching large areas like complete car parks [8]. Using canines to search undercarriages requires well experienced handlers to supervise and monitor the progress of the search. From a safety perspective, the trainer would be expected to operate further away from any explosives than a VUIM search, but as they have to monitor the canine, they may be close enough to be in danger.

Within the last 5 years multiple in-road undercarriage imaging systems have risen to prominence and been used extensively at security checkpoints for secure facilities. Typically, these systems use multiple upward pointing cameras embedded into the road surface (or a removable 'speed bump') which images a car as it drives overhead. These systems, like Gatekeeper [4], typically generate a mosaic image of the vehicle undercarriage, which can be correlated against a database—allowing anomalies to be highlighted to security personnel. Besides being costly, these in-road imaging systems require a moving car to travel over them—as would occur at vehicle checkpoints and so, unlike the other techniques described, do not allow inspection of a stationary vehicle. Inspection of a stationary vehicle is an important factor for remote control and accelerometer triggered explosives.

Several mobile robot systems, notably iRobot's Packbot [6] and TARDEC's ODIS [5], have been used for undercarriage inspection in conflict areas including Iraq and Afghanistan. These mobile robots typically act as an extension to the use of mirror based inspections, where a user can tele-operate a robot from a safe distance and perform an online inspection of the undercarriage based on the live video feed. Such a system requires a high level of operator competency (involving driving the robot, monitoring a live video feed searching for anomalies and controlling a pan/tilt camera) and without a complete picture of the vehicle anomalies are more difficult to detect.

One robot which attempts to bridge the gap between generating a mosaic image for operators (as this is much easier to spot anomalies on compared to a live video feed) and maintaining a stationary vehicle, has been created by the Security Automation and Future Electromotive Robots (SAFER) initiative [7]. For their robot, researchers tele-operate a mobile robot beneath a vehicle and generate an image mosaic of the vehicle undercarriage using Fourier based image registration. Although this approach improves operator visibility, it still requires a skilled operator to drive the robot and has a relatively narrow field of view which would require multiple passes to scan the vulnerable areas of the vehicle.

Each of the surveyed techniques, although widely used and functional has key short-fallings, either in operator safety, the operator expertise required to operate or the requirement of a non-stationary car. The Vehicle Undercarriage Inspection Robot (VUIR) system described in the following section incorporates the portability and operator protection of the mobile robot solution with the ease of use and complete undercarriage view of the in-road camera solutions.

3 System Description

The system described in this section consists of a mobile robot which automatically navigates underneath a vehicle, uses a wide angle catadioptric camera to image the vehicle undercarriage and performs a template matching image registration operation to generate a mosaic image of the vehicle undercarriage. Besides initially placing the robot in the approximate starting pose (within ±1 m and ±15°) the only role the user plays is to review the mosaic image returned by the robot and identify anomalies.

A block diagram showing the electronic hardware used for the undercarriage imaging is shown in Fig. 1. Due to a lack of on-board processing power image data is analysed on a nearby computer. The current implementation uses off-line image processing using Simulink to generate the mosaic but for future implementations, a high speed wireless connection to the mobile robot (as shown in Fig. 1) would allow simultaneous mosaicing whilst the robot is moving. The operation of the system, with respect to image processing is described by the flow diagram in Fig. 2, where fundamental operations of Vehicle Acquisition, Undercarriage Imaging and Image Registration are shown to govern the operation of the system.

The VUIR, is a small differential drive robot with a baseline track width of 165 mm. An AVR Mega 2561 is the core microcontroller which controls the VUIR behaviour and interfaces to the computer using a 2.4 GHz wireless link. The motors driving the robot are 30:1 ratio geared DC motors producing up to 200 RPM on the output shaft. The motors are fitted with 120 Count Per Revolution (CPR) optical encoders giving a resolution of 27 μm per count. The robot is powered by a high-power density 11.1 V Lithium Polymer (LiPO) battery to give several hours of run time between charges.

The area of the operation, underneath vehicles, imposes several constraints for the design and construction of the robot. In Victoria (Australia), the state in which the authors reside, a legal minimum car height is mandated as 100 mm, hence the

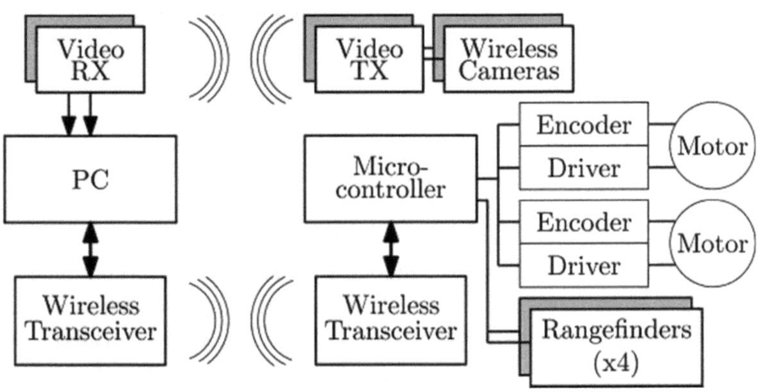

Fig. 1 Hardware block diagram

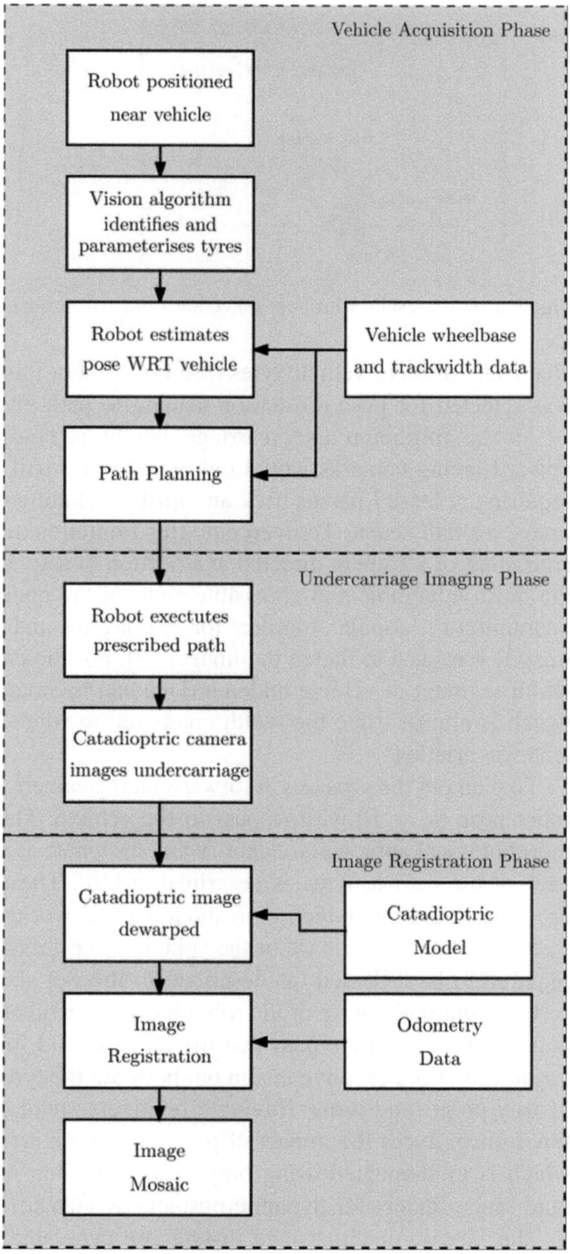

Fig. 2 System flow diagram

robot should be less than this height to ensure it can fit under a variety of vehicles. In addition, car tyres are dark objects which absorb light well, hence limiting the effectiveness of scanning laser range finders (or LIDAR) which were tested by the authors and found to give very inaccurate readings for detection and

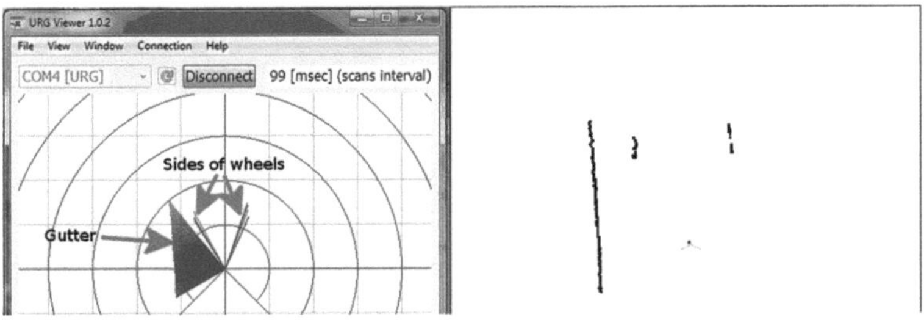

Fig. 3 LIDAR results with poor reflection from tyre surface

characterization of vehicle tyres (see Fig. 3). For this reason a vision based system was selected for pose estimation leading to path planning.

As the minimum undercarriage height is comparatively low, conventional upward facing cameras would only capture a small portion of the undercarriage, requiring at least 7 passes back and forth (including in and out between wheels) to image a small sedan. To overcome this limitation, a catadioptric imaging system, consisting of a camera directed at a portion of a spherical mirror, is used to image the most vulnerable and accessible areas of the undercarriage in only two passes. Scaramuzza's popular toolbox for calibration and re-projection of catadioptric images was used to flatten the images [9]. For the camera system used, a 650 mm width segment of vehicle undercarriage can be imaged from a distance of 15 cm, which is almost triple the width compared to when a slim camera with a 28 mm lens was trialled.

To plan out these passes, a forward facing camera is used to estimate the current robot pose (x, y, θ) with respect to the vehicle. The pose estimation step uses a morphological approach to identify and designate as fiducial points the positions of each of the vehicle tyres as described in [10]. These tyre fiducial points are then processed, along with their corresponding real-world coordinates (derived from the track width and wheelbase of the vehicle), to estimate the pose and hence the path required to be followed (as described in Ross et al. [11]).

Once the initial pose of the robot has been estimated, a path planning algorithm is used to pre-plan the path that the robot should follow. The wide angle camera negates the need to move in and out between the tyres, hence reducing the number of way points and turns. Having a relatively small (<10 m) route with relatively few turns reduces the impact of path integration errors [12], leading to a solution which is implemented using only dead-reckoning odometry for robot navigation (and range finders for dynamic obstacle avoidance).

The mosaic undercarriage images are generated using template matching, as some areas of the undercarriage have poor or repetitive features which may result in poor performance for feature based methods. To make the mosaic generation more efficient, odometry data is used to reduce the search space for which template matches are compared, a factor which decreases significant false match errors which can easily occur for the more homogeneous areas of the undercarriage. Using odometry data to aide in template matching does not suffer as badly as accumulated errors in path integration does in navigation as each frame is relative only to the next frame. In addition, it allows for rotationally invariant mosaic generation using the differential odometry counts between the two wheels to yield an orientation variable.

Gorsich et al. [7] rightly point out that the distance recorded by the odometry will correspond to a different number of pixels movement of the undercarriage image based on the undercarriage height. Intuitively, significantly more pixels will be shifted when imaging a low-profile sports car compared to a truck for the same ground distance travelled by the mobile robot.

Like Gorsich, the approach discussed in this chapter assumes images to be approximately co-planar, which for a single vehicle approximates to give reasonable results, although some higher areas will seems bunched up and lower areas stretched out. Two possible solutions are suggested for further research to solve this problem. Firstly, an upwards facing range-finder could be used to measure the undercarriage height (doubled with the functionality of detection of being under the vehicle). Using this range finder data a relation can be developed to scale the odometry data to be used as a search space based on the undercarriage height.

An alternate scheme to provide undercarriage height invariance would be to initially use the odometry data to generate a sufficiently large search space and dynamically adjust this Region of Interest (ROI) based on the difference between the odometry predicted template match and the actual template match region. Therefore, for relatively coplanar regions of the car the ROI would be sufficiently small and the odometry should predict the match position well. As the undercarriage height varies, the search space would increase and settle towards a new relation between odometry and pixel movement based on an increasing error in odometry prediction.

The mobile robot undercarriage imaging system described in this section is designed to autonomously image a vehicle undercarriage and provide the user with a mosaic image showing the complete vehicle undercarriage for inspection. Rather than requiring an expert operator to navigate the robot whilst monitoring a controllable camera, this approach allows for simple operation with a complete view of the vulnerable areas of the undercarriage.

4 Imaging Results

This section presents the results of the undercarriage imaging conducted with two small imitation IEDs planted on the vehicle. The imitation IEDs were blocks of wood which were covered in black tape and mounted with rare-earth magnets to attach to the vehicle undercarriage. The vehicle under test was a 2005 Toyota Corolla hatchback of which control images were captured for a visual comparison.

In Fig. 4 a segment of the mosaic assembled image is shown with the two imitation IEDs circled for the readers identification. The two IEDs have

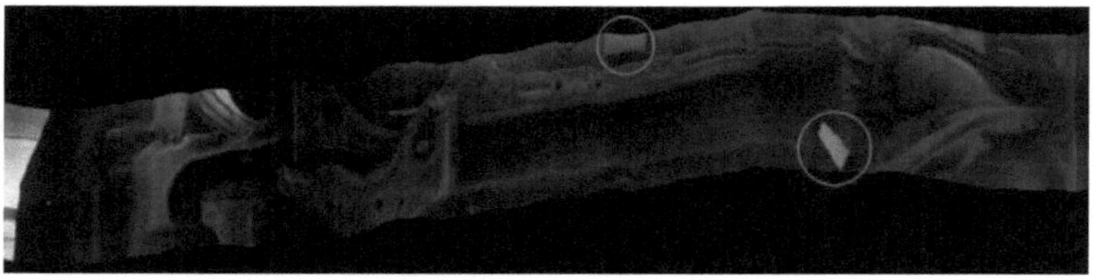

Fig. 4 Section of a scanned undercarriage which includes two (*circled*) simulated IEDs

dimensions of [148 × 69 × 35 mm (L × W × H)] and [120 × 45 × 35 mm (L × W × H)].

The robot currently traverses the path at a speed of 13 cm/s, a relatively slow speed to reduce vibrations and ensure a large amount of overlap for image registration. For a 7 m total path (for a small sedan), a total mission time of 55 s would be required, leading to the mosaicing of 1375 frames. To significantly decrease the scanning time a more robust robot with shock absorbers coupled with a higher frame-rate camera (currently 25 fps) could be used.

One major feature of the VUIR approach is that a mosaic image is generated to identify IEDs, as compared with the more difficult task of identifying them from a moving live video feed. There are four significant areas of improvement which could lead to better clarity for the generated mosaics.

Firstly, the catadioptric camera system uses a relatively low resolution (VGA) camera, which when flattened has a re-projection error in the order of ±2 pixels. Using a high definition camera would enable significantly better images to be captured—improving the final generated mosaic. Additionally, if a user wanted to more closely inspect an area of the vehicle, the software inspection package could produce the original non-flattened image frames which are coupled to segments of the mosaic image, yielding a higher quality image as shown in Fig. 5.

Secondly, the current system uses only ambient lighting, resulting in poor performance in very low light conditions and less detail for darker recesses of the vehicle undercarriage. Some rival commercial implementations have used LEDs behind frosted Perspex to provide a soft white light to evenly illuminate the

Fig. 5 High quality non-remapped vision of IED

undercarriage [13]. In addition, for some captured images, bright reflections towards the sides of the vehicle can skew the white balance resulting in a further loss of detail. One solution would be to use a polydioptric imaging sensor to minimise the reflections received which would have the additional advantage of inferring depth, and hence the odometry to pixel ratio, from the overlapping stereoscopic image regions.

Thirdly, the mosaics produced by the proposed system are closely coupled to the quality of the odometry used to generate the dynamic ROI. When odometry results are accurate, this dynamic ROI approach significantly reduces computational time, enables rotational invariance and increases template matching accuracy. When the odometry is poor, frames will become mis-aligned and result in a dis-jointed mosaic image. One possible solution would be to monitor the template matching alignment error (e.g. Sum of Squared Differences) and increase the size or adjust the orientation of the ROI when it rises above a threshold value. Alternately, different forms of odometry, like optical flow odometry implemented with optical mouse sensors could be used to improve accuracy [14].

Finally, the area most significantly *affected* by a lack of clarity (as clear from Fig. 6 is where the robot is turning a corner. The lack of clarity in cornering has three significant contributing factors; the non-alignment of the camera to the centre of rotation, odometry inaccuracies and frames with significant noise at the edges being copied over good frames previously laid down (as the flattened catadioptric images have more noise towards their edges). To improve the cornering performance the camera should be more closely aligned to the turning axis of the robot and higher quality images (coupled to accurate odometry) should be used.

In contrast to the other undercarriage inspection techniques previously discussed, the proposed technique keeps humans safe—with a mobile robot performing an inspection and compiling a mosaic image showing the easily accessible areas beneath the car undercarriage for review by an operator who can select higher quality (non-remapped) images of key areas of interest.

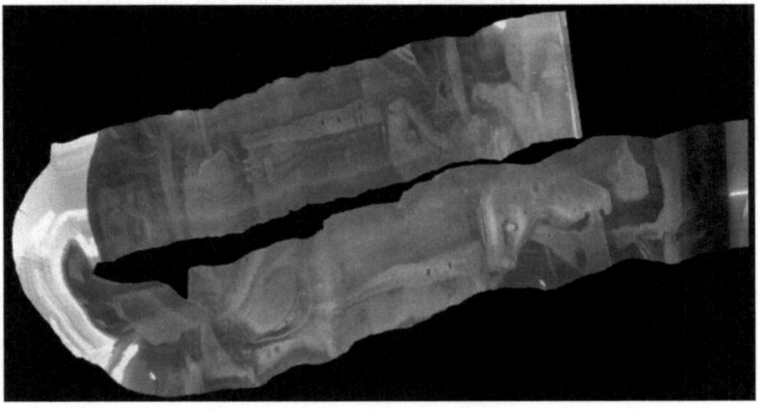

Fig. 6 Mosaic vehicle undercarriage result

5 Conclusion

This chapter has presented a novel approach to vehicle undercarriage inspection where a mobile robot fitted with a catadioptric imaging system is used to image a car undercarriage. The system keeps personnel away from potentially dangerous vehicles and provides a mosaic view of the car undercarriage whilst allowing users to manually inspect areas of interest in higher resolution detail. The portable and inexpensive mobile robot implementation performs acquisition and path planning resulting in minimal training required for personnel.

References

1. J. Frank Bolz, K.J. Dudonis, D.P. Schulz, *The Counter-Terrorism Handbook: Tactics, Procedures and Techniques*, 2nd edn. (CRC Press, Boca Raton, 2002), pp. 143–160. Chap. 8
2. M. Muradov, O. Allenova, Zelimkhan Yandarbiyev is blown up in Qatar. Curr. Dig. Russ. Press **7**(56), 11–12 (2004)
3. P. Dickson, J. Li, Z. Zhu, A. Hanson, E. Riseman, H. Sabrin, H. Schultz, G. Whitten, in *Mosaic Generation for Under Vehicle Inspection*. Sixth IEEE Workshop on Applications of Computer Vision (WACV) (2002), pp. 251–256
4. Gatekeeper, Inc: GKH-2011 AUVIS Product Information (2010)
5. M. Berkemeier, L. Ma, in *Discrete Control for Visual Servoing the ODIS Robot to Parking Lot Lines*. IEEE International Conference on Robotics and Automation (ICRA), (2005) pp. 3149–3154
6. B. Yamauchi, PackBot: A versatile platform for military robotics. Proc. SPIE **5422**, 228–237 (2004)
7. D. Gorsich, G. Gerhart, A. Koschan, D. Page, J.C. Ng, M. Abidi, SAFER under vehicle inspection through video mosaic building. Ind. Robot Int. J. **31**, 435–442 (2004)
8. R.J. Harper, J.R. Almirall, K.G. Furton, Identification of dominant odor chemicals emanating from explosives for use in developing optimal training aid combinations and mimics for canine detection. Talanta **67**(2), 313–327 (2005). Forensic Chemical Analysis
9. D. Scaramuzza, A. Martinelli, R. Siegwart, in *A Toolbox for Easy Calibrating Omnidirectional Cameras*. Proceedings of IEEE International Conference on Intelligent Robots and Systems (IROS) (2006)
10. R. Ross, J. Devlin, in *Vehicle Tyre Parameterisation using Binary Search Thresholding and Contour Fitting*. IEEE International Conference on Broadband Communications and Biomedical Applications, IB2COM2011 (2011)
11. R. Ross, A. Martchenko, J. Devlin, A 3-Degree of Freedom Binary Search Pose Estimation Technique. Under Review for: J. Mach. Vis. Its Appl. (2011)
12. J. Borenstein, H. Everett, L. Feng, D. Wehe, Mobile robot positioning: Sensors and techniques. J. Robot. Syst. **14**(4), 231–249 (1997)
13. G. Robotics, The ferret: Under vehicle inspection robot. Available: http://www.generalrobotics.us/ (2012)
14. R. Ross, J. Devlin, S. Wang, Towards refocused optical mouse sensors for outdoor optical flow odometry. IEEE Sens. J. **12**(6), 1925–1932 (2012)

Improving System Performance Via Reevaluation of Models

Holger H. Rapp

Abstract This work presents a real-time system consisting of a standard industrial arm, two cameras and a controlling PC that is able to catch and juggle a ping-pong ball thrown by a human. We briefly discuss the experimental setup and flow of information, but we focus on overcoming the two main problems: predicting the impact velocity and coping with the slow reaction time of the robot. We explain why this task is impossible to accomplish using the linear model for the flight path of the ball that is predominant in the robot juggling literature and we present a novel fluid-dynamical model. Experiments are presented comparing both models and showing that a better model bundled with a little more computation can drastically increase the systems intelligence and performance. This drastically reduces the demands on the robot's agility and speed.

1 Introduction

Many of the juggling tasks that humans can perform have been investigated for machines as well. A comprehensive overview is given by [1].

The subtask we are interested in is best described as paddle juggling: the idea is that a ball is kept in the air by repeatedly hitting it upwards. If the starting

Based on "A Ping-Pong Ball Catching and Juggling Robot: a Real-Time Framework for Vision Guided Acting of an Industrial Robot Arm", by Holger H. Rapp which appeared in the Proceedings of the 5th International Conference on Automation, Robotics and Applications (ICARA 2011). © 2011 IEEE.

H. H. Rapp (✉)
Institute of Measurement and Control Theory,
Karlsruhe Institute of Technology—KIT,
Engler-Bunte-Ring 21, 76131 Karlsruhe, Germany
e-mail: holger.rapp@kit.edu

G. Sen Gupta et al. (eds.), *Recent Advances in Robotics and Automation*,
Studies in Computational Intelligence 480, DOI: 10.1007/978-3-642-37387-9_15,

conditions—the ball's initial position and velocity—are precisely known this problem can be solved with an open loop controller and no sensors [2–4]. If there are some uncertainties in the starting conditions the task will require some kind of sensory input and therefore a closed loop controller. The work of Rizzi et al. is most influential in this area. Notably, [5] presents a paddle juggler that was remarkable for its time: a distributed system was used to interpret data from a stereo camera system to close the control loop of a custom built robot. The vision system and linear models notably constrain the starting conditions though, that means that the ball must be released in a very specific fashion. Therefore, the machine is unable to catch a ball thrown by a human.

Recently, the use of stock industrial robots has become viable and those systems have therefore moved into the focus of research. Nakashima et al. [6] uses a six degrees-of-freedom (DOF) industrial robot comparable to ours, but keeps the constraints on the starting conditions of the ball for similar reasons as prior work.

This chapter is based on a conference paper [7] and is meant as a supplement to the original work. While it is self contained, it cuts short on some of the topics elaborated in the paper and presents more discussion in other parts. The task that is to be solved can be summarized as follows: The system must be able to catch a ball thrown by a human towards it. The ball should then be juggled and kept bouncing on the racket.

The main challenge is solving the initial catching task when there are no constraints on the starting conditions of the ball. The subsequential juggling is then also handled because it is a simplified version of the catching. For the catching, the system must be able to cope with the uncertainty that comes with human involvement: each throw is different.

While solving the task, two obstacles were most difficult to overcome using the traditional approach with a linear model: the prediction of the impact velocity on the racket and the slow reaction of the robot itself. We use a standard industrial robot not designed for real time control. It therefore has a long dead time before movements are started and it offers no trajectory control. Only incremental movement commands are accepted, i.e. no direct control over the movement velocity or trajectory is possible.

The velocity on the other hand is the variable that determines the angle of the racket so that the ball will be reflected straight upwards. If this is estimated badly the ball will bounce in another direction and be out of reach for the robot—it will not be able to juggle it. The velocity must be predicted precisely.

Both problems can be solved simultaneously by replacing the linear model of the ball's flight via a nonlinear, fluid-dynamical model. We will present an online tracker for this model and compare it with an online tracker for the linear model and a bundle adjustment method over all measurements. We will see that the fluid-dynamical model allows for earlier movement of the robot, more precise impact velocity prediction and cheaper calculation than the globally optimized bundle adjustment method.

2 Description of the Framework

2.1 Hardware

The experimental setup is depicted in Fig. 2. It consists of the robot itself, the industrial PC doing the inverse kinematics for the robot movement and the controlling PC which runs the software part of our framework. The sensors of the system are connected to this PC: two standard industrial cameras which acquire images at 60 Hz (Fig. 1).

More detailed information about the individual parts have been provided in [7].

2.2 Software

The basic flow of information is as follows: first, the ball needs to be detected and associated in the images of both cameras. The ball position in world coordinates can then be found by triangulation. The positions for each frame together with a fixed desired height over ground z_{impact} is then fed into a predictor for the impact

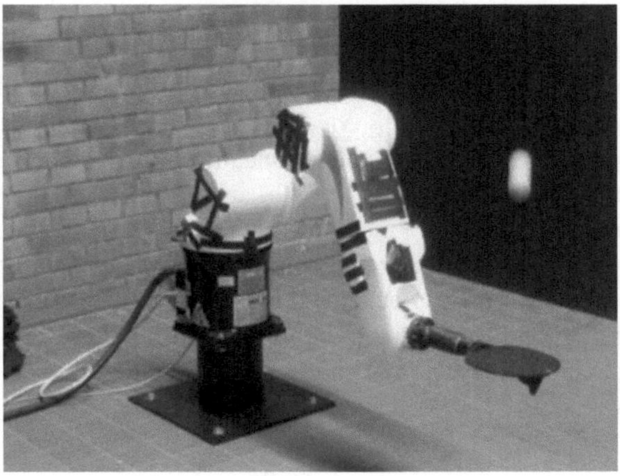

Fig. 1 The robot used in the experiments juggling a ball

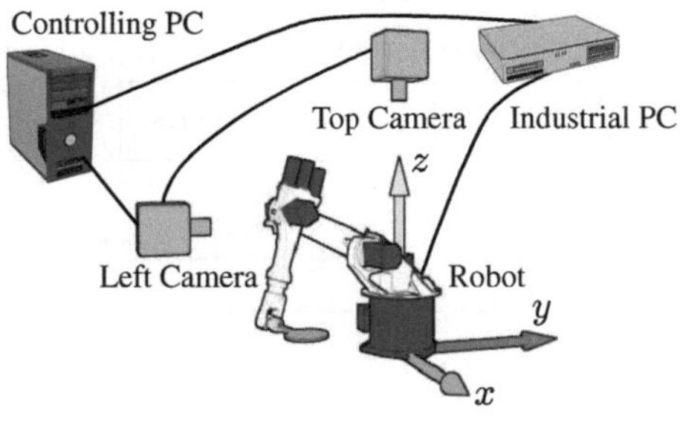

Fig. 2 Experimental setup and origin of work space

parameters of interest: the impact position $\boldsymbol{x}_{\text{impact}}$, the impact velocity $\boldsymbol{v}_{\text{impact}}$ and the time t_{impact} at which the ball will cross the height z_{impact}.

The individual components are discussed briefly in the next paragraphs. We will skip over the details of the image processing and only explain the controlling components of the system. Interested readers can find more information in [7].

The detailed flow of information is depicted in Fig. 3. As soon as a new image pair is acquired from the cameras, the images are processed and the ball is searched in each of them. If the ball is detected in each image, the two pixel coordinates are used to triangulate the current ball position $\boldsymbol{b}$ in space. This ball position is then passed on to a predictor as a new measurement. This predictor uses a model (see Sect. 3.1) to infer the impact position and impact time of the ball. This information is fed to two distinct controllers—max impulse and vertical reflection—which outputs the setpoint for the robot movement: a Cartesian position $\boldsymbol{x}_{\text{setpoint}}$ and two angles defining the normal of the racket's surface $\boldsymbol{\varphi}_{\text{setpoint}}$. The sample-and-hold element waits until its inputs have converged. It will then issue one movement command to the robot which acts back on the ball by hitting it, effectively closing the control loop. A PID element corrects the output angles of the vertical reflection controller and ensures the stability of juggling.

2.2.1 Control Components

Independent of the model or method used, the predictor outputs an estimate of the three impact parameters: the impact position $\boldsymbol{x}_{\text{impact}}$, the impact velocity $\boldsymbol{v}_{\text{impact}}$ and the impact time t_{impact}. $\boldsymbol{x}_{\text{impact}}$ is used directly to position the racket. The other two parameters are used to determine the normal vector for the racket expressed through the two Euler angles $\boldsymbol{\varphi} = (b, c)$ and the point in time when the robot should start to move upwards to give a maximum impulse in z direction onto the ball.

For the vertical reflection, it is assumed that the ball makes an ideal elastic collision with the racket and is therefore reflected. The ball arrives with a velocity $\boldsymbol{v}_{\text{impact}}$ and we want it to be reflected straight upwards: its direction after reflection should be $\boldsymbol{d} = \boldsymbol{e}_z$. We can use the reflection law

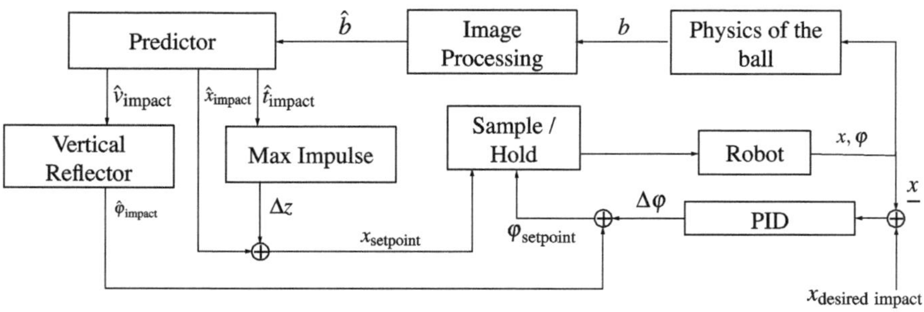

Fig. 3 Control diagram

$$\boldsymbol{d} = \boldsymbol{v}_{\text{impact}} - 2(\boldsymbol{v}_{\text{impact}}^T \boldsymbol{n})\boldsymbol{n} \tag{1}$$

to derive the required normal vector $\boldsymbol{n} = (n_x, n_y, n_z)^T$ of the racket. For this, we multiply (1) from the left with $\boldsymbol{v}_{\text{impact}}^T$ and solve for $\boldsymbol{v}_{\text{impact}}^T \boldsymbol{n}$, choosing the sign of the solution so that $\boldsymbol{n} = \boldsymbol{e}_z$. Re-substituting this in (1) and solving for $\boldsymbol{n}$ results in

$$\boldsymbol{n} = \frac{\boldsymbol{d} - \boldsymbol{v}_{\text{impact}}}{\sqrt{2(||\boldsymbol{v}_{\text{impact}}|| - \boldsymbol{v}_{\text{impact}}^T \boldsymbol{d})}}. \tag{2}$$

The robot expects the setpoint orientation in Euler angles, therefore the component must output the two corresponding angles that lead to this normal vector. They can be calculated from $\boldsymbol{n}$ via $b = \arctan n_x/n_z$ and $c = \arctan n_y/n_z$.

However, vertical reflection is not enough: the ball loses kinetic energy due to friction and deformation each time it collides with the racket. It is therefore mandatory to put energy into the ball to keep it airborne. Since we cannot control the velocity of the robot's movements directly, we send a movement command at a precise time so that its movement velocity is at its maximum value when the ball collides with the racket. The delay T_0 between movement command and achieving maximum velocity was determined in off-line experiments beforehand. Knowing T_0, we give the robot an upwards movement command by an amount of Δz at time $t_{\text{impact}} - T_0$. The robot will then hit the ball with its maximum velocity. After a constant delay ΔT, the command to move down again is issued to make sure that the robot is ready for the next bounce. The output of the maximum impulse component can therefore be written as

$$\Delta z(t) = \Big[\sigma\big(t - (t_{\text{impact}} - T_0)\big) - \sigma\big(t - (t_{\text{impact}} - T_0 + \Delta T)\big)\Big]\delta z, \tag{3}$$

with σ being the Heavyside step-function and δz being the constant amount that the racket should move upwards, in our case $\delta z = 0.05$ m.

Given the impact velocity prediction is accurate, the two components guarantee that the ball is reflected straight upwards at the first contact of the ball with the racket. There are some sources of deviations though: the racket might be installed with a slight offset in the angles compared to the robot coordinate system, the dead time of the robot varies slightly from movement command to movement command, the vertical reflection component neglects angular velocity and the deformability of the ball and all the parameters are only estimations. This leads to instability: the robot would loose the ball after a few bounces because it would leave its control space. We want the robot to always juggle the ball where it can reach it easily and where its movements are as fast as possible.

To achieve this, a PID controller is used that corrects the angles outputted by the vertical reflection component. Its input is the difference between the desired impact position $\boldsymbol{x}_{\text{desired impact}}$ in the middle of the working area of the robot and the estimated impact position $\boldsymbol{x}_{\text{impact}}$. The integral part of the controller guarantees that the real impact position gets closer and closer to the setpoint with each consecutive bounce.

3 Prediction of the Trajectory

Given N measurements and a desired impact height z_{impact}, we want to predict the impact position $\boldsymbol{x}_{\text{impact}}$, impact velocity $\boldsymbol{v}_{\text{impact}}$, and impact time t_{impact} of the ball on the racket.

We first have to choose a model that describes the physical conditions of the throw (see Sect. 3.1), then we have to decide on a mechanism to predict the impact state of the model given some observations (see Sect. 3.2). The combination of model and predictor should provide accurate predictions with N being as small as possible.

3.1 Models

We investigated two distinct models. The next section introduces the linear model which is commonly used in the literature. It is analytically tractable but unable to capture the physics. The following section discusses a fluid-dynamical model that incorporates all the dominant effects on the movement of the ball.

3.1.1 Linear Model

A point mass m is assumed for the ballistic motion. The only force that acts on the mass is the gravitational force $\boldsymbol{F}_{\text{g}} = m(0, 0, -g)^T = -mg\boldsymbol{e}_z$ which results in a linear—in fact constant—differential equation describing the motion. This model is only suitable for slow velocities and high masses m, as than air drag and other nonlinearities can be ignored.

The flight path of the ball is a parabola with the initial position $\boldsymbol{x}_0 = (x_0, y_0, z_0)^T$ and initial velocity $\boldsymbol{v}_0 = (v_{0_x}, v_{0_y}, v_{0_z})^T$:

$$\boldsymbol{x}(t) = \boldsymbol{x}_0 + \boldsymbol{v}_0 t - \frac{1}{2} g t^2 \boldsymbol{e}_z \quad \text{and} \quad \boldsymbol{v}(t) = \boldsymbol{v}_0 - g\boldsymbol{e}_z t \tag{4}$$

Solving $x_z(t_{\text{impact}}) = z_{\text{impact}}$ for t_{impact} gives either one or two solutions in the future. If there are two solutions, we are always interested in the later time because this is the one where the ball will come down on the racket:

$$t_{\text{impact}} = \frac{1}{g}\left(v_{0_z} + \sqrt{v_{0_z}^2 + 2g(z_0 - z_{\text{impact}})}\right). \tag{5}$$

Substituting this into (4) directly yields the desired information for the impact parameters.

This model has the advantage of being analytically tractable. However, for a flying ping-pong ball the assumption of no interaction with the surrounding air is incorrect: friction is acting on the very light ping-pong ball in non neglectable ways.

3.1.2 Fluid-Dynamical Model

Therefore, we present a novel model which takes the effect of friction into account. The dominant effects are the decelerating force $\boldsymbol{F}_\mathrm{d}$ due to air drag and the Magnus effect $\boldsymbol{F}_\mathrm{M}$, which changes the flight direction of the ball due to its rotation. We will discuss both effects in this order.

The Reynolds number for our case is high, we can safely assume the border case of turbulent flow in all cases. This allows us to assume the following equation for the air drag:

$$\boldsymbol{F}_\mathrm{d} = \frac{1}{2}\rho A\, C_D |\boldsymbol{v}|^2 \frac{\boldsymbol{v}}{|\boldsymbol{v}|}. \tag{6}$$

Herein $A = \pi r^2$ is the cross section of the ball that is hit by the flow, $\boldsymbol{v}$ is the velocity of the ball, C_D is the drag coefficient which depends on the geometry of the flying object and is between 0.1 and 0.5 for a spherical object and ρ is the density of air.

If a sphere with radius r rotating with angular velocity $\boldsymbol{\omega}$ is thrown through a medium with velocity $\boldsymbol{v}$, the flow velocity of the medium is different for its top and bottom part. This corresponds to a difference in static pressure at the two sides of the sphere and leads to a buoyant force of

$$\boldsymbol{F}_\mathrm{M} \approx 2\rho C_M \pi r^2 (\boldsymbol{\omega} \times \boldsymbol{v}) \tag{7}$$

given that $|\boldsymbol{\omega}|r \ll |\boldsymbol{v}|$ which is the case here. This effect is called Magnus force. C_M is a dimensionless number called Magnus factor that depends on the Reynolds number and on the geometry and surface roughness of the object in the flow. For our experiment with a reasonable smooth sphere, we expect $C_M < 0.5$.

Combining the Magnus force $\boldsymbol{F}_\mathrm{M}$ and the air drag $\boldsymbol{F}_\mathrm{d}$ together with the gravitational force $\boldsymbol{F}_\mathrm{g}$ we get the complete model's differential equation

$$\boldsymbol{F} = m\ddot{\boldsymbol{x}} = \boldsymbol{F}_\mathrm{g} + \boldsymbol{F}_\mathrm{d} + \boldsymbol{F}_\mathrm{M} = -g\boldsymbol{e}_z - \frac{1}{2}\rho A\, C_D |\boldsymbol{v}|^2 \frac{\boldsymbol{v}}{|\boldsymbol{v}|} - 2\rho C_M \pi r^2 (\boldsymbol{\omega} \times \boldsymbol{v}). \tag{8}$$

This model describes the flight curve of our ping-pong ball physically correct and exhaustively, but it has the disadvantage of high complexity: $\boldsymbol{\omega}$ is changing significantly from throw to throw but cannot be measured and must therefore be estimated from the measurements of the ball's position. C_D and C_M can be assumed to be constant in our case, but cannot be directly measured either.

3.2 Predictors

Each of these models needs to be fitted to measurements and predicted forward in time to get the impact parameters. For our experiments we chose to compare a bundle adjustment approach and a sequential estimation approach of both models.

The bundle adjustment of the fluid-dynamical model was used as ground truth as we will discuss in the next section. The other combinations were benchmarked against this ground truth.

The sequential estimation was done using the Kalman filter. The linear model can be used directly for the Kalman filter while adding a system variance that accommodates for the nonlinearities in the flight curve of the ball. This enhances the prediction performance by allowing the state to change slightly in a nonlinear way. Another advantage is that the initial state and variances can be learned from training data; this effectively incorporates prior knowledge about the throwing velocity. For tracking the fluid-dynamical model, we used a nonlinear extension to the Kalman filter known as the unscented Kalman filter [8]: the basic idea is to interpret state and covariances of the filter as a Gaussian approximation to the correct probability density function (PDF) and to translate this through the non-linear model. This is done via the unscented transform: one chooses enough points from the PDF to fully describe a Gaussian distribution and translates those through the model. The translated points describe a Gaussian approximation to the a-posteriori probability.

A major problem for our case is that the Kalman filter can only track system states, i.e. the current parameters of the model. To predict the impact time t_{impact} with the current system state we trace the model forward in time until we reach the impact height z_{impact}. This can be done analytically for the linear model. For the fluid-dynamical model one has to use a differential equation solver with event handling that can integrate the current state forward in time and report any zero crossings in the z component. The found impact time t_{impact} is then used to predict the system state at impact.

4 Experiments

The two models and the two prediction approaches were compared in experiments. Since the precise time, speed and position of the impact could not be measured directly, a ground truth needed to be generated from the available measurements. As we will detail in 4.1, this has been done by doing a bundle adjustment of all measurements of a throw using the fluid-dynamical model. This leaves three combinations to be benchmarked: linear model with bundle adjustment, linear model with Kalman filter and fluid-dynamical model with Kalman filter.

4.1 Generation of the Ground Truth

We used 746 distinct throws for our experiments. All throws were conducted by humans and there were 14 different throwers. For the model evaluation, we used all throws, even those not caught by the robot. We investigated the measurements till the first impact on the racket happened, because all consecutive bounces are easier as the horizontal velocities are very small if the vertical reflection worked properly on the first bounce.

Since the ball was only measured at discrete time intervals, we cannot guarantee that we have really seen the ball at the point of impact. Therefore we have to make an estimation of the actual position of impact on the racket. This is done by fitting the fluid-dynamical model to all measurements of the throw using a Levenberg-Marquardt algorithm. The best parameters can than be used to find the state at impact height z_{impact} which represents the ground truth for the evaluations.

The remaining problem is the estimation of the values C_D and C_M, which need to be determined before fitting individual throws can be properly done. For this, all throws were taken into account. The fluid-dynamical model was fitted to each individual throw while treating the initial ball position $\boldsymbol{x}_0$, velocity $\boldsymbol{v}_0$, angular velocity $\boldsymbol{\omega}_0$, C_D and C_M as unknowns to be varied. This gave us samples from a distribution over C_D and C_M which can be seen in Fig. 4. The many outliers come from throws that do not fit the model: for example balls that were only rolling on the floor. To cope with the outliers, we fitted a students-t distribution to this data, which gives a sane first mode of $C_M = 0.1418$ and $C_D = 0.4903$. These values have been used to pin C_M and C_D. The ground truth was generated by refitting all throws without C_M and C_D in the parameter set and predicting the impact position. Figure 5 shows the first eight bounces of a sample throw. The colored circles are the measurements of the ball positions, each color represents one bounce. The black line is the fitted fluid-dynamical model for the first bounce and the black circle is the ground truth for the impact position.

4.2 Results

We used the following scheme for the benchmark: for each throw, we fed the predictor one measurement at a time and let it predict the impact parameters. We

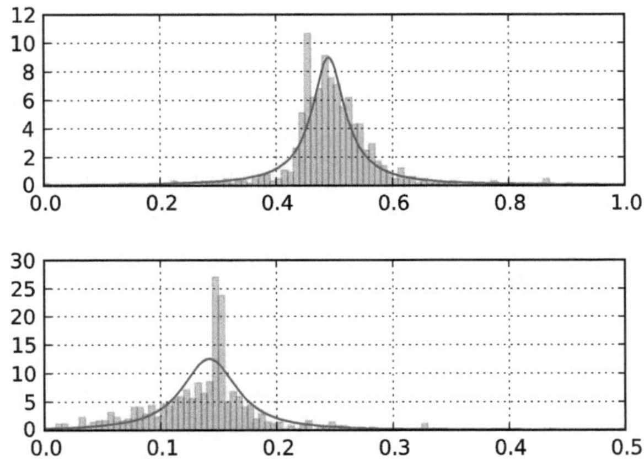

Fig. 4 Value–normalized frequency. Determination of constants C_F (*upper*) and C_M (*lower*)

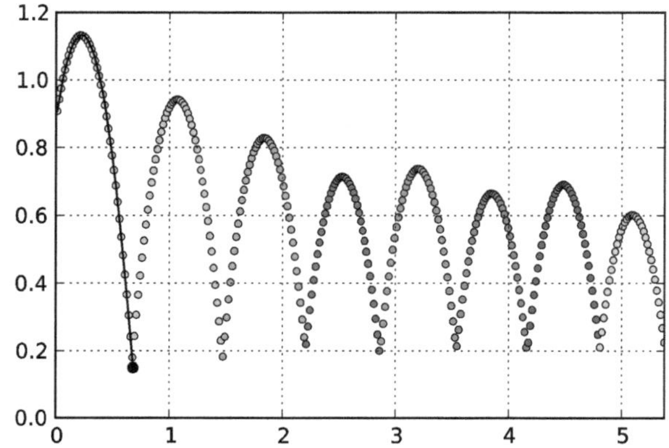

Fig. 5 $t[s] - z(t)[m]$. First bounces of a sample throw

denoted the root squared-error (RSE) in the impact position RSE $\Delta\boldsymbol{x}_{\text{impact}}$, the RSE in impact velocity RSE $\Delta\boldsymbol{v}_{\text{impact}}$, and the absolute error in impact time $|\Delta t_{\text{impact}}|$. Errors that were off by more then 5 units (m for position, ms for velocity and s for time) were discarded as outliers. In this way we got a mapping from the number of measurements used for a prediction N to a set of prediction errors $\{\text{RSE } \Delta\boldsymbol{x}_{\text{impact}}, \text{ RSE } \Delta\boldsymbol{v}_{\text{impact}}, |\Delta t_{\text{impact}}|\}$. The initial state and covariances for the Kalman filters were learned by partitioning the sets of throws into four sets of 149 throws and one set of 150 throws and using leave-one-out cross validation.

The result of this testing scheme for N over RSE $\Delta\boldsymbol{x}_{\text{impact}}$ and RSE $\Delta\boldsymbol{v}_{\text{impact}}$ can be seen in Fig. 6. Depicted are mean (solid line), median (dashed line) and one standard deviation around the mean (shaded areas).

For the linear model, we notice that the full bundle adjustment using the current and all past measurements in the predictor performs worse than the Kalman approach. This is remarkable as it uses much more information for the prediction and takes significantly longer to compute. This can be explained by the system variance in the Kalman filter which allows nonlinear deviations from the model and underlines how bad the linear model really captures the reality.

The fluid-dynamical model converges faster on the correct impact position than the linear model even though it has more parameters to estimate from the measurements. The earlier convergence allows for earlier robot movement. The main improvement however is in the impact velocity prediction: the linear model's performance with predicting the impact velocity is not sufficient to catch the initial throw by a human: the ball bounces out of the robots reach immediately. The convergence of the fluid-dynamical model is fast enough to make an excellent prediction in time and allows the system to excel at the task we formulated in the first section of this chapter.

For $N = 15$ the t-test was used to test the hypothesis that the RSE/absolute error of the linear model with Kalman tracker is smaller or equal to the one of the fluid-dynamical model. The test rejected the hypothesis for each impact parameter ($\boldsymbol{x}_{\text{impact}}$, $\boldsymbol{v}_{\text{impact}}$, and t_{impact}) for a confidence level of 95 %.

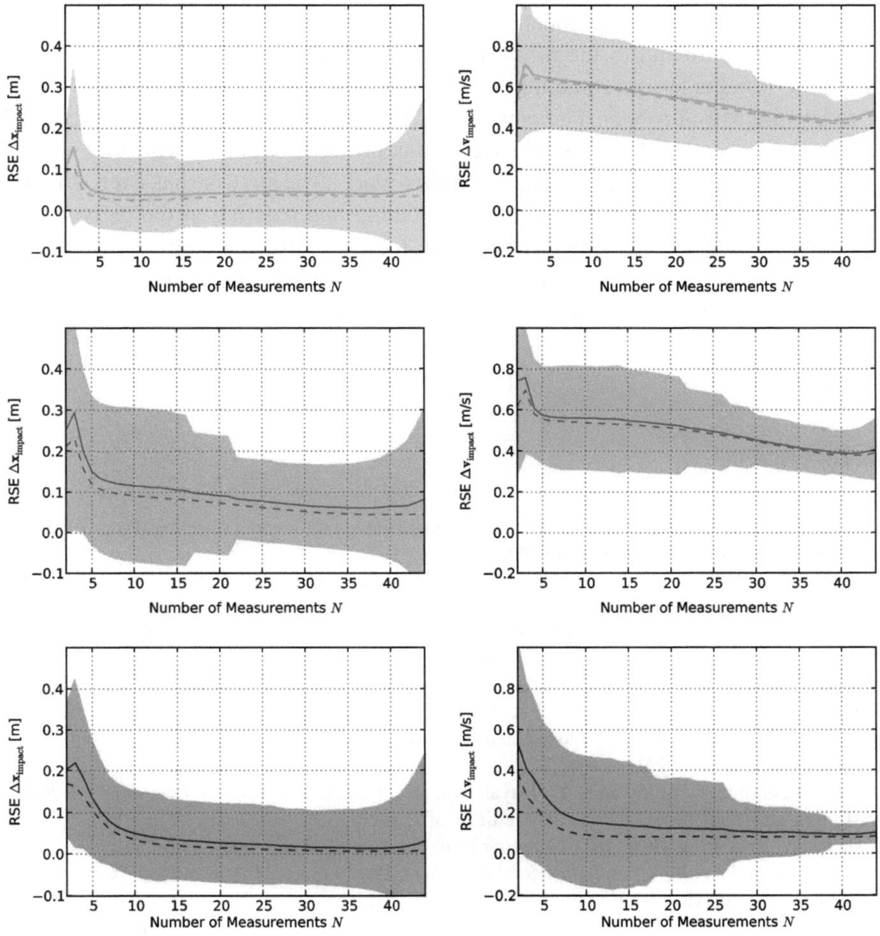

Fig. 6 Prediction performance comparison of impact position (*left*) and impact velocity (*right*). *Top* linear model with bundle adjustment. Middle: linear model with Kalman predictor. *Bottom* fluid-dynamical model model with unscented Kalman predictor

5 Summary

This paper described a real-time framework for a ping-pong ball catching and juggling industrial robot arm. The hardware setup is all off-the-shelve comprised of two cameras, a PC and a small industrial robot. The algorithms employed proved to be fast and stable enough for the task at hand.

We have discussed a linear model that is widespread and usually used in juggling tasks and a novel fluid-dynamical model which captures the nonlinearity of the physics better. It is remarkable how this simple change of model helped overcoming the obstacles defined in the first section: it predicts the impact speed very precisely which leads to small errors in the calculation of the angle of the racket and at the same needs much less measurements to do so. This allows to send a movement command to the robot much earlier and gives it more time to react. Therefore it helps to overcome the second obstacle as well: the slow reaction time of todays industrial robots. The capabilities of the system can best be appreciated when seen in action, that is why a video was provided together with this work[1].

We have two take-away points from this work. The first is that even difficult real-time controlling tasks like tracking, catching and reacting to the physics of a fast moving ball are possible with todays industrial robots. The second is that shortcomings in hard- or software can sometimes be overcome by reevaluating and adjusting one's model of reality. This point is poignantly summarized as "Brains, not brawn".

References

1. S. Schaal, C.G. Atkeson, Open loop stable control strategies for robot juggling. Proc. IEEE Int. Conf. Robot. Autom. **3**, 913–918 (1993)
2. M. Buehler, Robotic Tasks with Intermittent Dynamics, PhD thesis, Yale University, New Haven, 1990
3. T.L. Vincent, in *Controlling a Ball to Bounce at a Fixed Height*. Proceedings of American Control Conference (1995), pp. 842–846
4. P. Reist, R. D'Andrea, in *Bouncing an Unconstrained Ball in Three Dimensions with a Blind Juggling Robot*. Proceedings of ICRA 2009, pp. 1774–1781, IEEE 2009
5. A.A. Rizzi, L.L. Whitcomb, Distributed real-time control of a spatial robot juggler. IEEE Comput. **25**, 1224 (1992)
6. A. Nakashima, Y. Sugiyama, Y. Hayakawa, in *Paddle Juggling of One Ball by Robot Manipulator with Visual Servo*. Proceedings of IEEE ICARCV (2006)
7. H. Rapp, in *A Ping-Pong Ball Catching and Juggling Robot: A Real-Time Framework for Vision Guided Acting of an Industrial Robot Arm*. Proceedings of Automation, Robotics and Applications (ICARA), 2011, pp. 430–435, IEEE 2011
8. S. Julier, J. Uhlmann, H. Durrant-Whyte, in *A New Approach for Filtering Nonlinear Systems*. Proceedings of the American Control Conference, vol. 3, pp. 1628–1632 (American Automatic Control Council, Evanston, IL, 1995)

[1] http://www.mrt.kit.edu/rappweb/pprobot_supplementary.mp4 (65 MB)

Smart Floor with Learning Capability for Mobile Robot System

Soo Hyeok Kang, Yong Ho Kim, Byung-Cheol Min, Soon-Geul Lee, Jinung An and Dong Han Kim

Abstract In this chapter, we suggest a new information space concept called "smart floor". It is the floor storage of specific information designed for route guidance of a mobile robot. A mobile robot can reach its goal position using the information saved in smart floor as well as automatically update the information for a changing environment. We constructed smart floor using RFID tag packaging and fabricated a mobile robot mounted with a passive RFID tag in ultra high frequency (UHF) bandwidth. The passive RFID tag is inexpensive and does not require a power supply. The primary information stored in smart floor is direction-values and Q-values. These values are the resultant Q-learning method that is employed to guide a mobile robot form an arbitrary position to the goal position. This research will contribute to the development of various applications such as mobile robot applications.

Based on Implementation of Smart Floor for Multi-Robot System, by Soo Hyeok Kang, Yong Ho Kim, Eun Jin Lee, Soon-Geul Lee, Byung-Cheol Min, Jinung An, and Dong Han Kim which appeared in the Proceedings of the 5th International Conference on Automation, Robotics and Applications (ICARA 2011).

S. H. Kang · Y. H. Kim · D. H. Kim (✉)
Department of Electronics and Radio Engineering,
Kyung Hee University, Gyeonggi, Seoul, Korea
e-mail: donghani@khu.ac.kr

B.-C. Min
Department of Computer and Information Technology, Purdue University, IN,
West Lafayette, USA

S.-G. Lee
Department of Mechanical Engineering, Kyung Hee University, Gyeonggi, Seoul, Korea

J. An
Pragmatic Applied Robot Institute, DGIST, Daegu, Korea

G. Sen Gupta et al. (eds.), *Recent Advances in Robotics and Automation*,
Studies in Computational Intelligence 480, DOI: 10.1007/978-3-642-37387-9_16,

1 Introduction

Currently, mobile robots are being used as service robots, providing convenience and service mixed with the ubiquitous sensor network (USN) environment. In addition, a research concerning location recognition and simultaneous localization and mapping (SLAM) is being conducted [1, 2].

The representative research focuses on location recognition using an RFID system [3]; active RFID tags are placed on the floor and track the path when a mobile robot with an RFID reader and antenna passes over the tags. The active RFID tags are difficult to integrate into the floor since they require a power supply and are fairly expensive. Having a poor recognition ratio and short location recognition, an RFID 13.56 MHz reader is not suitable for applications in a real environment. Other researches concentrate on location recognition of a mobile robot using an infrared sensor system [4]. The authors installed several infrared sensors on the ceiling and tracked the path of the mobile robot; this is an example of global vision where it would be impossible for a robot to accomplish its mission when it navigates outside of the recognition area.

We describe the development of smart floor using RFID tag packaging and the fabrication of a mobile robot mounted with a passive RFID system using UHF bandwidth. The results of path planning through Q-learning are stored in smart floor and used for maneuvering the mobile robot. The mobile robot is able to update the stored navigation values as it moves. The method we suggest in this chapter can be applied wherever an RFID tag can be embedded into a material.

2 Smart Floor

2.1 Smart Floor Concept

Living in the area of ubiquitous technology allows people to acquire information regardless of time or place and affords many types of conveniences. Due to the human effort about obtaining information, the concept called intelligent space or smart space was created [4–6]. The study of fulfilling a robot's mission has been doing in the intelligent space now [4].

We suggest a new information space concept, called smart floor, which can be applied to workspace for a mobile robot, shown in Fig. 1. Smart floor provides the robot with various types of information for route guidance. Smart floor has three primary strengths.

First, explicit communication is unnecessary with smart floor. In general, the transmitter/receiver interface must be stable for mutual communication. If either the transmitter or receiver does not operate properly, then the necessary information cannot be exchanged. However, since a mobile robot can obtain specific

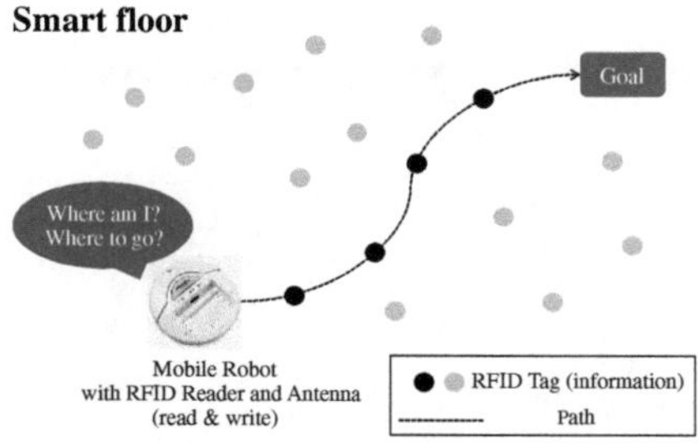

Fig. 1 The concept of smart floor and its implementation

information at specific locations on the smart floor, the explicit communication required is no need. This feature is called location-based information system.

Second, accurate position recognition is unnecessary with smart floor. If we are not aware of the exact destination, a navigation device helps in providing an alternative navigation route. Similar to a navigation device, the robot is able to reach the desired destination using the information provided by smart floor.

Third, it is feasible to run multi-agent learning with smart floor. Efficiency can be remarkably ameliorated when multiple robots work together on one task. However, in this operation of using multiple robots, it is required to communicate and cooperate with each other. On smart floor, when one robot leaves information on the floor, other robots can obtain the information and complete their tasks using the information. Unnecessary communication and information acquisition can be eliminated because only the robot at the specified location receives information in the location-based information system.

2.2 Smart Floor Implementation

The smart floor is designed by burying RFID tags into the floor. Route guidance and location information for fulfilling a mobile robot's task are stored in the RFID tags. Since the RFID tag packaging is buried in the floor, performance evaluation should be preceded by an analysis of the information acquisition rate of the buried RFID. Thus, we measured the acquisition distance of RFID tags in various materials.

RFID tag packaging is an experiment intended to measure the acquisition distance of the RFID tag in diverse packaging environments. We chose passive RFID tags for the following reasons: they do not require a power supply, they are inexpensive, and they can be permanently used. Prior to this research effort, we had measured the acquisition distances of passive RFID tags in various materials [7]. For the experiment, we placed the tags on the top of the tile, shown in Fig. 2.

General purpose passive RFID tags have the characteristics shown in Table 1 and have memory storage for 16 figures of 64 bits. The information for smart floor route guidance is stored in the memory. As shown in Fig. 3, a structure of data consists of coordinates, Q-value, direction-value, and tag ID in order. We

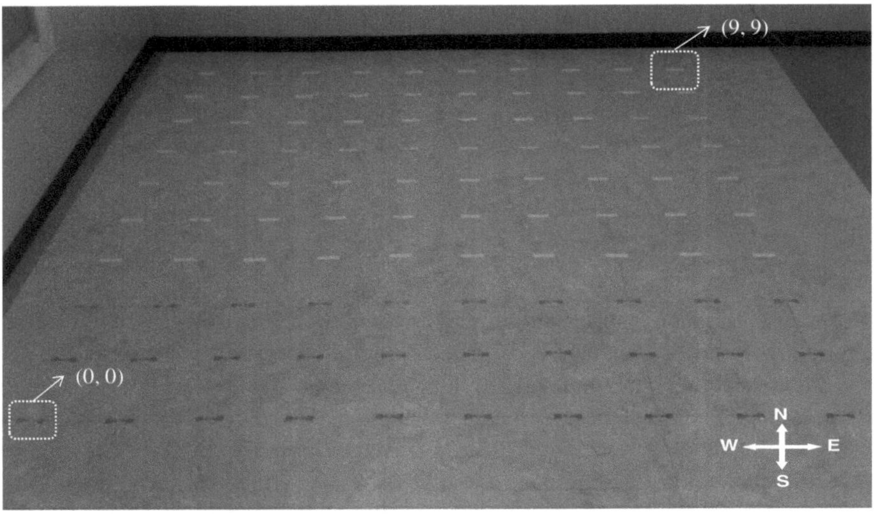

Fig. 2 Smart floor

Table 1 Typical passive RFID tag specifications

Property	Specifications
Protocol	EPC Class 1 Gen 2
Operational frequency range	860 to 960 MHz
Memory	240bit EPC/64bit TID (Tag ID)

examined the design of the smart floor composed of passive RFID tags in this section. The next section discusses the probability of creating a capable mobile robot with a passive RFID system in the next section.

2.3 Passive RFID System for Smart Floor

The RFID system of smart floor consists of a computer, a RFID reader and antenna, and mobile robot, as shown in Fig. 4. The computer issues navigation commands to the mobile robot by using the tag information received from the RFID reader and antenna. The mobile robot with this passive RFID system is then capable of recognizing the location and can read or write the tag information into smart floor.

The RFID reader used in our research operated in the 900 MHz UHF bandwidth; this is the most commonly used bandwidth and possible to detect objects moving at approximately 100 km/h [8]. As shown in Fig. 5a, we selected the MKUR-300 from Minerva [9] as the RFID reader, and mounted it on the mobile robot. Table 2 lists the detailed specifications of the MKUR-300.

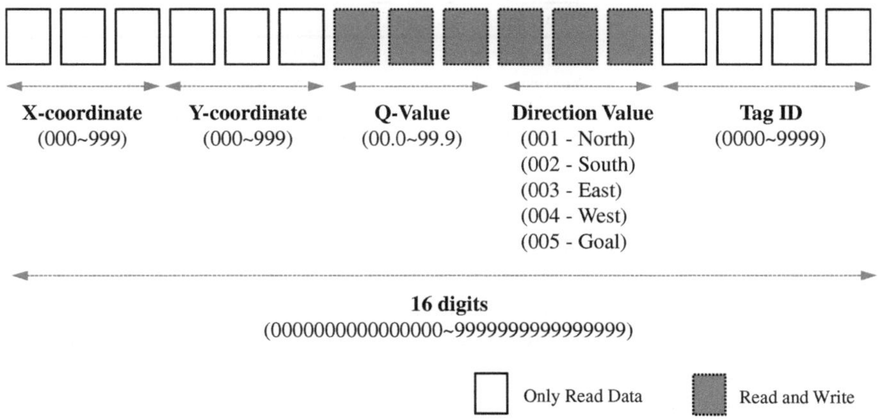

Fig. 3 Data structure of a RFID tag

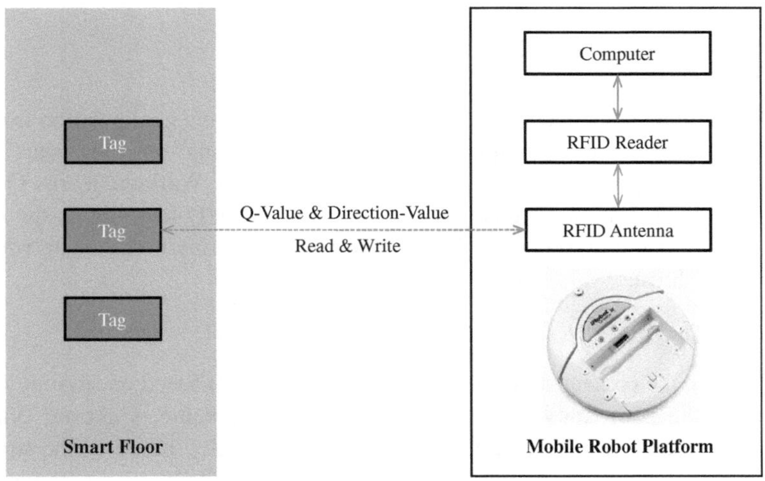

Fig. 4 RFID system in smart floor

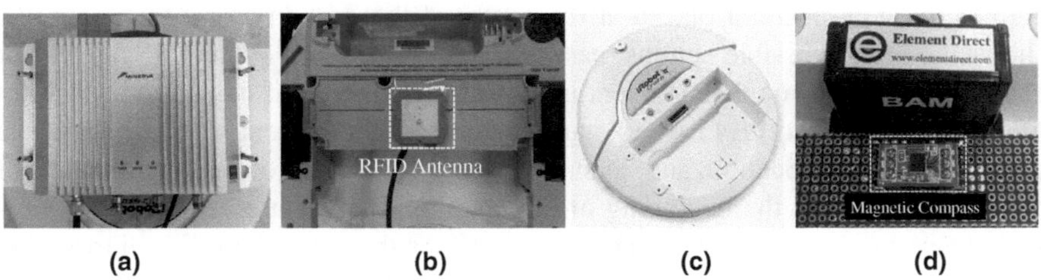

Fig. 5 Passive RFID system. **a** RFID Reader. **b** RFID antenna. **c** Mobile robot. **c** Mobile robot

Table 2 MKUR-300 specifications

Property	Specifications
Frequency band	910.4–913.6 MHz
Air protocol	EPC Class 1 Gen 2, ISO 18000-6C
ANT Port	4 Port (RX/TX Common Type)
Connection	RS-232

We chose a low-profile antenna, which supports the 900 MHz UHF bandwidth. Figure 5b shows the RFID antenna attached on the soleplate of the mobile robot. Figure 5c shows 'Create robot', the mobile robot manufactured by the iRobot Corporation, that was used for this research. This mobile platform supports various sensors and extension modules. We added a magnetic compass sensor on the BAM module in order to measure the direction angle; the BAM module supports Bluetooth communication, as shown in Fig. 5d.

3 Path Planning Using Q-learning

A path should be planed to allow a mobile robot to reach the goal position from the current position. We accomplished this via Q-learning on the smart floor. Q-learning, an algorithm which was firstly suggested by Watkins, learns Q-function that is a function of state-action [10]. We performed Q-learning on the lattice space as path planning on smart floor and used 1-step Q-learning update which is the simplest form.

$$Q(s,a) \leftarrow Q(s,a) + \alpha[r_{t+1} + \gamma \max Q(s_{t+1}, a_{t+1}) - Q(s_t, a_t)] \tag{1}$$

In Eq. (1), the present state chooses a path (movement) based on strategy. In the next state, the path (movement) having the largest Q-value is chosen. Various paths (movements) will be attempted with a low possibility, based on the strategy. In Q-learning, paths (movements) are chosen independently from the present strategy of a specific moment. And then, the ideal value function is directly estimated by learning the paths (movements) value function. Figure 6 shows the path to the goal position placed at the center of the grid environment. Figure 6a shows the set of possible actions: east (E), west (W), south (S) and north (N), for each state. We assume that an agent is subject to defined grid-world, so it does not move the outside world. First, an agent determines the present state, which is the column and row location and then chooses a single action among the set of possible actions. At the beginning of the learning, the agent extensively explores state-space by the selection of various actions. As the agent gains familiarity with the state-space through the learning process, it learns to select the suitable action for reaching the goal position. As shown in Fig. 7, the agent will obtain a '1' which is a compensation value if it reaches the goal position. If the agent attempts

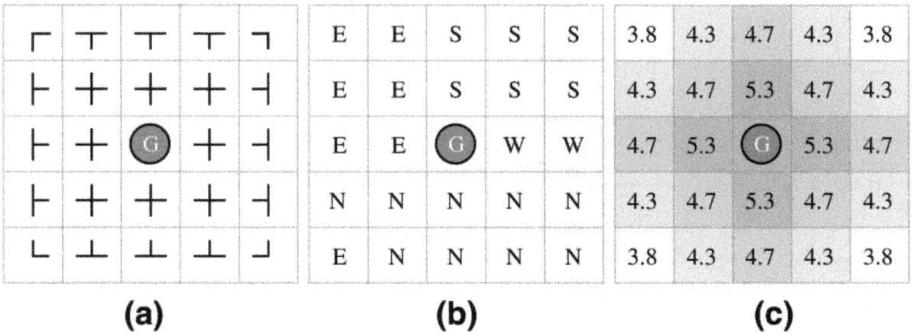

Fig. 6 The shortest distance to G in the grid-world. **a** Possible actions. **b** Optimal strategy. **c** Q-value

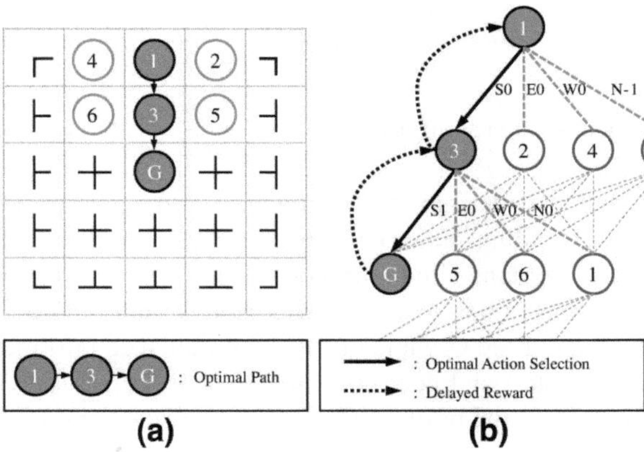

Fig. 7 Selection of the best action and backup process of reward value. **a** Optimal path. **b** Backup diagram

an action that exits the lattice space, it will obtain '−1'. Any other actions of the agent will be then given the compensation value of '0'.

As shown in Fig. 7a, if the initial state is 1, then the optimal route becomes 1 → 3 → G. Figure 7b shows the diagram used in choosing a chain of actions from state 1 to G. The numbers expressed with the actions indicate a compensation value.

4 Experiments

We discussed generating and saving the direction value and Q-value at the corresponding lattice space in smart floor, though Q-learning in the previous section. The Q-value dictates the proper movement based on the highest value compared to

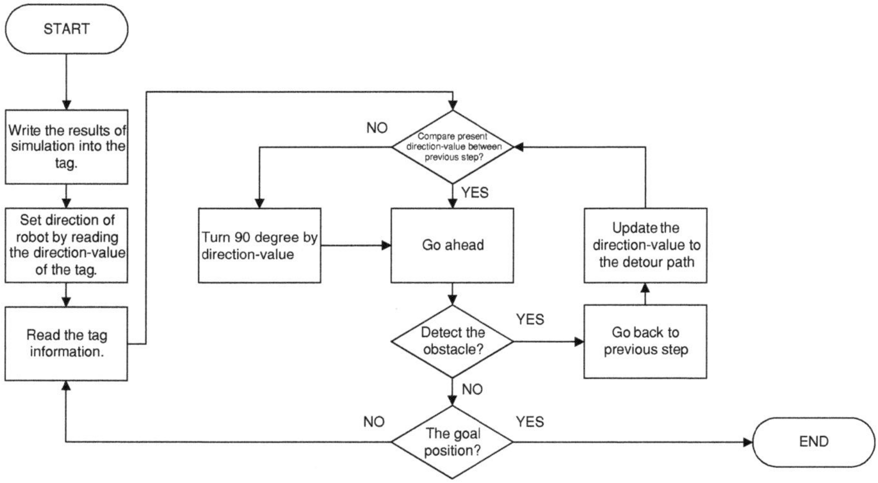

Fig. 8 Flowchart for a mission in smart floor

the present value, and provides navigation guidance to the goal position. Figure 8 shows the process of fulfilling the mobile robot's mission on smart floor.

We designed the Q-learning simulator to save the necessary data for a robot to fulfill its mission on smart floor. We used Microsoft Visual Studio 2010 to make the simulator and print out the Q-value of the route guidance information for each case, after defining the start and goal positions.

For the first case of the goal position (7, 5), we run fist the simulator to save the Q-values and direction-values information on smart floor. Figure 9 illustrates the

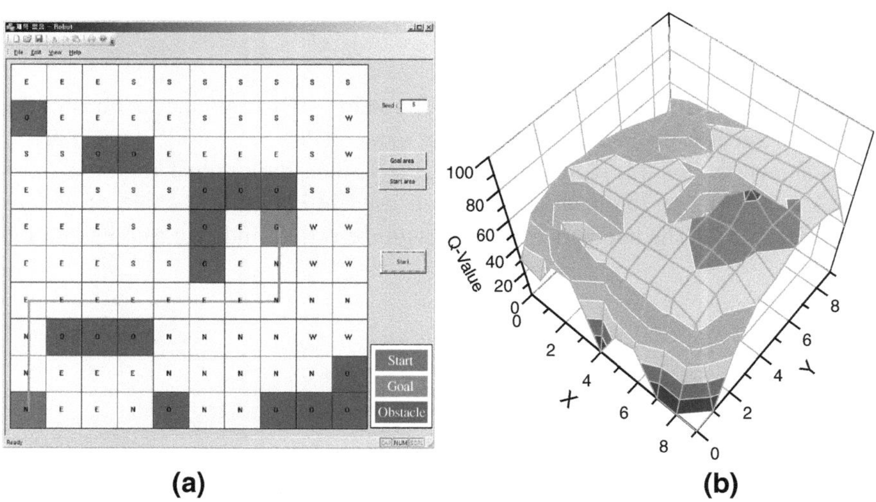

Fig. 9 Simulation results for the first case: the goal position (7, 5). **a** Path to goal position. **b** Q-value for each state

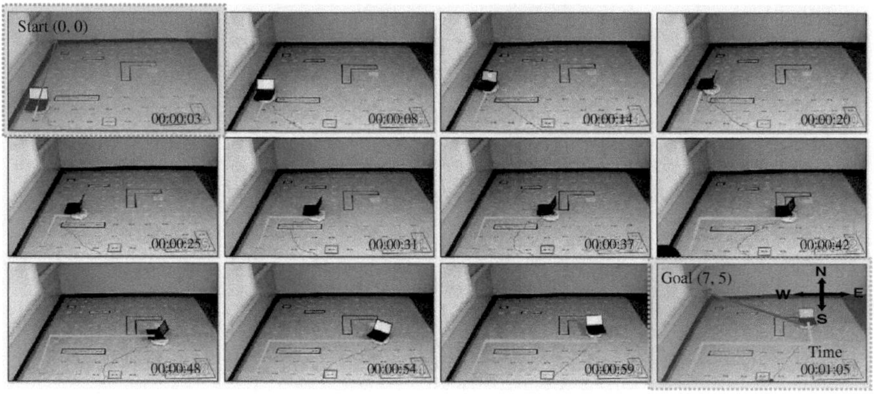

Fig. 10 Experimental results for the first case: goal position (7, 5)

paths generated from the starting location (0, 0) to the goal positions. After that, we run the mobile robot on smart floor. As a result, the robot could successfully reach the goal position by moving in a "Read", "Move", "Stop" order, on the tag space which had the interval of 30 cm between each to each. Not only that, there were no collisions with obstacles surrounded by black on smart floor. Figure 10 shows the experimental results.

Figure 11 shows the second case where the robot should take a detour route in order to avoid an unexpected obstacle on the robot's pre-planned path. We located the arbitrary obstacle at the (0, 3) location, to see whether the robot choose a detour route or not.

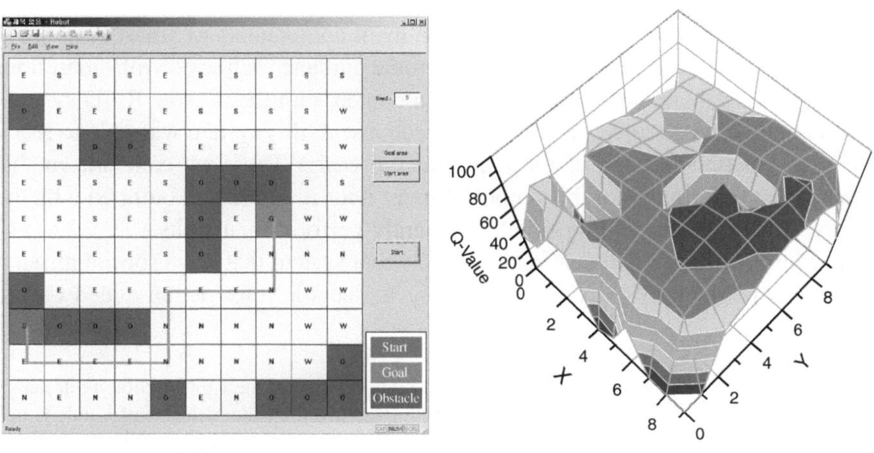

Fig. 11 Simulation results for the second case: detour path for the goal position (7, 5). **a** Path to goal position. **b** Q-value for each state

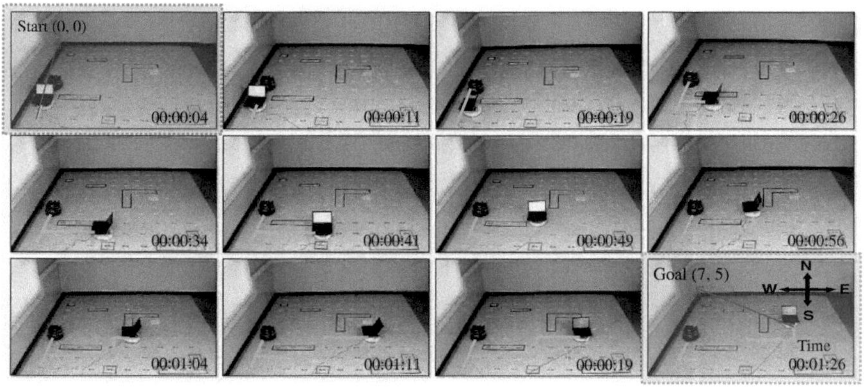

Fig. 12 Experimental results for the second case: the goal position (7, 5)

As a result, as shown in Fig. 12, the mobile robot first followed the route by relying on the values of the direction guiding data stored on smart floor. However, when it detected the obstacle by a bumper sensor, the robot instantly stopped at (0, 3). Then, the robot rewrote the direction-value of RFID tags and could finally reached the goal position (7, 5). Figure 12 shows the traces of the robot moving towards the goal.

5 Conclusions and Future Works

In this chapter, we proposed the new concept of an information space system called smart floor. First of all, we introduced the main components of smart floor such. And then, we showed various examples of route guidance achieved with smart floor by introducing the mobile robot platform mounted with the passive RFID system. Consequently, we verified that the mobile robot with this system was able to reach the goal position in each of the experimental cases. Thus, we conclude in this chapter that the strength of this proposed smart floor is that robots outfitted with the passive RFID system will be able to efficiently perform their tasks.

In future work, we will deal with a multi-robot system that has a special mission to clean a particularly dirty area according to specific information added on the tag.

Acknowledgments This research was supported by a grant from the Construction Technology Innovation Program (CTIP) (No. 10CCTI-A050948-04) funded by the Ministry of Land, Transportation and Maritime Affairs (MLTM) of the Korean government, and the Industrial Strategic Technology Development Program (No. 10035544-2010-01) funded by the Ministry of Knowledge Economy (MKE), Korea. Also, This work was supported by the DGIST R&D Program of the Ministry of Education, Science and Technology of Korea (11-BD-01).

References

1. A. Kleiner, J. Prediger, B. Nebel, in *RFID Technology-based Exploration and SLAM for Search and Rescue*. International Conference on Intelligent Robots and Systems. Beijing, China, 9–15 Oct 2006
2. Hahnel, D, in *Mapping and localization with RFID technology*. Internatlomi Confsmnce on RobDtics and Automation. New Orleans. LA (2004)
3. B.S. Choi, J.W. Lee, J.J. Lee, in *Localization and Map-Building of Mobile Robot Based on RFID Sensor Fusion System*. Conference on Industrial Informatics. Daejeon, Korea, 13–16 July 2008
4. E.M. Gorostiza, J.L. Galilea, F.J. Meca, D.S. Monzú, F.E. Zapata, L.P. Puerto, Infrared sensor system for mobile-robot positioning in intelligent spaces. Sensors **11**, 5416–5438 (2011)
5. S.J. Cho, in *On Interface for Information Acquisition Using RFID of Smart Space*. Proceedings of KFIS Fall Conference 2004, vol. 14, Issue 2, (2004)
6. X. Wang, J.S. Dong, C.Y. Chin, S.R. Hettiarachchi, in *Semantic Space: an infrastructure for smart spaces*. IEEE Computer Society 2004, vol. 3 (2004), pp. 32–39
7. S.Y. Nam, D.H. Kim, Evaluation of RFID System for Location Based Services in the Building. The Institute of Electronics Engineers of Korea, vol. 48, Issue 1, (2011)
8. http://www.rfidjournalkorea.com/news/articleView.html?idxno=8226
9. http://www.minerva.co.kr/contents/product/mkur_300.asp?a_site=3_1&sub_flash=2
10. C.J.C.H. Watkins, Q-learning. in *Machine Learning*, vol. 8 (Kluwer Academic Publishers, Boston, 1992), pp. 279–292

High Level Sensor Data Fusion of Radar and Lidar for Car-Following on Highways

Michael Schnürmacher, Daniel Göhring, Miao Wang and Tinosch Ganjineh

Abstract We present a real-time algorithm which enables an autonomous car to comfortably follow other cars at various speeds while keeping a safe distance. We focus on highway scenarios. A velocity and distance regulation approach is presented that depends on the position as well as the velocity of the followed car. Radar sensors provide reliable information on straight lanes, but fail in curves due to their restricted field of view. On the other hand, lidar sensors are able to cover the regions of interest in almost all situations, but do not provide precise speed information. We combine the advantages of both sensors with a sensor fusion approach in order to provide permanent and precise spatial and dynamical data. Our results in highway experiments with real traffic will be described in detail.

1 Introduction

The interest in autonomous cars has grown in recent years, as they provide new insights for general robotic systems in areas like safety, machine learning and environmental perception [1]. This work is based on Radar/Lidar Sensor Fusion for Car-Following on Highways, by Daniel Göhring, Miao Wang, Michael Schnürmacher, and Tinosch Ganjineh which appeared in the Proceedings of the

M. Schnürmacher (✉) · D. Göhring · M. Wang · T. Ganjineh
Freie Universität Berlin, Institut für Informatik, Arnimallee 7 14195 Berlin, Germany
e-mail: m.schnuerm@fu-berlin.de

D. Göhring
e-mail: drgoehring@zedat.fu-berlin.de

M. Wang
e-mail: miao.wang@fu-berlin.de

T. Ganjineh
e-mail: ganjineh@inf.fu-berlin.de

G. Sen Gupta et al. (eds.), *Recent Advances in Robotics and Automation*,
Studies in Computational Intelligence 480, DOI: 10.1007/978-3-642-37387-9_17,

One key aspect of driving autonomous cars is the detection of obstacles in order to ensure safety and to prevent collisions with other traffic users. The main challenge here is to keep track of the followed vehicle even at big distances, winding roads, and high speeds. The application also demands for a precise estimation of other cars' position and velocity. The approach presented here fuses data from lidar scanners and from a long range radar.

There exist a broad spectrum of sensors (including radar, ultrasonic, lidar, camera, etc.) that are suitable for object detection. Unfortunately, data from one specific sensor is usually not enough to infer all information needed for complete autonomous driving under all possible conditions. For that reason sensor data fusion has been widely used in order to construct more reliable and robust systems. A review of state-of-the-art techniques and problems in multisensor data fusion is given in [2]. The fusion of camera and lidar data is one of the most popular approaches and has been studied, e.g., in [3]. An approach for obstacle detection and tracking combining stereo vision and radar is presented in [4].

The fusion of lidar and radar data has also been studied earlier. The combination is promising, because the sensors are complementary in many aspects [5]. In [6] the fusion is used to make the system more robust against adverse weather conditions, whereas [7] focuses on improving the reliability of object-lane association and the reduction of false alarms (phantom obstacles). In [8] a pre-crash system on the basis of lidar-radar data fusion is developed.

In car-following applications it is mandatory to have continuous and precise position and velocity information about vehicles driving ahead. Lidar scanners do not measure dynamic information whereas radar sensors give only rough position information. The data fusion approach presented here integrates both measurements into a common state estimation process. The result is a precise and permanent position and velocity estimation.

The remainder of this chapter is structured as follows. Section 2 describes the sensors and their configuration on our testing car. Section 3 describes the fusion architecture. Data association, out-of-sequence measurements and state estimation are common challenges in any data fusion approach, which is covered in Sects. 4 and 5. Section 6 describes how the car uses the estimated data for velocity planning. Experimental results during a test run on public highways are given in Sect. 7. Finally, Sect. 8 draws conclusions and discusses future work.

2 Sensor Configuration

We use our autonomous test car "MadeInGermany" for experiments, depicted in Fig. 1. The car is equipped with lidar scanners, radars and cameras for obstacle detection. For this work we use six ibeo Lux lidar scanners and a TRW radar sensor for sensor data fusion. The coverage of these sensors is illustrated in Fig. 2.

Fig. 1 Autonomous car "MadeInGermany" (*MiG*)

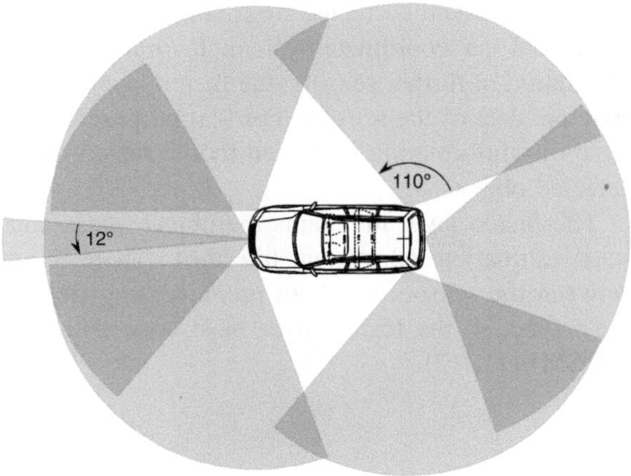

Fig. 2 Six lidars (110°) and one front radar (12°)

Each lidar sensor has an horizontal field of view of 110°, however for car-following the front three scanners are relevant. The radar sensor has a relatively smaller horizontal field of view of 12°.

Both sensors have a similar range of up to 200 m (depending on the material and inclination angle of the reflecting surface), which makes them ideal for ACC-like applications on highways.

Lidar scanners use light pulses to measure distances to other objects. Our lidar sensors have a scan frequency of 12.5 Hz and are configured to make measurements precisely each 0.25°. The distance resolution is 0.04 m. Radar sensors use radio waves to measure distances and velocities by using the Doppler effect. Their distance precision is comparable to lidars but its angular resolution is relatively bad.

By fusing the sensor data we want to exploit the advantages of both sensors. The radar has a narrow field of view and tends to lose the followed car especially in curves. Furthermore, the angular resolution of the radar does not suffice to localize other objects relative to the road lanes in all situations. All these problems are handled by integration of lidar data. In addition, dynamics of other vehicles is required. Velocity estimates from lidar data must be inferred by tracking detected objects. These estimates are relatively imprecise and should be improved by integration of radar measurements (radars only measure relative velocities, so the overall tracking process is still important).

3 Data Fusion Architecture

We use a centralized fusion architecture that is divided into sensor layer and fusion layer (see [9, 10]). This is illustrated in Fig. 3.

The sensor layer is implemented separately for each sensor. First, the data is time stamped and corrected from potential delay and jitter. It is then brought into the global car coordinate system. In order to make the system flexible for the integration of further sensors, the fusion layer abstracts as far as possible from the concrete data of the sensors. For that purpose, a set of features plus precision and reliability information is needed by the fusion modules to associate and track the detected objects.

The fusion layer maintains the object tracks and handles data association and state estimation (Bayesian filtering). There is no hardware synchronization between the sensors and both preprocessing steps are not guaranteed to take the same time, so the fusion layer must cope with Out-Of-Sequence Measurements (OOSM) [11, 12].

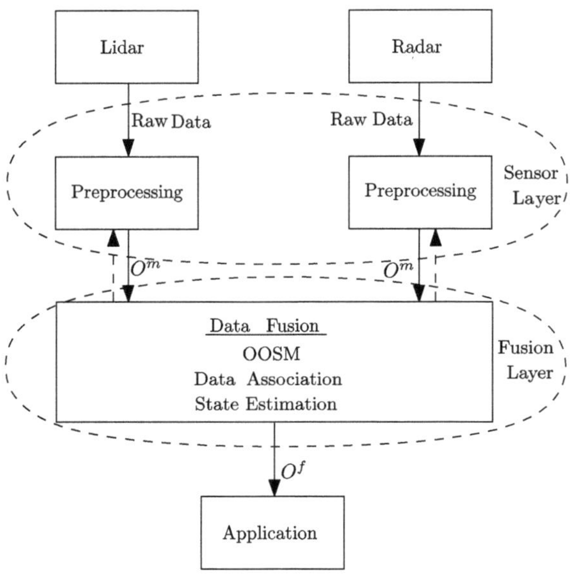

Fig. 3 The fusion architecture is separated into sensor layer and fusion layer. After data preprocessing, the measured object data with their standardized feature set (O^m) is passed to the fusion modules. Here, the objects are tracked and the resulting filtered object hypothesis (O^f) are sent to the application modules. Strong hypothesis can be fed back to the preprocessing modules in order to improve challenging tasks like e.g. clustering

4 Out-of-Sequence Measurements

When younger measurements appear later than older measurements in a system, they are said to be out-of-sequence. OOSMs are usually hard to integrate into a filtering process [13]. We use buffering and measurement reprocessing to handle the OOSM problem. It is easy-to-implement and finds an optimal solution in the sense that no data is lost and the latest information is immediately processed and passed on (not held back). As [12] correctly points out, the approach is memory (buffering) and time (reprocessing) consuming. In our case, this aspect is negligible because the fusion region and thus the number of potential fusion objects is relatively small. A short summary of solutions for the OOSM problem is given in [12] and [11] discusses different synchronization strategies to minimize the worst case latency.

Our approach is illustrated in Fig. 4. The data fusion module maintains a track T for each detected object. It keeps a history of the filtered object state ($O^f_{t_i}$) at different time frames t_i together with the (unfiltered) object data ($O^m_{t_i}$) computed in the sensor specific modules at the same time. It holds that $t_i \leq t_j$ for $d \geq i > j \geq 1$, where d is the depth of the history of the track. The filter prediction and update steps for the tracks are as follows: $O^f_{t_{i+1}}$ is predicted to time stamp t_i, which is then $\hat{O}^f_{t_i}$. This temporal object is updated with the measurement $O^m_{t_i}$ which then becomes the new filtered state $O^f_{t_i}$.

When new objects are added to the fusion module, they are first associated to the tracks. In each track we search for the state estimation that is closest to the new object but not older (binary search on the time stamps). The state is predicted to the new time frame and can now be compared to the object. Such an object always exists except if the track was initiated with objects only from one sensor. In that case the latest estimate is used for comparison iff the time difference is not too big, otherwise the track is not matchable. A consensus between different potential associated tracks is found using a nearest neighbor approach. The total history is then reprocessed from the new pair onwards by a chain of predict and update steps.

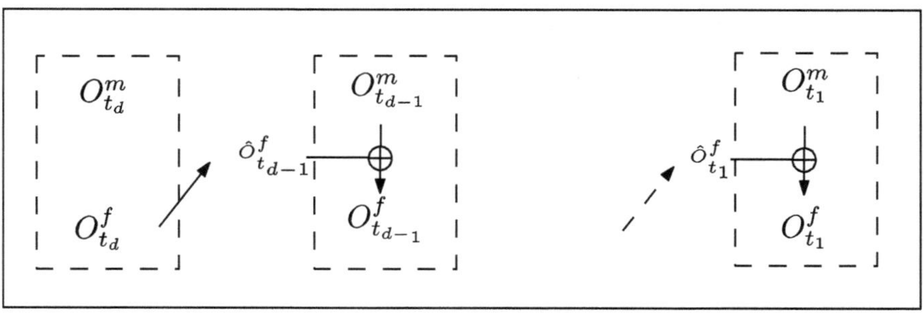

Fig. 4 A track maintains a history for each obstacle hypothesis

After each iteration, the history of each track can be cleaned from unnecessary data. It holds that data from the same sensor arrive in-sequence. Let t_{lidar} and t_{radar} be the time stamp of the latest lidar and radar object data of the track and let dt_{old} be some aging time threshold. Then define

$$t_{old} = \max(\min(t_{lidar}, t_{radar}), \max(t_{lidar}, t_{radar}) - dt_{old}).$$

It is save to clean the history from all pairs that are older than t_{old}.

5 State Estimation

We use a constant acceleration model to describe the filtered states of each track. It consists of its position p_x and p_y, its velocity v_x and v_y, and its acceleration a_x and a_y. In [9] a fusion algorithm is presented that uses two models that are chosen depending on the information quality provided by the respective sensor. In addition to the point model it implements a bicycle model that is used when information about the object's orientation is available (from geometric data when the objects are close enough). When following a vehicle, only its rear is scanned from both sensors, so the geometric information does not suffice to fine tune the orientation of the car and is thus determined by dynamic information only. Nevertheless, the approach is considered essential for future applications, especially when cars are turning tightly in dense traffic and thus orientation information is indispensable.

A Kalman Filter [14] is used to implement the motion model. The state is approximated by a Gaussian distribution with mean $x = (p_x, p_y, v_x, v_y, a_x, a_y)$ and covariance matrix P.

A track is created for each new object that enters the fusion module and that cannot be associated to any already maintained track. The initial state and covariance values of the Kalman Filter depends on the sensor. Each track passes through an confirmation phase before it can be sent to the application modules: A track must have three consistent processed objects to be confirmed (e.g. velocity consistency). The track state estimation consists of a cycle of prediction and measurement update steps.

State prediction is a linear transformation of the current state estimate $\hat{x} = Fx$. The state transition matrix F is set such that $(\hat{p}_x, \hat{p}_y) = (p_x, p_y) + (v_x, v_y)dt$, $(\hat{v}_x, \hat{v}_y) = (v_x, v_y) + (a_x, a_y)dt$ and $(\hat{a}_x, \hat{a}_y) = (a_x, a_y)$. The prediction adds an error to the new estimation due to the inaccuracy of the motion model. One major error is due to the fact that we do not model turning maneuvers and another error is caused by cars changing their acceleration. The first is expressed by assuming a maximal angular velocity ($\dot{\alpha}_{max}$) and the latter by assuming a maximal acceleration divergence ($\dot{a}_{max}$). Given these values the process noise covariance matrix can easily be computed.

After the state prediction, $\hat{x}$ is updated with a new measurement (see measurement fusion models in [15]). The measurement uncertainty is different for both sensors in terms of position and velocity. This is described for lidar and radar in Sects. 5.1 and 5.2, respectively.

5.1 Lidar Preprocessing

A lidar scan contains many scan points, of which usually more than one belong to the same object. It is advantageous to integrate all points that are reflected by the same object into a common cluster, because we assume that each followed track is associated to exactly one object in the real world. Clustering, like data association, is also a challenging task. For similar reasons common clustering problems like over- and under-segmentation are easier to handle for our application, because vehicles tend to keep bigger distances when they move fast and on wider lanes. We use a distance based cluster algorithm with an adaptive threshold. Two scan points are associated to the same cluster if their relative distance is smaller than some distance threshold. The threshold is increased for higher velocities and greater distances to the sensor. In order to make the clustering more robust against over-segmentation, we search for convex shapes in each cluster and split them when our convexity assumption is violated.

A lidar measurement is only a position measurement. We take the center of gravity of all points in an object cluster as the measurement of the object's position. This is reasonable for the car-following application, as we only see the object from the rear. When objects are seen from different perspectives, we would use many different parts of the object as position measurement, which leads to inaccurate velocity and acceleration values.

As discussed before we must also compute a precision measure for each measurement (measurement covariance). As described in Sect. 2 the lidar readings (distance and angle) are very accurate. Unfortunately, these parameters cannot be used as values for the covariances of the measurements. First, the measurement position is computed from many of these readings, which also do not always hit the same part of the object. Second, the measurements are transformed into a common global coordinate system before they arrive at the fusion layer. Both add an additional error to the measurement position.

5.2 Radar Preprocessing

Instead of raw data our radar sensor provides a list of already filtered objects. Each object consists of its position (defined by its distance to the sensor and relative angle) and of its velocity relative to the sensor.

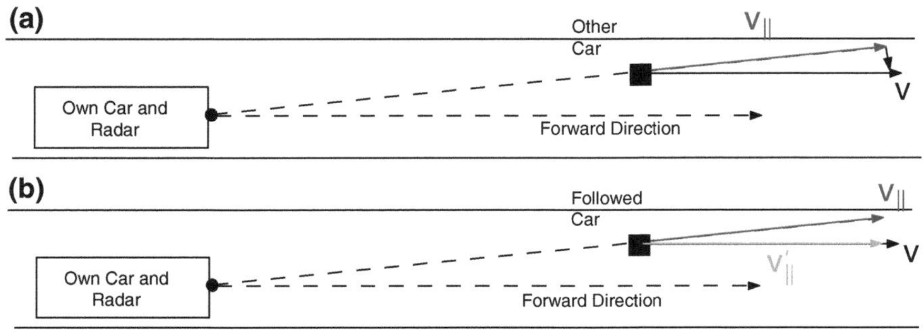

Fig. 5 Radar-velocity-measurement: **a** v is the actual velocity of the followed car, $v_{\|}$ the one measured by the radar sensor **b** and $v'_{\|}$ is the one we assume

The object position information from the radar sensor is directly used as position measurement.

In contrast to the lidar, the radar directly provides information about the object's velocity by using the Doppler effect. This velocity is more precise than the velocity inferred from a sequence of lidar position measurements. Unfortunately, this information is incomplete as only the velocity component relative to the sensor is measured. This can lead to the following type of problem: In Fig. 5, the radar detects an object that drives in front of us on the same lane but is slightly shifted laterally. Moreover, their forward direction is the same. The true velocity of the object is denoted by v. The radar only measures v's component relative to our car ($v_{\|}$). The effect is that the object seems to leave the lane, although it actually does not. In order to prevent this, we only use $v'_{\|}$, which is the component of $v_{\|}$ that is parallel to the forward direction of the own car. This is a tolerable approximation because the radar aperture angle is very small and by integrating additional information from the lidar sensor the perpendicular component can be inferred (although less precise).

6 Velocity Planning

A behavioral module is in charge of generating a set of trajectories along the road network given a specific mission destination. A low-level controller then determines a safe target speed for which brake/acceleration are computed. The detected obstacles are used to evaluate the trajectories for possible lane change maneuvers. If the optimal choice is to follow a vehicle, then the obstacle position and velocity information are used to keep a safe distance.

In particular, each obstacle contour point is checked for collision with the trajectory augmented by the car's dimension. A majority of obstacles and their corresponding contour points are usually located off-road including buildings, trees, and pedestrians. Off-road obstacles are categorized as such and disregarded to optimize collision checks.

All relevant obstacles are stored in a spatial kd-tree [16] to test containment within the trajectories' domain. If interfering obstacles are found along the trajectory they are evaluated by a sigmoidal function over distance to the obstacle:

$$e_{collision} = \lambda * \frac{1}{1 + \exp(\mu_1 \, d_{obst} - \mu_2 \, d_s)}, \tag{1}$$

where d_{obst} denotes the longitudinal distance between car and obstacle and d_s a safety distance to be maintained. The symbols λ, μ_1 and μ_2 denote tuning parameters.

6.1 Target Speed for Car-Following

If an obstacle has been detected on the planned trajectory, we want the car to match the speed to the obstacle. A simple approach is to use the obstacle's distance and speed and to assume a constant acceleration (constant negative acceleration for braking). We can calculate the desired current velocity v_0 with

$$v_0 = \sqrt{2a_{brake}(s_0 - s_E + s_{safe}) + v_E^2}, \tag{2}$$

where s_0 is the car's current position, s_E the position of the obstacle, v_E the velocity of the obstacle, s_{safe} a safety distance and $a_{brake} = -2\,\mathrm{m/s^2}$ a given constant negative acceleration. For a static obstacle, v_E is set to 0. Since we assume a highway scenario we only consider the longitudinal component v_E of the velocity of the vehicle in front of us.

The velocity function over position for a dynamic object and a given braking acceleration is depicted in Fig. 6. Given the distance to an obstacle and the obstacle's velocity one can calculate the desired velocity of the car. A given safety distance would shift this function to the left.

7 Experimental Results

The precision of the position and velocity measurements of the sensor data was described by covariance matrices in Sect. 5. Their values result from experiments on the unfiltered data, which are described Sect. 7.1. Other approaches exist to determine the covariance matrices for Kalman filters by automatically learning [17]. However, this is difficult in our case because a necessary ground truth is hard to obtain.

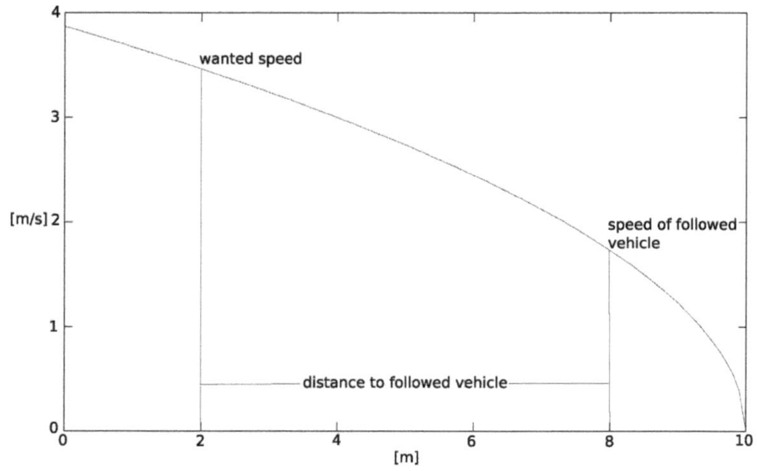

Fig. 6 Velocity function over position. The resulting car velocity is calculated from the current car position, the followed car's position and the followed car's velocity

Car-following on highways with data from lidar or radar only has some crucial limitations. Our approach improves the object lane localization (over radar), extends the field of view in curves (over radar), computes complete object velocity estimations (in contrast to incomplete information from radar), and improves the precision of the velocity estimates (over lidar). Section 7.2 quantifies our results about the velocity precision improvement with real data from a car-following scenario on a German highway.

7.1 Measurement Precision

This section describes the computation of the measurement covariance matrices from experimental data. We separate the position and velocity measurements in two independent parts: The dimension in measurement direction $\|$ (which is approximately the forward direction of the car, because we scan the car in front of us) and the dimension perpendicular to this direction. We only compute the precision information from $\|$ because the values of the perpendicular dimension are less meaningful. It often occurs that the position measurement is biased with respect to the true center of the preceding car, as radar and lidar data is not always reflected from the same position of the car. Because the perpendicular direction of the velocity measurement taken by the radar is not provided, the values are set to a high uncertainty.

We use data from a following scenario on a straight road (no curves) to compute $\sigma(p_{\|}^{lidar})$, $\sigma(p_{\|}^{radar})$ and $\sigma(v_{\|}^{radar})$. The principal components of the position

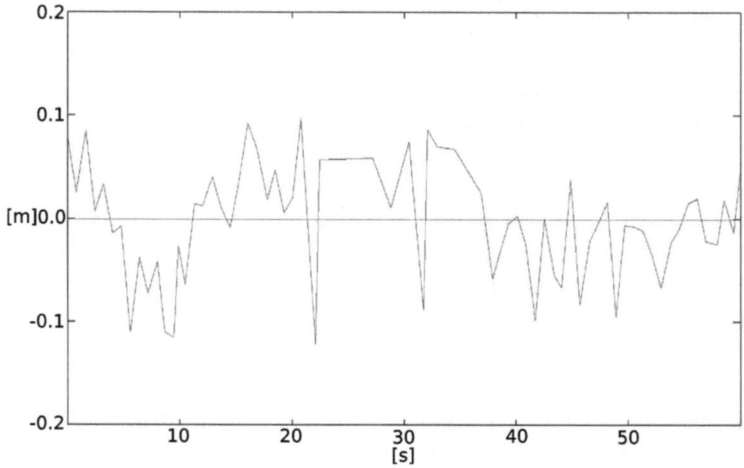

Fig. 7 Error of distance measurements from the lidar along the || component with respect to the estimated ground truth

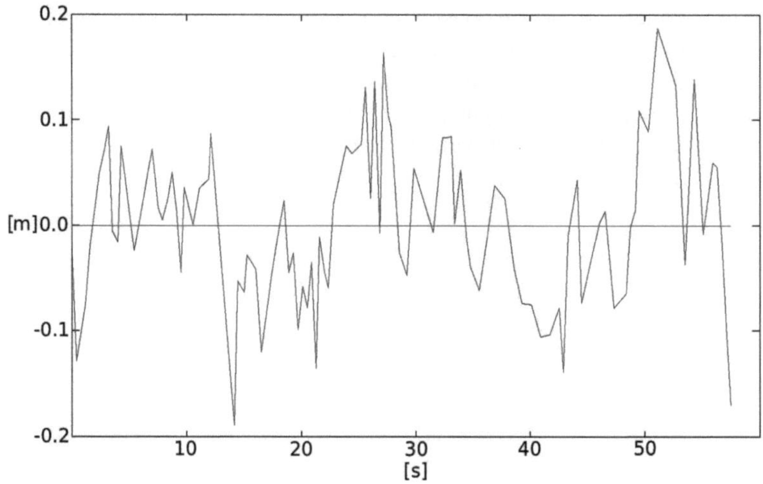

Fig. 8 Error of distance measurements from the radar along the || component with respect to the estimated ground truth

data are computed with PCA. The component with the biggest variance gives us ||. We project the position values on || and plot them against their time stamp. The ground truth is estimated by fitting a least squares polynomial to the data. We subtract this polynomial from the data. The results are $\sigma(p_{\|}^{lidar} \approx 0.052$ (Fig. 7) and $\sigma(p_{\|}^{radar}) \approx 0.074$ (Fig. 8). The same can be done for the radar velocity measurements: $\sigma(v_{\|}^{radar}) \approx 0.022$.

7.2 Velocity Precision

This section shows the improvement in velocity precision of a followed vehicle by applying the data fusion approach.

The *Automobil-Verkehrs- und Übungs-Straße* (AVUS), a highway passage of Autobahn A115 and former motor racing circuit in the south-western districts of Berlin, Germany, was chosen as testing course. Data was recorded of an autonomous drive of "MadeInGermany" in real traffic. We do not analyze the whole run but use an extract of the data (a time slice of about one minute), in which three different cars were followed at various distances and at a speed of about 23 m/s. The chosen track contains two straight parts and a curve.

Figure 9 shows the estimated velocities of the followed vehicles for lidar data only (top) and for data from lidar and radar processed with the fusion algorithm.

The ground truth for velocity measurements is again obtained by a piecewise least squares polynomial fit. As in Sect. 7.1 the order of the polynomial is chosen such that it best fits the expected function (without overfitting) by visual judgment. Three regions (R1–R3) are analyzed separately, as depicted in Fig. 9. Each of the region corresponds to a different followed vehicle. Note that the artificial ground truth adds an error to the analysis, but we considered it adequate enough to be significant.

Results are shown in Fig. 10. It can be seen that the mean velocity estimation in both cases was very close for all regions. We could also improve the velocity precision in all regions significantly. The smallest improvement was made in R3 (factor of about 3.93). The improvement in R1 and R2 in contrast are much bigger (R1: 7.36, R2: 6.73).

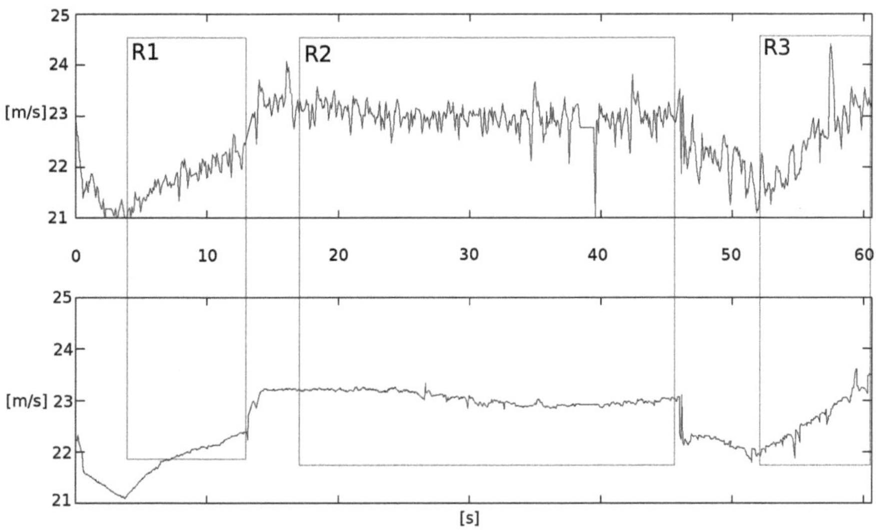

Fig. 9 Comparison of the velocity precision of tracked lidar objects (*top*) with objects from fusion of lidar and radar data (*bottom*)

	μ_{lidar} $\left[\frac{m}{s}\right]$	μ_{fusion} $\left[\frac{m}{s}\right]$	σ_{lidar} $\left[\frac{m}{s}\right]$	σ_{fusion} $\left[\frac{m}{s}\right]$	σ_{lidar} / σ_{fusion}
R1	21.800	21.863	0.162	0.022	7.36
R2	23.036	23.121	0.249	0.037	6.73
R3	22.382	22.538	0.381	0.097	3.93

Fig. 10 Results of the velocity variance analysis

8 Conclusion and Future Work

We presented a real-time efficient radar/lidar obstacle fusion approach, combining the advantages of both—accurate and highly available position estimation with lidar and precise velocity estimation with radar. The described algorithms have been tested on our autonomous car and led to a comfortable following behavior on the German Autobahn in real traffic conditions. In experiments we increased the precision compared to the lidar velocity and also achieved a good position and velocity estimation whenever radar data were unavailable, thus compensated the narrow field of view of the radar sensor.

Future work will focus on further integration of rear orientated blind spot radar to implement a safe and robust lane changing behavior. Further, a multi-object tracking and fusion system for road intersections, maybe with occlusions will be part of our future research.

Acknowledgments The authors wish to thank the German federal ministry of education and research (BMBF).

References

1. M. Buehler, K. Iagnemma, S. Singh, *The DARPA Urban Challenge: Autonomous Vehicles in City Traffic, Volume 56 of Springer Tracts in Advanced Robotics* (Springer, New York, 2009)
2. B. Khaleghi, A. Khamis, F.O. Karray, S.N. Razavi, *Multisensor Data Fusion: A Review of the State-of-the-Art*. Information Fusion, 2011
3. N. Kaempchen, K.C. Fuerstenberg, A.G. Skibicki, K.C.J. Dietmayer, *Sensor Fusion for Multiple Automotive Active Safety and Comfort Applications* (2004), pp. 137–163
4. A. Ewald, Multisensor obstacle detection and tracking. Image Vis. Comput. **18**, 389–396 (2000)
5. J. Dickmann, F. Diewald, M. Maehlisch, J. Klappstein, S. Zuther, S. Pietzsch, S. Hahn, M Munz, *Environmental Perception for Future Integrated Safety Systems* (2009)
6. B. Yamauchi, *All-weather Perception for Man-portable Robots using Ultra-wideband Radar* (2010), pp. 3610–3615
7. C. Blanc, L. Trassoudaine, Y. Le Guilloux, R. Moreira, *Track to Track Fusion Method Applied to Road Obstacle Detection* (2004)
8. M. Skutek, Ein PreCrash-System auf Basis multisensorieller Umgebungserfassung. PhD thesis, Technische Universitaet Chemnitz, 2006

9. M. Darms, P. Rybski, C. Urmson, An adaptive model switching approach for a multisensor tracking system used for autonomous driving in an urban environment. **2009**, 521–530 (2008)
10. H. Winner, S. Hakuli, G. Wolf, Handbuch Fahrassistenzsysteme. Vieweg+Teubner and GMV Fachverlag GmbH, 2009
11. N. Kaempchen, K. Dietmayer, in *Data Synchronization Strategies for Multi-sensor Fusion*. Proceedings of the IEEE conference on intelligent transportation systems (2003), pp. 1–9
12. W. Wang, X. Huang, M. Wang, Out-of-sequence measurement algorithm based on gaussian particle filter. Inf. Technol. J. **9**, 942–948 (2010)
13. Y. Bar-Shalom, M. Mallick, H. Chen, One-step solution for the general out-of-sequence-measurement problem in tracking. *Aerospace* (2002)
14. R.E. Kalman, A new approach to linear filtering and prediction problems. Trans. ASME-J. Basic Eng. **82**, 35–45 (1960)
15. J. Gao, C. Harris, Some remarks on kalman filters for the multisensor fusion. Inf. Fusion **3**(3), 191–201 (2002)
16. B.C. Ooi, *Spatial kd-Tree: A Data Structure for Geographic, Database* (1987)
17. P. Abbeel, A. Coates, M. Montemerlo, A.Y. Ng, S. Thrun, in *Discriminative Training of Kalman Filters*. Proceedings of robotics: science and systems (2005)

Closed-Loop Control of Patient Handling Robots

Christian Ziegler and Jörg Franke

Abstract Patient handling robots are increasingly employed to enable a flexible positioning of the patient for diagnostic and therapeutic purposes. Due to the greatly differing robot loads because of varying patient weights especially serial kinematics are no longer able to keep up with the constantly increasing medical demands for positioning accuracy. In this chapter a low-cost measuring system for a permanent integration into the workspace of a patient handling robot is introduced which can measure the pose of a patient couch with high accuracy. This enables closed-loop control of the patient couch. In tests on a robot system an average positioning error of 0.12 mm was achieved for a diversity of medically relevant poses.

1 Introduction

The fields of application for robots in medicine could be continually expanded in recent years. The robots employed are used to position the therapy or diagnostic systems flexibly in relation to the patient or for patient handling. From a medical point of view a high accuracy of the robots is one of the most important requirements. A periodic control of the positioning accuracy and, if necessary, measures to increase the accuracy of the positioning systems are therefore indispensable. [1, 2] In this context the exact positioning of the patient via a patient handling robot for diagnostic or therapeutic purposes poses a particular challenge.

C. Ziegler (✉) · J. Franke
Institute for Factory Automation and Production Systems (FAPS),
Friedrich-Alexander-University Erlangen-Nuremberg,
Egerlandstr. 7-9 91058 Erlangen, Germany
e-mail: ziegler@faps.uni-erlangen.de

J. Franke
e-mail: franke@faps.uni-erlangen.de

G. Sen Gupta et al. (eds.), *Recent Advances in Robotics and Automation*,
Studies in Computational Intelligence 480, DOI: 10.1007/978-3-642-37387-9_18,

Very high requirements have to be met for the accurate positioning especially in radiation therapy. To this end the patient's tumor has to be positioned in the so called isocenter, a fixed point inside the room on which the treatment beam is directed. The tumor tissue can then be irradiated with a previously minutely planned radiation. To ensure that all tumor cells are killed and the radiation level for the surrounding healthy tissue is kept at a minimum the pose of the patient couch has to be adjusted accurately. In Fig. 1 (left) an exemplary patient handling robot is shown [3, 4].

To attain high accuracy, usually robot systems which are customized by the manufacturers or specialized companies are utilized. As these require intricate calibrations, typically before delivery to the customer, they are considerably more expensive than conventional systems. For varying robot loads the state of the art positioning accuracy of these patient handling systems is 0.5 mm. In some cases the positioning error is much higher and can be up to several millimeters, especially in vertical direction [6]. Additionally, they have to be inspected from time to time and if necessary re-calibrated. There have been standardized methods for years now for these procedures. In most cases the kinematic chain of the robots is closed via measuring systems such as cameras or laser stripe sensors. Afterwards the kinematic model of the total system is numerically optimized [7, 8] based on the measurements. Another feasible approach is the selection of the most suitable joint configuration [9, 10]. There is evidence that the task of accurately positioning patient handling robots can be better solved by a room integrated measuring system. In this way, after an initial calibration of the system in the treatment room,

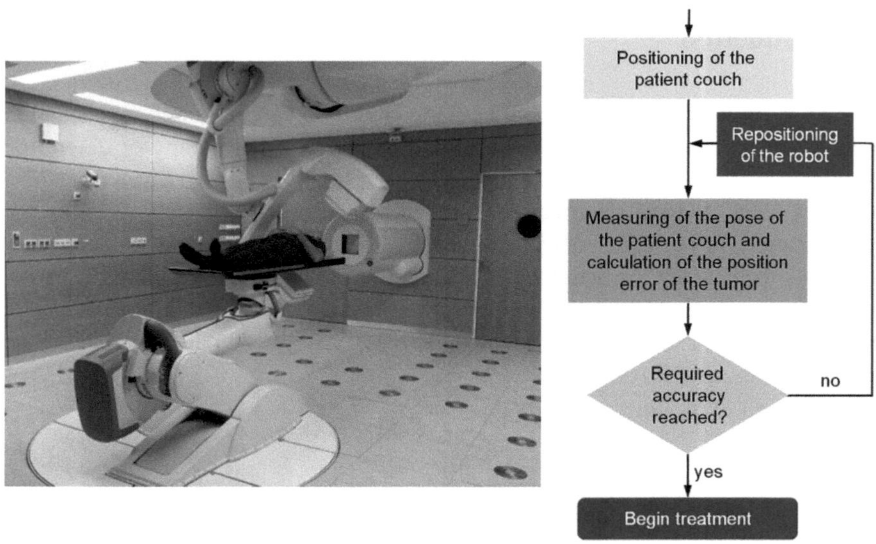

Fig. 1 *Left* Patient handling robot with six degrees of freedom for tumor positioning in particle therapy based on a Kuka KR 240 [5]; *Right* Basic concept for closed-loop control of the patient couch

the current pose of the end-effector respectively the patient couch can be determined at any time. Based on these measurements the pose of the robot's end-effector is controlled in-the-loop if necessary, see Fig. 1 (right).

As a consequence of the described inadequacies of existing calibration systems as well as the advantages of a pose measuring some main research goals can be deduced. The achievable tumor positioning accuracy should be less than 0.5 mm for all relevant poses. For economic reasons the costs for hardware components should not exceed the costs for a high accuracy robot calibration ($\approx$\$5000–\$10000). Furthermore, the system components may not influence the treatment workflow negatively. Easy and safe access to the patient couch has to be ensured at all times [1, 11].

2 Stereo Camera System for Pose Control

In recent years there have already been increased research endeavors to utilize room integrated measuring systems for closed-loop control of robots. By now various measuring systems are available on the market which can in principle realize the positioning respectively pose control with the required accuracy. Often summed up under the term "Large-Scale Metrology" and designed for measuring tasks in big operating volumes, these systems are typically optical and mobile [12, 13]. So called laser trackers cover a comparatively broad range of application areas and have a very high measuring accuracy. The use of additional equipment enables the coverage of all six degrees of freedom of the device under test [3]. A different approach is pursued with the Indoor-GPS. Here signals in the form of laser planes are emitted by transmitters that are then analyzed by the receivers via triangulation. To determine all six degrees of freedom of an object the position of multiple receivers can be measured simultaneously [14]. In robot system solutions camera-based concepts have been implemented primarily in recent years. Particularly mobile measuring systems based on infrared stereophotogrammetry could be established in several industrial and clinical application areas [15, 16].

Though in principle the described available measuring systems are sufficient to determine the current pose of the patient couch with adequate accuracy, due to the very high initial costs of at least several 10 T\$ the mentioned measuring systems are to be judged as uneconomical for the intended permanent use at only one machine. This necessitates the development of a novel measuring system based on cost-effective sensor systems. For this at first suitable cost-effective sensor systems have to be identified. The diagram in Fig. 2 summarizes the main selection criterions for the intended use for common measurement systems as well as for sensors. Beside the reachable accuracies and the costs, the illustration shows the amount of degrees of freedom (DoF) that can be measured with good performance. Ultimately systems which realize the measuring via comparatively low priced imaging sensors can be classified as particularly qualified to meet the requirements. Besides conventional cameras, sensors on the basis of the laser triangulation principle should be

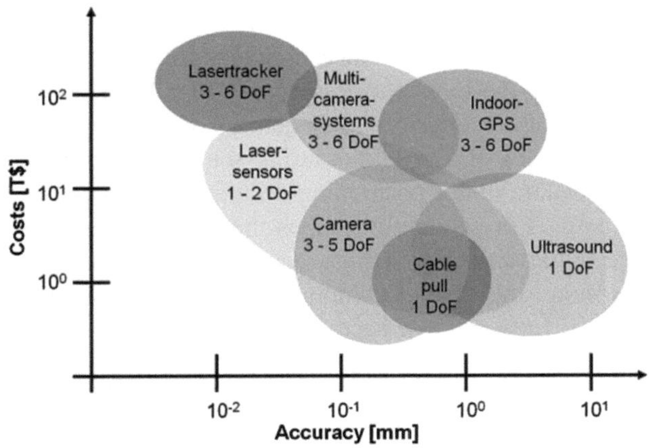

Fig. 2 Because of relatively low hardware costs and the ability to cover several degrees of freedom (DoF) cameras have high potential for use in cost-effective measurement applications

mentioned most notably. Given the trend towards higher resolution CCD- and CMOS-chips at a simultaneous decline in prices, further enhancements are to be expected in the coming years in terms of measuring resolution and sensor accuracy.

In the end a two camera system that is sunk into the floor underneath the isocenter and angled at the patient couch is favored, see Fig. 3 (left). The position analysis is realized by markers which are attached to the patient couch. Based on the position information of at least three markers, the pose of the patient couch is calculated. For that markers can be attached to the bottom side of the patient couch widely spread in form of a matrix. Hence, the measuring volume can be reduced significantly, as only markers directly below the isocenter inside the field of view of the cameras have to be analyzed. Therefore the pose of the patient couch is determined directly underneath the tumor so that there is just a minimal distance between the tumor and the measured pose. Thus, all relevant deviations that add up inside the kinematic chain (including the deflection of the patient couch) are determined and can be compensated by a closed-loop control of the robot.

After the position of the tumor on the patient couch is determined via imaging—computed tomography is frequently used—the patient couch respectively the tumor is moved into the first irradiation pose by the uncalibrated robot. Figure 3 (right) shows the process of determination of the pose error vector of the patient couch as described in the following. Initially each of the two aligned and calibrated cameras takes an image of the markers attached to the bottom side of the couch. Both images are then searched for the identifiable data matrix codes. From the set of codes found in both images at least three markers are chosen and their position is determined. For this purpose for each of the circles assigned to the chosen data matrix codes an edge detection of its elliptic projection is carried out. Based on this, the corresponding image coordinates of the circle centers are calculated with sub-pixel accuracy. In a next step the exact position of the circle centers in the room is identified stereoscopically on the basis of the corresponding image

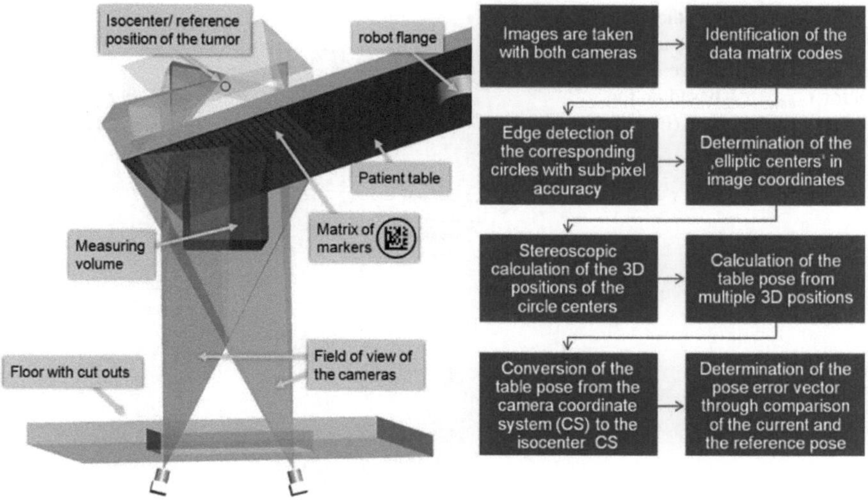

Fig. 3 *Left* Camera and marker configuration of the stereoscopic measuring system; *Right* Process of the pose error determination

coordinates. From that the pose of the couch can be determined in the coordinate system of the camera based on the positions of three markers. By means of the transformation matrix gained from a preceding camera calibration the pose of the couch respectively the tumor in the isocenter coordinate system can be calculated. Through a comparison of the determined actual pose with the mathematically determined target pose the pose error can then be calculated. If the pose error exceeds an acceptable threshold, adjustment values for each degree of freedom are sent to the robot control. After the robot has adjusted the axes, a new pair of images is taken and the pose error is determined again. If the required pose accuracy of the patient couch is reached the treatment can begin.

3 Calculation of the Pose Error Vector

To determine the pose error vector $T_{tumor,ideal}$—that is needed to control the robot—several transformation matrices have to be known. Figure 4 illustrates all essential coordinate systems and the corresponding transformations. The following equation shows the computation of $T_{tumor,ideal}$:

$$T_{tumor,\,ideal} = T_{tumor,\,couch} * T_{couch,\,camera} * T_{camera,\,iso} * T_{iso,\,ideal}$$

For the conducted experiments on the test cell the transformation $T_{tumor,couch}$ between the patient couch coordinate system and the tumor coordinate system was determined on a 3D measuring machine. In clinical applications $T_{tumor,couch}$ is defined through the usage of immobilization devices or can be measured e.g. by

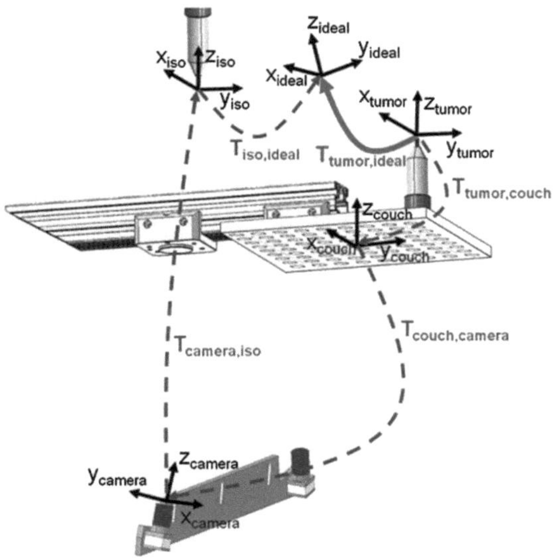

Fig. 4 Depiction of the required transformations for calculation of the pose error vector $T_{tumor,ideal}$ based on the test set-up of Sect. 4 (Fig. 6)

using 3D X-ray imaging. As described in Sect. 2, the transformation $T_{couch,camera}$ is calculated based on the images from the stereo camera system. As the transformation $T_{camera,iso}$ between the camera coordinate system and the isocenter coordinate system is constant, it only has to be determined once during installation of the camera system. The last required transformation $T_{iso,ideal}$ can be derived from the individual treatment plan of the patient. During our experiments and also in a lot of medical applications only the orientations of the ideal coordinate system changes and the positions of the two coordinate systems stay constant in or nearby the isocenter.

4 Experiments

To be able to make a statement concerning the accuracy of the measuring system first of all a test set-up was realized in which the relative pose between a 400 × 400 mm sized marker plate (13 × 13 markers with Ø = 20 mm) and the camera system could be adjusted precisely. The stereo camera system consists of two identical 5 megapixel cameras (½″-CMOS-chip). Both cameras have lenses with a focal length of f = 12 mm. They are mounted on the machine bed of a materials testing machine at a distance of 500 mm (stereo basis) and aligned to the marker plate which is attached to the linear positioning system of the machine, see Fig. 5. The distance between the cameras and the marker plate can then be freely adjusted via the positioning system.

In a first series of tests the accuracy of the measured pose was determined for multiple distances. The diagrams in Fig. 5 show the results of the experiments.

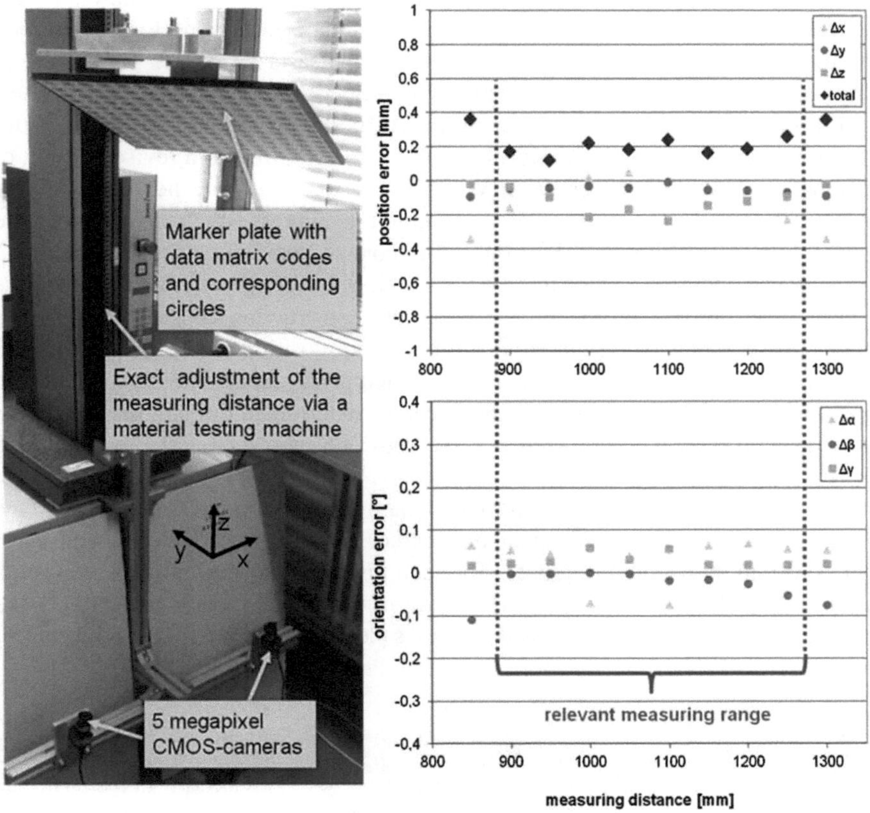

Fig. 5 Test set-up for measuring the accuracy of the stereo camera system *(left)* and measured position and orientation errors *(right)*

The three dimensional positioning error inside the relevant measuring range from 900 to 1250 mm remains fairly constant at a value of 0.2 mm. Only above and below the required altitude range the measuring accuracy deteriorates slightly. The measured angle errors stay below 0.1° inside the relevant measuring range. In a further series of tests the reproducibility of the measurements was examined. For this purpose ten measurements each were taken of different poses of the marker plate and subsequently analyzed stereoscopically via the described method. A comparison of the measuring results shows that the maximum deviations between measurements of the same pose occur in vertical direction. In the worst case the highest determined difference between the minimum and maximum value was 0.038 mm. Though this value is small in comparison to the acceptable position error it can be assumed that through analysis and average determination of multiple stereoscopic images a slight improvement of the measuring result could be achieved. Concerning the rotations a maximum deviation of 0.045° between two measurements was recorded.

The preliminary examinations on the materials testing machine show that the accuracy of the developed stereo camera system is sufficient for closed-loop control of the patient couch inside the relevant measuring range. To be able to draw further conclusions concerning the suitability of the described approach a test cell with an integrated stereo camera system is set up, see Fig. 6. For a better sensor resolution lenses with a focal length of f = 16 mm and a revised design of the markers are used. Figure 7 shows an image sample from the stereo camera system. In this illustration the configuration of the markers which are used for pose calculation is highlighted. For the pose control experiments a conventional uncalibrated industrial robot replaces the patient handling robot and a substitute construction the patient couch. The substitute construction consists of a marker plate on which pins representing the current tumor position can be mounted. Via a verification system (two vision systems installed perpendicularly to each other) the position error of the pose control system can be determined through measuring the Euclidean distance between a pin at a fixed position representing the isocenter and the tumor pin, see Fig. 6.

To evaluate the closed-loop control two main series of tests were conducted. One with a shorter tumor pin (120 mm above marker plane) and one with a longer tumor pin (190 mm above marker plane). For each tumor pin several poses were tested. Table 1 lists the results of the pose control of the patient couch respectively the substitute construction for the tested poses. As the abort criterion for the pose control

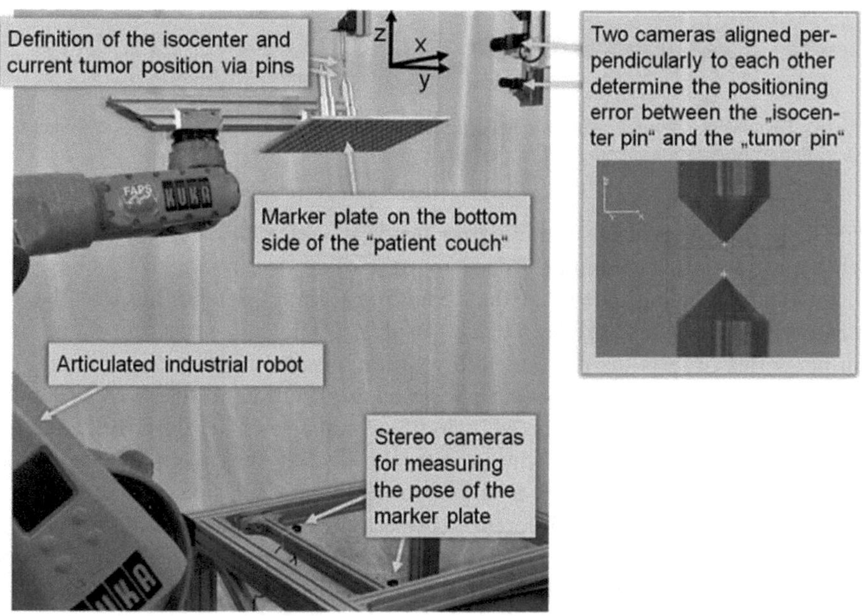

Fig. 6 *Left* Test set-up to determine the accuracy of the closed-loop control of an industrial robot; *Right* Edge detection to determine the position error

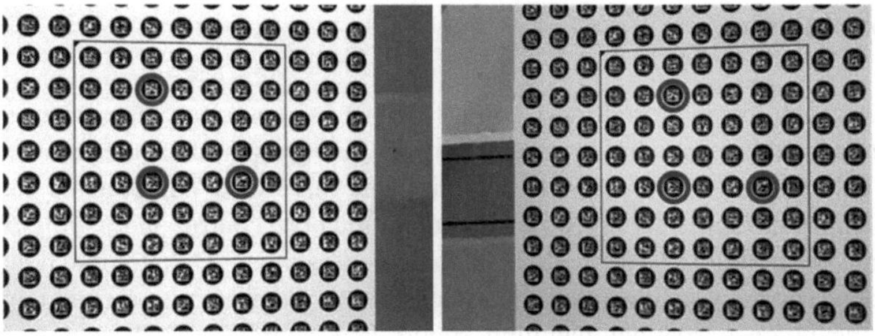

Fig. 7 The sample shows corresponding images from the stereo camera system; the highlighted markers are used for pose calculation

Table 1 Measured tumor position errors after pose control of the robot for tumor positions 120 mm (*left*) and 190 mm (*right*) above the maker plane

Rotation around…		Average position error (mm)	Maximum position error (mm)	Rotation around…		Average position error (mm)	Maximum position error (mm)
No rotation		0.08	0.12	No rotation		0.09	0.12
x-axis	5°	0.09	0.11	x-axis	5°	0.12	0.16
	10°	0.09	0.13		10°	0.07	0.11
	15°	0.10	0.12		15°	0.14	0.21
y-axis	5°	0.12	0.16	y-axis	5°	0.05	0.09
	10°	0.15	0.20		10°	0.10	0.17
	15°	0.20	0.22		15°	0.21	0.25

experiments a maximum length of 0.1 mm of the tumor error vector $T_{tumor,\ ideal}$ was defined. Depending on the robot pose an average of six to ten control loops were necessary to reach the abort criterion. Usually the accuracy requirement of 0.5 mm was fulfilled after one or at most two control loops. For each pose five positioning processes were conducted. The average tumor positioning error of all tests is 0.12 mm. The highest measured positioning error is 0.25 mm and occurs at a tilt angle of the substitute construction around the y-axis of 15° with the longer pin. To avoid major deformations or displacements of the tumor or target volume, 15° are usually the limit tilt angle in clinical treatment.

To determine whether the positioning error is depending on the accuracy of the robot an additional weight was attached to the flange of the robot. This led to an additional deviation of slightly above 0.5 mm compared to the unloaded state. Subsequently a pose control for three poses (no rotation; 10° around x-axis; 10° around y-axis) was conducted in the described manner. The resulting average positioning error for this is 0.03 mm above the comparison values without the additional load.

5 Conclusion

The approach of a closed-loop control via a stereoscopic measuring system described in this paper holds several advantages compared to the conventional robot calibration. Integrating all software and hardware components into the treatment room ultimately allows for an automatic pose control and tracing process that can be monitored by medical technical personnel. Due to the non load-dependent accuracy achievable by the measuring system it can be conducive to a higher accuracy of patient handling robots and thereby to a higher treatment quality. The use of cost-effective components (hardware costs $\approx$ $3000) furthermore allows for an economical system requiring fewer expenses than robot calibration. Because of the modular concept of the measuring system the basic idea of the closed-loop control of positioning devices can also be transferred to other applications.

Acknowledgments This work was partially supported and funded by Siemens Healthcare and the German Federal Ministry of Education and Research (BMBF), project grant No. 01 EX 1012 A.

References

1. C. Ziegler, J. Franke, in *A Cost-Effective Stereo Camera System for Online Pose Control of Patient Handling Robots*. Proceedings of IEEE 5th International Conference on Automation, Robotics and Applications (ICARA), (2011) pp. 459–464, Dec 6–8
2. H.T. Russell et al., Medical robotics and computer-integrated surgery, in *Springer Handbook of Robotics*, 1st edn., ed. by B. Siciliano, O. Khatib (Springer, Berlin Heidelberg, 2008)
3. C. Herrmann et al., in *Positioning and Synchronization of Industrial Robots*. Proceedings of International Conference on Indoor Positioning and Indoor Navigation (IPIN), (2010), Sep 15–17
4. C.J. Karzmark, C.S. Nunan, E. Tanabe, *Medical Electron Accelerators* (McGraw Hill, New York, 1993)
5. K. Herrmann, in *Enhancing Flexibility and Precision—Robots for Patient Positioning and Imaging*. Proceedings of PTCOG, (2007) May 18–23
6. M.A. Meggiolaro, S. Dubowsky, C. Mavroidis, Geometric and elastic error calibration of a high accuracy patient positioning system. Mech. Mach. Theor. **40**(4), 415–427 (2005)
7. L. Beyer, J. Wulfsberg, Practical robot calibration with ROSY. Robotica **22**(5), 505–512 (2004)
8. C.S. Gatla et al., An automated method to calibrate industrial robots using a virtual closed kinematic chain. IEEE Trans. Rob. **23**(6), 1105–1116 (2007)
9. J. Hollerbach, W. Khalil, M. Gautier, Model identification, in *Springer Handbook of Robotics*, 1st edn., ed. by B. Siciliano, O. Khatib (Springer, Berlin Heidelberg, 2008)
10. Y. Veryha, J. Kurek, Application of joint error mutual compensation for robot end effector pose accuracy improvement. J. Intell. Rob. Syst. **36**(6), 315–329 (2003)
11. J. Franke, C. Ziegler, J. Merhof, Workflowoptimierung im Umfeld medizinischer Großgeräte. Prof. Process **3**, 26–28 (2010)
12. W.T. Estler et al., Large-scale metrology—an update. Ann. CIRP **51**(2), 587–609 (2002)
13. M.J. Puttock, Large-scale metrology. Ann. CIRP **27**(1), 351–356 (1978)

14. Schmitt R et al, in *Performance Evaluation of iGPS for Industrial Applications*. Proceedings of International Conference on Indoor Positioning and Indoor Navigation (IPIN), (2010), Sep 15–17
15. R. Elfring, M. de la Fuente, K. Radermacher, in *Accuracy of Optical Localizers for Computer Aided Surgery*. Proceedings of IFMBE, vol. 25/VI (2009), pp. 328–330
16. R. Schuetze et al., in *OptoPose—a Multi-Camera System for Fast and Precise Determination of Position and Orientation for Moving Effector*. Proceedings of 9th Conference on Optical 3D measurements, (2009), Jul 1–3

A Modular Hierarchical Control Scheme for Mobile Manipulation

Kelvin Gong and Allan I. McInnes

Abstract The addition of a mobile base to a robotic manipulator greatly extends the workspace of the manipulator, but introduces complex control problems involving coordination of base and manipulator motion. We describe a modular, hierarchical control scheme for a mobile manipulator, designed to coordinate motion of the manipulator and base to maintain various performance metrics. We demonstrate the effectiveness of our control scheme by developing a controller that maintains the stability and manipulability of a nonholonomic base with a 6 degree-of-freedom manipulator while executing manipulation tasks. We demonstrate the modularity of our control scheme by showing how the controller can be extended to avoid obstacles without requiring redesign of the rest of the controller. Simulation results show the controller completing a task involving multiple end-effector targets, avoiding simple obstacles, and maintaining stability and manipulability within desired limits.

Based on "A Hierarchical Control Scheme for Coordinated Motion of Mobile Manipulators", by Kelvin Gong and Allan I. McInnes which appeared in the Proceedings of the 5th International Conference on Automation, Robotics and Applications (ICARA 2011).

K. Gong (✉) · A. I. McInnes
Electrical and Computer Engineering, University of Canterbury,
Private Bag 4800, Christchurch 8140, New Zealand
e-mail: kelvin.gong@pg.canterbury.ac.nz

A. I. McInnes
e-mail: allan.mcinnes@canterbury.ac.nz

G. Sen Gupta et al. (eds.), *Recent Advances in Robotics and Automation*, Studies in Computational Intelligence 480, DOI: 10.1007/978-3-642-37387-9_19,

1 Introduction

Mobile manipulators are becoming an increasingly popular area of robotics research, and are also of great interest for both commercial and scientific applications. A mobile manipulator consists of a robotic manipulator mounted onto a mobile platform of some kind. The mobile platform dramatically extends the workspace of the manipulator, as it is no longer fixed in one position. A simple approach to managing a mobile manipulator is to treat it as two separate systems, either moving the base while keeping the manipulator fixed, or keeping the base stationary and using manipulator as if it had a fixed base. The difficulty that mobile manipulator research seeks to address is the coordinated control of the base and manipulator together. This is challenging, because the dynamics of the manipulator and mobile platform not only differ, but also interact. The addition of the mobile base also introduces redundancy into the system, which a controller needs to solve, and, with most mobile bases, nonholonomic constraints that the controller must work within. Researchers have developed various control schemes to tackle the challenge of mobile manipulation [1–3]. However, these tend to be monolithic control laws that are complicated to design and implement, and require redesign of the entire controller if new behaviour is required. In contrast, we are interested in developing a controller that is built from components which are well-understood in industry, and that can be modified without requiring redesign of the entire control scheme.

As an example of the kinds of application we have in mind, consider a scenario in which a human operator is controlling a mobile manipulator to perform some task. Ideally, to reach a particular goal location the operator should be able to simply direct the end-effector to the target position without having to worry about the complexities of base motion or base/manipulator interactions. The base should automatically follow the end-effector, while maintaining a pose that maximizes manipulability. Once at the target point, the operator may wish to pick up a load, but in doing so may tip over the mobile manipulator. In this case the controller should be able to automatically reconfigure itself, and move the base and manipulator together to increase the stability while minimizing movement at the end-effector during the reconfiguration period. If the operational context changes, and modifications to the way the manipulator reaches goal locations are required, we would expect to be able to reuse most of the existing control infrastructure.

To enable applications such as the scenario just described, we propose a modular, hierarchical control scheme for mobile manipulation. More specifically:

- We outline a design for a simple, yet effective, control scheme (Sect. 3) that integrates classical control methods for separate mobile bases and manipulators using a high-level state-based controller to provide coordinated movement. The responsibility of the high-level controller is to move the mobile manipulator to place the end-effector at target positions, while at the same time ensuring that combined system maintains desired levels of stability and manipulability.

- We demonstrate the effectiveness of the control scheme (Sect. 3.4) using a simulation of a multi-target manipulation task. The controller produces smooth motion between target positions, using a combination of base and manipulator movement to produce monotonically decreasing position errors in response to each change in target position. The stability and manipulability of the system stay within acceptable limits throughout the motion.
- We show how the controller can be extended with additional features (Sect. 4), in this case an obstacle avoidance controller, without requiring redesign of the entire control system. The obstacle avoidance controller drives the mobile manipulator around simple obstacles while still achieving the other controller goals.

We begin by providing some background (Sect. 2) on the manipulator model we have assumed for the design and simulation of our controller, and the stability and manipulability metrics used within the controller. We close by comparing our controller to related ideas (Sect. 5) developed by other authors.

2 Background

As a platform for testing our controller design we use a model of a mobile manipulator that consists of a nonholonomic mobile base with a 6 degree-of-freedom manipulator mounted at its geometric centre. The basic control scheme uses a pair of performance metrics that measure the stability and manipulability of the mobile manipulator. The model and metrics are described in the following sections.

2.1 Model

The nonholonomic mobile base model is of the differential drive type, which is common in research and industry. As shown in Fig. 1, the base is assumed to be rectangular, and moves in the plane. Here G denotes the global coordinate frame, L denotes the local coordinate frame centered about the robot, and R denotes the robot body frame. Rotations in the anti-clockwise direction about the z-axis are positive.

The equations of motion for the base model (Eqs. 1 and 2) describe the translational and rotational motion of the base.

$$m\ddot{x}_R + c_1\dot{x}_R = F, \tag{1}$$

$$I\ddot{\theta}_L + c_2\dot{\theta}_L = T, \tag{2}$$

where x_R is the robot x coordinate, θ_L is the robot heading relative to the local axis, F and T are the force and torque generated by the wheels, m is the mass of the robot, I is the inertia of the robot about its z-axis, and c_1 and c_2 are damping constants. Force generated by the wheels induces linear acceleration, while an imbalance in wheel forces induces angular acceleration. Nonholonomic constraints are built into the model, as the mobile base can only move linearly in the direction it is heading.

The manipulator model is a 6 degree-of-freedom (6DOF) manipulator based on the well-known PUMA-560 [4] design. The size and geometry of the manipulator has been simplified to reduce the complexity of the model, as the main focus of the research is on the control scheme. We use a standard kinematic model [5] for the manipulator motion. The local coordinate system of the manipulator is the same as the base-fixed coordinate system. Thus, as shown in Fig. 1, the origins of both the base and manipulator models coincide, and a manipulator base joint angle of 0 degrees means that the manipulator is pointing along the base's x-axis.

2.2 Performance Metrics

2.2.1 Stability

Our stability metric is a measure of how close the robot is to tipping over. Stability is computed as a ratio of the sum of moments about the robot body X or Y axis, and the maximum counter-moment produced by the weight of the robot. The total moment, M_a, about a body axis, $a = X$ or $a = Y$, is

$$M_a = \sum_{l \in Link} m_l g d_{al}, \tag{3}$$

where m_l is the mass of link l, g is the acceleration due to gravity, and d_{al} is the straight line distance between the axis a and the centre-of-mass of link l.

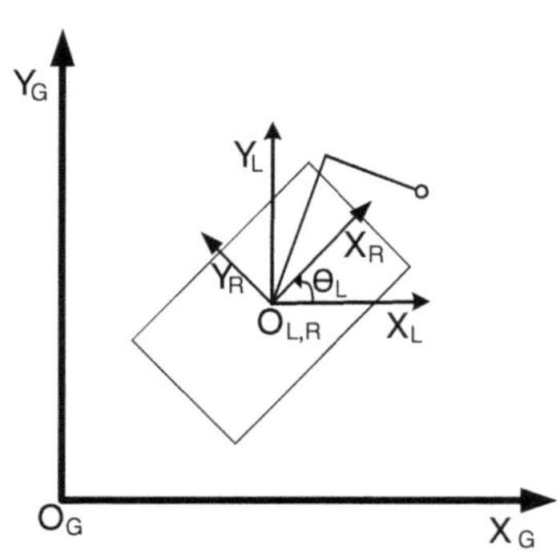

Fig. 1 Mobile base coordinate system definition

The maximum counter-moment is dependent on the geometry of the robot and the total mass of the system. The centre-of-mass of the mobile base is assumed to be in the middle of the mobile base, and so is equidistant for each of the parallel edges of the robot base. The maximum counter-moment possible about a base edge corresponds to the entire weight force of the robot acting relative to that edge. The maximum counter-moment, M_{ct_a}, about a body axis, $a = X$ or $a = Y$, is

$$M_{ct_a} = \frac{d}{2} F_w, \tag{4}$$

where F_w is the total weight force of the mobile manipulator through the centre-of-mass, and d is the width or the length of the mobile base.

The stability measure about an axis is then defined as the ratio of the maximum counter-moment to the net moment generated by the manipulator about a body axis:

$$Stability_a = \frac{M_{ct_a}}{M_a} \tag{5}$$

This can also be viewed as a ratio between the distance from the centre of the base to the edge, and the distance from the centre of the base to the current manipulator centre-of-mass. If this ratio is less than 1, then the centre-of-mass lies beyond the edge of the robot, and tip-over will occur. Thus an absolute stability value of greater than 1 is considered stable, with the sign of $Stability_a$ indicating the net direction of the moment about the respective axis.

2.2.2 Manipulability

Manipulability is a measure of a manipulator's ability to exert forces in all directions. As our measure of manipulability we use a standard manipulability metric [6], defined as:

$$\eta = \sqrt{| J(q)J(q)' |} \tag{6}$$

where q and $J(q)$ are, respectively, the joint angles and Jacobian matrix of the manipulator. This manipulability metric gives an indication of how far the manipulator is from singularities. If, as is commonly done, we exclude the wrist joints from the manipulability calculations and only take into account the base, shoulder and elbow joints, then the manipulability metric reduces to be directly proportional to the sine of the elbow joint angle. This is because an outstretched or completely folded manipulator effectively reduces the manipulator DOF by one since it will be no longer be able to move in one of its axes.

When working with a mobile manipulator, the base can improve the manipulability of the system, but only if all of the objectives are known beforehand. Otherwise the preferred direction to point the mobile base is essentially a guess. As we are focusing on objectives that are unknown to the controller until they are given by a human operator, the manipulability of the base is not considered. We instead seek to control the manipulability of the manipulator relative to the current base location.

3 Initial Controller Design

Our basic mobile manipulation controller provides for manipulation task execution while ensuring that the combined motion of the base and manipulator does not result in the robot tipping over, and using the mobility of the base to position the manipulator in ways that improve its manipulability. Rather than designing a single control law that achieves all of these goals, we use a hierarchy of controllers (Fig. 2) with separate responsibilities. In addition to simplifying the controller design task by splitting it into more tractable sub-problems, the hierarchical structure of the controller allows us to treat the component controllers as modular behaviours that can be modified or replaced without having to change the rest of the control system.

The overall controller is structured as three levels:

1. A low-level controller consisting of the control methods for moving the base and the arm to reach the target position.
2. A set of high-level controllers that determines target positions for the low-level controllers based on the required target points, the stability and manipulability metrics, and other robot goals.
3. A top-level arbitrator to decide what to do.

An overview of the low-level controllers, and the workings and design of the high-level controllers and top-level arbitrator, are described in the following sections.

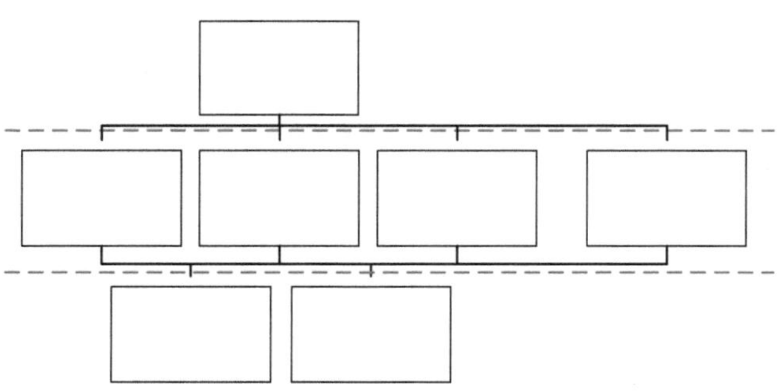

Fig. 2 Hierarchical structure of the controller

3.1 Low-Level Controllers

The robot base is controlled using two PID controllers. One controller manages forward motion as a function of the distance to the target position. The other controller manages turning motion as a function of the heading error. The resultant control actions are then combined to compute the commands to the left and right wheels. Using separate controllers lets us more precisely control the rate of turn relative to forward motion, and even rotate on the spot when necessary. We have opted to use PID controllers because they are well-understood in industry, are simple to implement, and do not require an accurate system model to perform well. However the modular nature of our control scheme makes it easy to replace the PID controllers with more sophisticated controllers if that proves necessary or advantageous.

The manipulator is controlled using a pseudo-controller that receives a set of joint targets, and controls the joint velocities such that all joints reach their final targets at the same time. The use of a pseudo-controller to imitate controller outputs allows us to quickly alter the performance characteristics of the manipulator to evaluate how it can affect the high level controllers and the overall system performance.The separation between the high-level and low-level controllers again allows a different manipulator controller to be substituted for the current one as necessary.

3.2 High-Level Controllers

The objective of the high-level controllers is to decide where to move the manipulator and base together to achieve a target position, whilst still maintaining the stability and manipulability measures within the margin required. This objective is achieved using three sub-controllers: a target position controller, a manipulability controller, and a stability controller. The top-level arbitrator determines which sub-controller is to be executed based on the relative importance of the behaviour provided by each controller, and the sensed state of the robot. Our control scheme can thus be viewed as a kind of behaviour-based controller [7], which has proven to be a successful way of structuring robotic controllers for unknown and dynamic environments. The design and implementation of the three sub-controllers are described in more detail in the following sections.

3.2.1 Target Position Controller

The position controller determines how to approach a predefined target point. The main consideration here is whether the target point is within range of the manipulator. If the target point is out of range, the controller sets the base to move closer. If the target is within range, the inverse kinematics are calculated for the manipulator to reach the target point. If an inverse kinematic solution is not found, the base is set to move closer. A default pose is set during the base move operation.

3.2.2 Manipulability Controller

Manipulability can only be improved by driving towards the target if the arm is out-stretched, or away from the target if the arm is folded. We define four quadrants around the base to determine the movement direction that will produce the fastest improvement in manipulability. As shown in Fig. 3, each quadrant covers a 90° range that is ±45° of the base-fixed axes. An arm position in one of the quadrants results in a movement in the corresponding direction to improve manipulability.

We define the threshold for improving the manipulability metric as a soft threshold, which means that the controller can allow the manipulability to drop below the threshold in some situations. A soft threshold allows the manipulator to still attempt to reach the target location while the base starts to adjust manipulability.

3.2.3 Stability Controller

The robot can become unstable and tip over about either the X or Y body axis. Due to the nonholonomic nature of the base we adopt a slightly different approach to ensuring stability depending on which axis is approaching the stability threshold:

- If tip-over is imminent about the Y-axis, the base moves forwards or backwards to reduce the moment load and improve stability.
- If tip-over is imminent about the X-axis, the base cannot directly move left or right due to the nonholonomic constraints. Instead, the controller rotates the base while keeping the arm in its original position. This transforms the X-axis

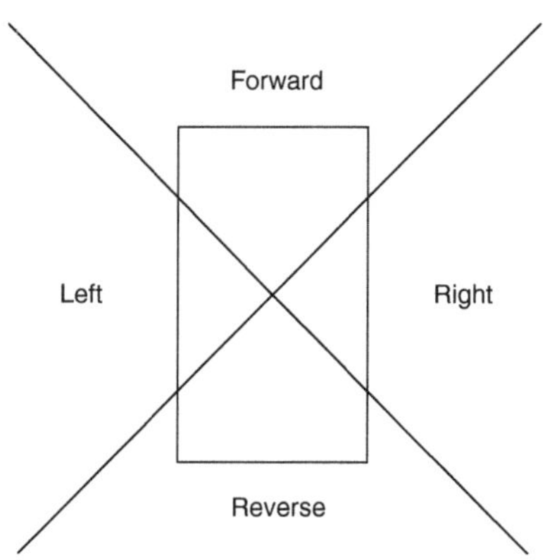

Fig. 3 Manipulability movement sectors

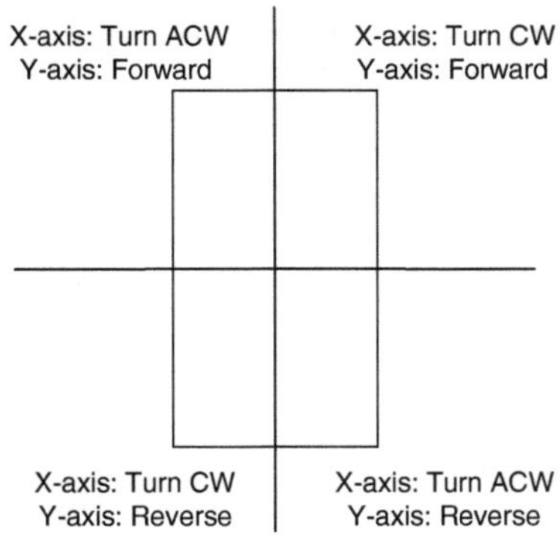

Fig. 4 Stability quadrants and corresponding controller responses

stability problem into a Y-axis stability problem, which can then be handled as above.

As Fig. 4 shows, the controller design splits the robot into quadrants, with a different movement direction resulting depending on where the stability problem is occurring.

The consequences of breaching the stability threshold can be potentially fatal for the mobile manipulator. Therefore, unlike the manipulability controller, we define the threshold for the stability metric as a hard threshold. If the threshold is reached, the manipulator stops trying to complete the current objective and addresses the stability issue with the base. The threshold is set to be slightly larger than the stability limit, which is 1, to provide some margin before the tip-over point is reached.

3.3 Arbitrator

The top-level behaviour arbitrator decides which sub-controller to execute depending on the state of the robot. In our present design, the arbitrator can be thought of as a flat state machine. Such state machines are common in industrial systems, and provide an easy method of assigning conditions for selecting appropriate behaviours. For more complex problems the flat state machine could be replaced by a hierarchical state machine or some other suitable arbitration system.

Figure 5 shows a state diagram for the controller, including the conditions for each state transition. At present we have treated the relative priority of different states as a compound condition on certain transitions. Thus the state machine in Fig. 5 prioritizes addressing the stability as most important, and includes an additional requirement on transitions into the manipulability state that requires stability to be acceptable before manipulability can be addressed. We have also added hysteresis conditions to the stability and manipulability thresholds to reduce the frequency of state oscillations going in and out of those respective states.

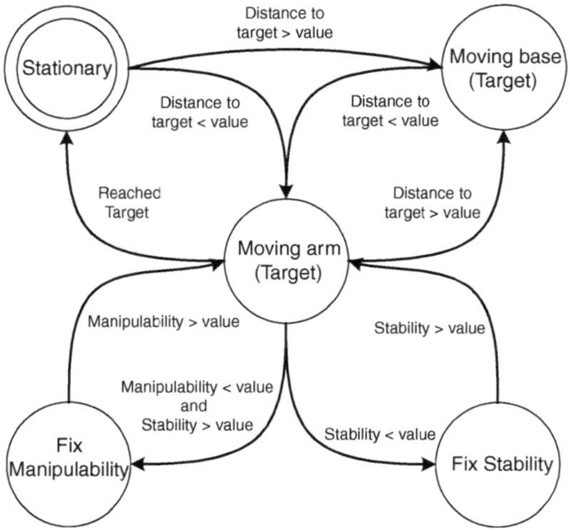

Fig. 5 State diagram of the state machine and the necessary conditions to cause transitions

3.4 Controller Performance

To examine the effectiveness of our controller we have developed a controller and nonholonomic mobile base model using Simulink [8], and combined it with a manipulator model created using Corke's Robotics Toolbox [5]. As a demonstration of the controller's behaviour, we have simulated a scenario in which the mobile manipulator is directed to reach a series of waypoints with the end-effector. A load in the form of a mass was placed at the end-effector to highlight stability issues as the manipulator reaches for the waypoints, and the waypoints were placed such that stability and manipulability problems could arise if the controller was not working correctly. During the simulation the controller does not know all the waypoints in advance, and is only given the next waypoint once it has successfully reached the current waypoint. This simulates a human operator carrying out an arbitrary task by telling the controller where to go, with the controller having to react operator commands instead of generating a plan *a priori*. The following controller parameters were used for the simulation:

- Manipulability threshold at 0.6 (with hysteresis at 0.9).
- Stability threshold at 1.25 (with hysteresis threshold at 1.6).

Figures 6, 7, 8, 9 and 10 show the results of the simulation. As Fig. 8 shows, the manipulability results are as expected. As the manipulator is in range of the desired waypoint, it reaches out to reach the waypoint, which reduces the manipulability. The transition from moving the base to reaching out to the target can be seen in the controller state results (Fig. 10). The soft threshold used by the manipulability controller is also evident, with the manipulability on many occasions dropping below the 0.6 threshold while the controller is working to improve it.

The stability results (Fig. 9) show that at no time during the simulation was the threshold breached. The stability controller was triggered a few times during the

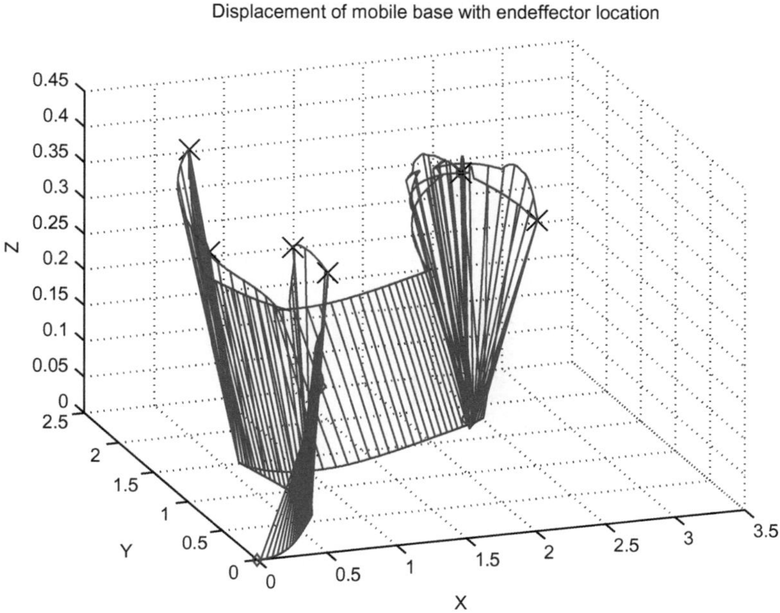

Fig. 6 Path of the end-effector and base. Connecting lines represent positions of the base and end-effector in the same time stamp. 'X' marker show the waypoint locations

second half of the simulation, and the hard threshold used meant that as soon as the threshold was reached the manipulator stopped the current task to improve the stability. Small changes in the manipulability measure during adjustments to improve stability are expected, as the manipulator and base move together.

In general the relative error between the end-effector location and the target waypoint monotonically decreases as the mobile manipulator attempts to reach the waypoint. Performance is good when the manipulability controller is triggered, with the error barely changing during these periods, as seen when comparing Figs. 8 and 10. When the stability controller is triggered there are periods when the error increases, in some cases—such as at 22 s—by a large amount. However, these interludes are short, and the end-effector always converges on all of the waypoints.

4 Extending the Controller

A modular controller design allows the controller to be adapted at a later stage to perform new tasks without having to redesign the entire controller. This modularity extends to all levels in the hierarchy, from the top-level down to the low-level. For example, a different arbitrator can be substituted to replace the state

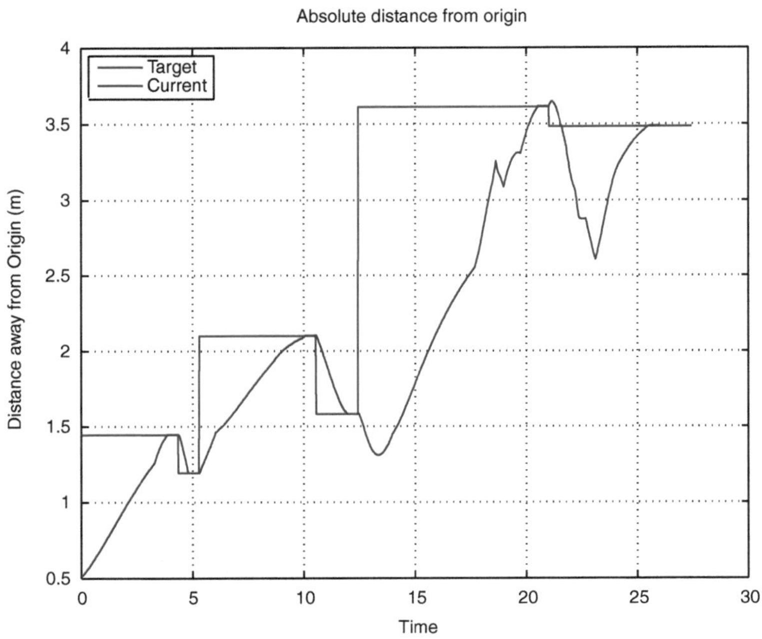

Fig. 7 Comparison of the distance from the origin to the target and end-effector locations showing the error during the simulation. Each new *horizontal target line* represent a new waypoint

machine, additional high level controllers can be added to increase functionality, and the base PID controller can be replaced to achieve new performance characteristics. To demonstrate the modularity of the controller design we show how the controller can be extended to incorporate a new high-level controller that performs obstacle avoidance for the mobile base, a common task in the field of robotics.

4.1 Obstacle Avoidance Controller

For this example we assume the obstacles are uniformly shaped circular columns, and also define a circular area big enough to enclose the mobile base. This simplifies and limits the obstacle avoidance problem to be two dimensional, with a goal of preventing the two circles from colliding. Collisions can be checked by comparing the distance between the centres of both circles to the sum of the radius of both circles. As long as the sum of the radii is less than the distance between the centre of both circles there are no collisions. We also define a buffer distance between the two circles, such that once the mobile base enters the buffer zone obstacle avoidance will be triggered. The buffer can be adjusted to reflect how

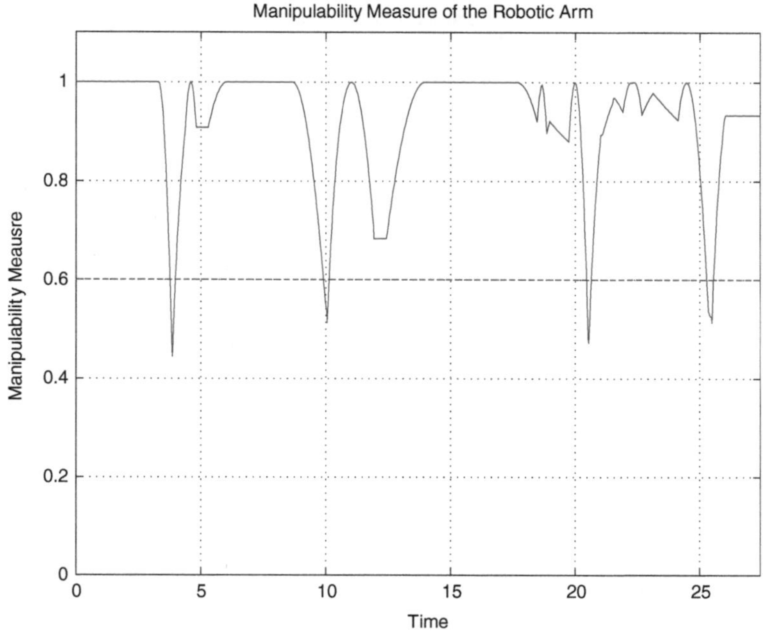

Fig. 8 Manipulability measure during the simulation. *Dashed line* shows the threshold

closely the robot should be allowed to approach obstacles. The buffer is also an indication of the maneuverability of the mobile base, as within the buffer distance the mobile base must be able to complete the necessary turning maneuvers to steer around the obstacle.

To fit within our controller framework, the obstacle avoidance controller must output destination points in the global reference frame. Therefore, the obstacle avoidance controller is designed to compute temporary waypoints that will take the mobile base around the obstacle. A temporary waypoint is calculated based on the current vector between the mobile base and the obstacle. The waypoint is placed with a constant offset along a vector perpendicular to the robot-obstacle vector and tangent to the circle enclosing the mobile base. The direction of the robot-obstacle vector changes as the mobile base drives around the obstacle, and a new temporary waypoint is recalculated as the robot moves. Thus the robot traces out a circular path around the obstacle until the obstacle is cleared, i.e., the heading to the desired waypoint differs substantially from the heading to the obstacle. Figure 11 shows a typical obstacle avoidance scenario, including the temporary waypoints–indicated by crosses–computed as the mobile base drives around the obstacle.

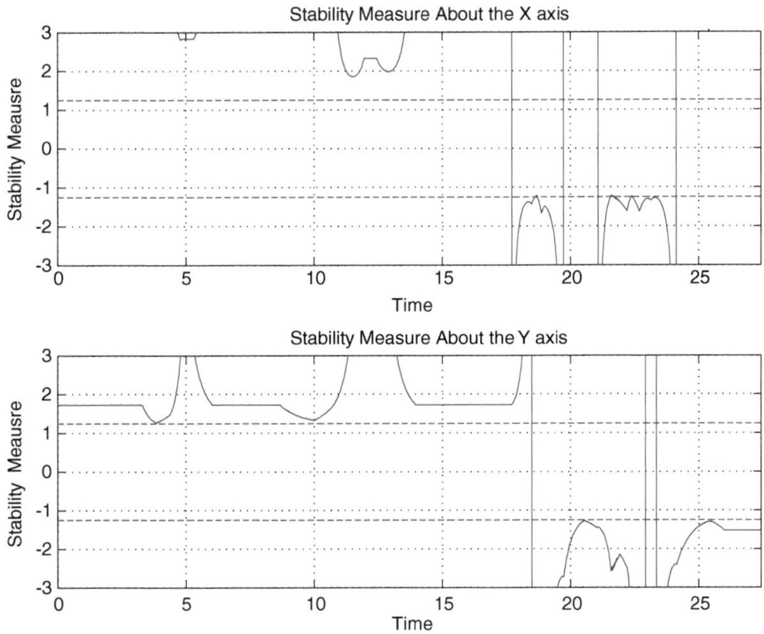

Fig. 9 Stability measure for both axes during the simulation. *Dashed line* shows the threshold

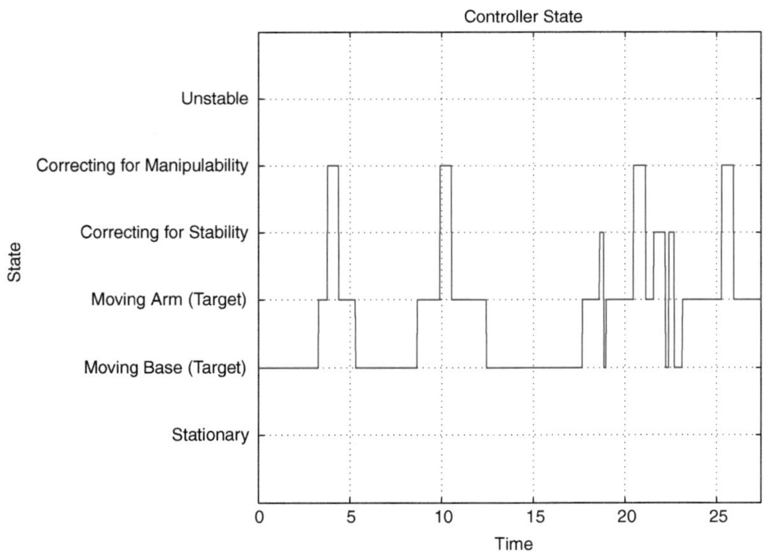

Fig. 10 Controller states during the simulation

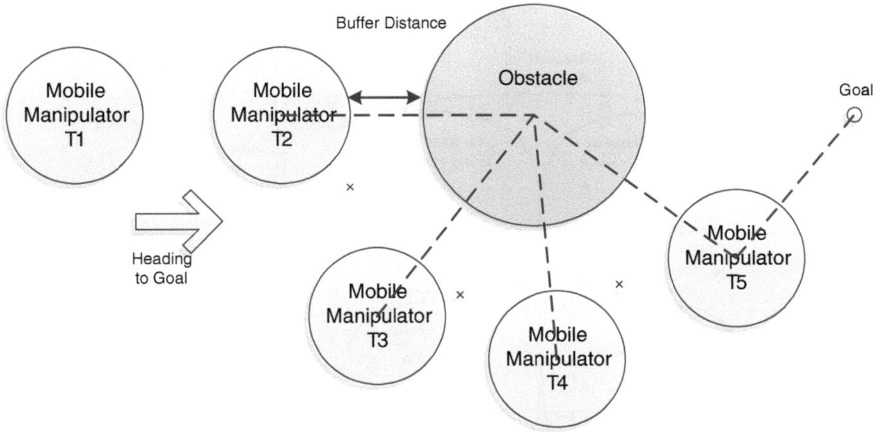

Fig. 11 Steps for the operation of the obstacle avoidance controller

4.2 Revised Arbitrator

To integrate the obstacle avoidance controller into the existing controller we define a new obstacle avoidance state within the arbitrator (Fig. 12), along with conditions to enter and leave the new state. To enter the obstacle avoidance state the mobile base has to be within range, and also heading towards, an obstacle. To leave the state the mobile base must be out of range or heading away from the obstacle. We assume that the robot only encounters obstacles while approaching a target point, and that the buffer zone between the mobile base and any obstacle is large enough to ensure that collisions with obstacles will not occur when the robot is adjusting its position to improve manipulability or stability. However, additional transitions between the obstacle avoidance state and the other states could be added if necessary.

Integration of the obstacle avoidance controller does not require changes to any of the other high-level controllers, or any modification of the low-level controllers. The obstacle avoidance controller thus becomes just another high-level controller that the arbitrator can select based on feedback from the surroundings and the desired objective. The output of the obstacle avoidance controller, in the form of waypoints, is executed by the low-level controllers in exactly the same way as the commands from other high-level controllers.

4.3 Controller Performance

Figure 13 shows a simulated scenario in which the mobile manipulator attempts to reach two waypoints: the first located on the far side of an obstacle, and the second back at the starting position. The obstacle has a radius of 0.2 m and the mobile

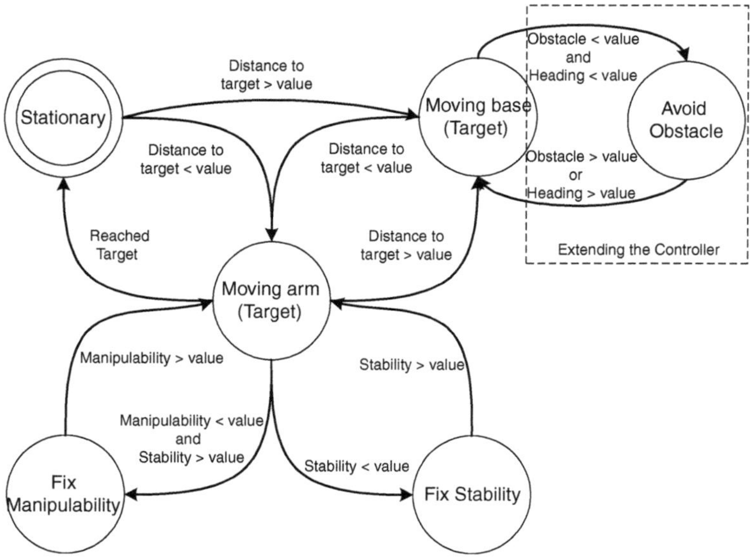

Fig. 12 State diagram of the state machine with the added obstacle avoidance controller

base circle has radius of 0.3 m. The buffer distance is set to 0.3 m. The mobile manipulator successfully moved around the obstacle to reach the targeted way-points without hitting the obstacle. A distance of roughly 0.3 m was maintained between the obstacle and the mobile base until the obstacle was cleared. There are two long time periods in which obstacle avoidance is executed, as the robot manoeuvres around the obstacle to reach the first goal, and then returns to the starting position to reach the second goal. Outside of the obstacle zone the robot behaves as previously, producing a smooth approach to each goal while maintaining stability and manipulability.

5 Related Work

Several researchers have explored controllers for coordinated motion of mobile manipulators. Many of these, such as Yamamoto and Yun [3], Seraji [9], Bayle et al. [1], and Fruchard et al. [2], focus on developing a single, often complex, feedback control algorithm to perform the manipulator tasks and maintain manipulability. In contrast, our controller is made up of a hierarchy of simpler controllers, which eases the design of individual controllers and allows the controller to be extended with new capabilities. Furthermore, none of the monolithic feedback controllers of which we are aware address the problem of maintaining platform stability.

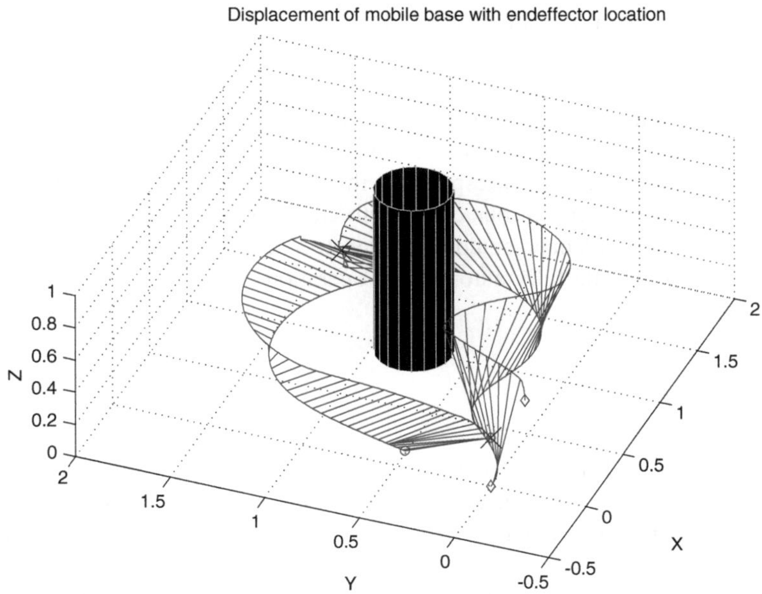

Fig. 13 Path of the end-effector and mobile base when reaching target with obstacle in the way

Huang et al. [10] approach the mobile manipulation problem through a motion planning scheme. This requires knowledge of both the workspace and the objective before the algorithm can be used. Their approach thus differs from our control scheme, which is aimed at reacting to objectives that are unknown to the controller until they are given. The zero moment point stability measure used by Huang et al. is comparable to ours. However, they define manipulability as a preferred pose that the controller attempts to maintain. Although their controller maintains stability, unlike our controller theirs does not compensate for external disturbances such as a load. Therefore they are limited to control the stability to within a small region, which translates to a smaller range of motion for the manipulator.

Iagnemma et al. [11] investigated the kinematic reconfigurability of the mobile manipulator system to increase stability in rough terrain. They take advantage of the ability of the mobile base platform to alter its shoulder joint angles, allowing it to tilt and change the wheel-base dimensions to help in improve stability. We have used a fixed mobile base platform, and therefore rely more heavily on the positioning of the manipulator to improve stability.

Hamner et al. [12] combine coordinated control of the mobile manipulator with autonomous operation in a factory assembly setting. They address manipulability, but do not consider tip-over stability. This is justified since their robots operate in a known factory environment, and use a heavy-duty mobile base.

Chen et al. [13] propose a hierarchical controller based on neural networks that is capable of identifying and compensating for the coupled dynamics of the mobile base and manipulator. They do not explicitly set out to maintain manipulability or stability, but to learn and identify the dynamics of the system. This could be a useful complement to a controller such as ours.

The hierarchical control structure we have implemented resembles parts of a Three-Layer Architecture [14], which is a common structure for autonomous robot controllers. Our sub-controllers correspond to the skill layer of a Three-Layer system, while the arbitrator corresponds to the sequencing layer. The deliberator or planning layer is missing in our control structure, as it is a purely reactive controller. However, a planner could be added to our controller to generate target waypoints, and provide fully-autonomous operation.

6 Conclusion

We have introduced a modular, hierarchical control scheme capable of coordinating motion for a mobile manipulator. The main goal of the controller is to control the mobile manipulator such that the stability and manipulability are always maintained at acceptable, if not optimal, levels. The hierarchical control system consists of low-level controllers that are responsible for the movement of the base and manipulator separately, and a collection of higher-level controllers that coordinate the low-level controllers to execute manipulation tasks while keeping the performance metrics within specified bounds. The effectiveness of the control scheme was tested in simulation by tasking the controller with reaching a series of waypoints with the end-effector while carrying a load.

The hierarchical approach to controller design simplifies the design and implementation process, and also allows additional sub-controllers or performance metrics to be added. We have demonstrated the ease with which the basic controller can be modified by adding a simple obstacle-avoidance controller to the system, and showing through simulation that it integrates well with the original controller design. Future work will include examining the impact of different threshold settings on the performance of the controller, exploration of extending the controller to handle energy management, and integration of a planning module.

References

1. B. Bayle, J.Y. Fourquet, M. Renaud, in *A Coordination Strategy for Mobile Manipulation*. Proceedings of the international conference on intelligent autonomous systems (IAS 2000) (2000), pp. 981–988
2. M. Fruchard, P. Morin, C. Samson, A framework for the control of nonholonomic mobile manipulators. Int. J. Robot. Res. **25**, 745–780 (2006). doi:10.1177/0278364906068374
3. Y. Yamamoto, X. Yun, Coordinating locomotion and manipulation of a mobile manipulator. IEEE Trans. Autom. Control **39**(6) (1995)
4. P. Corke, B. Armstrong-Hélouvry, A meta-study of PUMA 560 dynamics: a critical appraisal of literature data. Robotica **13**(3) (1995)

5. P. Corke, Robotics toolbox for Matlab (2008). Website, http://www.petercorke.com/Robotics_Toolbox.html
6. T. Yoshikawa, *Foundations of robotics: Analysis and control* (MIT Press, Cambridge, 1990)
7. R.C. Arkin, *Behavior-Based Robotics* (MIT Press, Cambridge, MA, 1998)
8. MathWorks, Simulink (2011). Website, http://www.mathworks.com/products/simulink/
9. H. Seraji, in *An On-line Approach to Coordinated Mobility and Manipulation*. Proceedings of the IEEE international conference on robotics and automation (ICRA '93) (1993), pp. 28–35
10. Q. Huang, K. Tanie, S. Sugano, Coordinated motion planning for a mobile manipulator considering stability and manipulation. Int. J. Robot. Res. (2000)
11. K. Iagnemma, A. Rzepniewskia, S. Dubowskya, P. P., T. Huntsbergerb, P. Schenker, Mobile robot kinematic reconfigurability for rough-terrain, in *Proceedings of SPIE Symposium on Sensor Fusion and Decentralized Control in Robotic Systems* (2000)
12. B. Hamner, S. Koterba, J. Shi, R. Simmons, S. Singh, An autonomous mobile manipulator for assembly tasks. Auton. Robots **28**(1) (2010)
13. Y. Chen, L. Liu, M. Zhang, H. Rong, Study on coordinated control and hardware system of a mobile manipulator, in *Proceedings of the 6th World Congress on Intelligent Control and Automation* (2006)
14. E. Gat, On three-layer architectures, in *Artificial Intelligence and Mobile Robots* (MIT Press, Cambridge, 1998)

Automated Test Case Generation for Industrial Control Applications

Reinhard Hametner, Benjamin Kormann, Birgit Vogel-Heuser, Dietmar Winkler and Alois Zoitl

Abstract The need for increasing flexibility of industrial automation system products leads to the trend of shifting functional behavior from hardware solutions to software components. This trend causes an increasing complexity of software components and the need for comprehensive and automated testing approaches to ensure a required (high) quality level. Nevertheless, key tasks in software testing include identifying appropriate test cases that typically require a high effort for (a) test case generation/construction and (b) test case modification in case of requirements changes. Semi-automated derivation of test cases based on models, like UML, can support test case generation. In this paper we introduce an automated test case generation approach for industrial automation applications where the test

Based on "Test Case Generation Approach for Industrial Automation Systems", by Reinhard Hametner, Benjamin Kormann, Birgit Vogel-Heuser, Dietmar Winkler, and Alois Zoitl which appeared in the Proceedings of the 5th International Conference on Automation, Robotics and Applications (ICARA 2011). © 2011 IEEE.

R. Hametner (✉) · A. Zoitl
Automation and Control Institute (ACIN), Vienna University of Technology (VUT), Gusshausstrasse 27–29/E376 1040 Vienna, Austria
e-mail: hametner@acin.tuwien.ac.at

A. Zoitl
e-mail: zoitl@acin.tuwien.ac.at

B. Kormann · B. Vogel-Heuser
Automation and Information Systems (AIS), Technische Universität München, Boltzmannstraße 15 85748 Garching, Germany
e-mail: kormann@ais.mw.tum.de

B. Vogel-Heuser
e-mail: vogel-heuser@ais.mw.tum.de

D. Winkler
CDL for Software Engineering Integration for Flexible Automation Systems, Vienna University of Technology (VUT), Taubstummengasse 11 1040 Vienna, Austria
e-mail: dietmar.winkler@tuwien.ac.at

G. Sen Gupta et al. (eds.), *Recent Advances in Robotics and Automation*,
Studies in Computational Intelligence 480, DOI: 10.1007/978-3-642-37387-9_20,

cases are specified by UML state chart diagrams. In addition we present a prototype application of the presented approach for a sorting machine. Major results showed that state charts (a) support efficient test case generation and (b) enable automated generation of test cases and code for industrial automation systems.

1 Introduction

Trends in industrial automation application development show a shift of functional behavior from hardware solutions to software components to increase flexibility with respect to frequent changing requirements and modified configurations. Additional added value of software components arise from decreasing project development time, cost, and accelerating development processes of industrial control and automation systems. Nevertheless, this trend leads to an increased complexity of software components and the need for effective testing approaches.

Nevertheless, experiences in the field of software engineering [1], e.g., database applications and web applications, showed that testing is an integral part of the development process, both in traditional processes as well as in agile control software development projects [2]. However, automating (parts of) test processes along the testing chain, e.g., test case generation, execution of test cases, analysis of results and reporting, is an important foundation for increasing test performance and product quality. Frequent test runs embedded within defined engineering processes, e.g., applying continuous integration and test, enable fast feedback on deliverables and the current development status for project monitoring purposes [3].

In addition, automation supported test case generation based on UML models (a) support flexibility of systems product development, (b) enable immediate response to changing requirements, and (c) enable automated code generation of functional behavior and test cases. Models, e.g., defined by the Unified Modeling Language (UML) [4], represent a real world setting and can be used as foundation for automated test case (and code) generation [5].

The generation of test cases based on available system knowledge is an important factor of complex automation systems, such as industrial automation systems. We investigated three approaches for test case construction/generation.

Manual approach—By using this approach the extraction of test case information from test specification has to be done manually. Therefore, the manual constructed test cases for further processing should be included in the control application as well. Manual test case construction, implementations, and modification require appropriate programming skills.

Semi-automated approach—The second approach which we called semi-automated approach is partly automated. First of all, test cases have to be specified by UML state chart (SC) diagrams. We realized a Java-based XML parser implementation, which extracts test case information from UML specification automatically. Then, these test cases have to be implemented for the control environment manually.

The third methodology is a *fully automated approach*. There is the need to get a set of executable test cases for applications in automation systems development projects. Different to the semi-automated approach the model-transformation method is a suitable method instead of the parser functionality to extract the test case information from their specification. With this work we focus on improving our existing semi-automated approach towards a fully automated model based testing infrastructure.

The remainder of this paper is structured as follows: Sect. 2 presents the research issues for the fully-automated test case generation approach. Section 3 summarizes related work on model-based testing with focus on industrial automation systems. The automated test case generation approach for industrial automation applications is explained in Sect. 4, followed by an example implementation of the suggested test case generation process in Sect. 5. Finally, Sect. 6 summarizes the findings, concludes and identifies further work.

2 Research Issues

Towards model-based development [6], this work proposes UML SCs as suitable model to provide the necessary data to generate a suite of test cases with a dynamic test case generation approach.

RI-1. Is a model-based test specification method sufficient for an automatic test case generation in the domain of industrial automation? Although the model driven design approach has shown benefits also in industrial automation [6] it has not been investigated if a model driven approach is also suited for the test process.

RI-2. Can test cases be automatically generated in a platform independent way? Industrial automation characterized by many different execution platforms. Having test cases in a platform independent way are beneficial as it allows to test equal behavior on different target platforms.

RI-3. Does the test case model (i.e., a model which includes all test cases) allow an automatic generation of executable test implementations for industrial control systems? The generated test case code needs to fulfill requirements like real-time execution and hardware access as well as limitations in processing power and memory of industrial control systems.

3 Related Work

This section summarizes related work on model driven development and gives an overview on model based or model driven testing in the field of software engineering. Finally an introduction and related work on testing automation systems is presented.

3.1 Model-Driven Development

Model-based testing [7] is a well-known approach for test case generation in the field of software engineering. Models can be derived from the system requirement specifications as foundation for automated test case generation [8, 7]. The specification of the models can be defined by UML diagrams. Different UML diagrams for the specification process are available, but only a subset of the diagram types is really suitable for the automation domain. A selection of suitable UML diagrams has been shown in [9, 10], and [11]. Panjaitan and Frey [11] proposed an approach for modeling IEC 61499 applications by UML and also the entire development process of distributed control systems. Therefore, they used several UML diagrams e.g., state chart-, component-, sequence-, activity-, use case-, and class diagrams for the model transformation process. Hussain and Frey [12] proposed a UML-based development process with an automated test case generation process. They automatically generate component diagrams from scenario specifications in terms of sequence diagrams. In contrast to [11], our approach deals with the information extraction for the automated test case generation with their execution environment and not the generation of the systems behavior itself from the UML specification. Zhang et al. [13] used UML for specifying IEC 61499 function blocks to use them for UML notations and supporting the FB paradigm for further applications.

3.2 Model-Driven Testing

Accurate examples for models to be used for testing purposes are Finite State Machines (FSM), formally defined in [14]. The FSM can be used to generate test cases fulfilling test coverage. There exists a set of desirable properties for testing of FSMs. Action Coverage is defined as the desirable property of executing every possible action at each state at least once. Chow [15] introduces Branch Coverage, Switch Coverage, Boundary-Interior Coverage, and H-Language as test coverage criteria. Branch Coverage traverses an FSM in order to visit each branch, so that the complexity of all possible paths reaching to infinity at worst can be reduced. Switch Coverage describes a branch-to-branch tuple, meaning that in and out branches of a state are covered by test sequences [16]. Boundary-Interior Coverage as described in [15] characterizes test sequences causing loops to be traversed once without additional iterations. H-Language is a similar approach for Boundary-Interior Coverage loop testing. Reachability Analysis is the process of computing the reachable states from the initial states [17].

3.3 Testing of Industrial Automation Systems

Model-based testing of industrial control applications aims at supporting processes for test case generation and soft commissioning. The models can be derived from the system requirement specifications, enabling automated code and test case generation from their models [8, 7]. The results of [9] are used for the development of the test case generation of this publication. Winkler et al. [18] shows a component oriented test approach for testing IEC 61499 function blocks. This component consists of separated functional, diagnosis, test methods.

Moser et al. [19] showed how ontologies can be used to represent both system and testing knowledge of typical production automation systems, and described an approach for automatically deducting test cases based on these ontologies. In this work, the key challenge is how to derive test cases (and test code) based on UML models, i.e., UML SC, for application in the automation systems development projects.

In contrast, models in the domain of software engineering development, function blocks (FBs) are common practice for modeling systems behavior of automation system applications. FBs encapsulate software components which are reusable for other control applications. This function block oriented paradigm is used in the IEC 61131-3 [20] and the IEC 61499 Standard [21] for industrial process measurement and control systems.

4 Automated Test Case Generation

This section presents the proposed test case generation approach for supporting the development process, testing control applications, and it is explained how the information of the test cases can be derived from UML specification. Function blocks (FBs) are a common practice for modeling system behavior in the automation domain, encapsulating software components for reusable applications.

We identified two different UML SC diagram models per control component: Plant Behavior Model and the Control Behavior Model for the *Test Case Specification*. Hegny et al. [2] used this approach for the generation of IEC 61499 applications and simulates plants behavior. Their concept for the simulation framework is based on the coupling of control application and the plant simulation application. Different to [22], we use these UML SC descriptions for the test specification of the machine.

The *Plant Behavior Model* describes the physical and mechanical behavior of a component or parts of a system, e.g., mechanical axis. Also timing information about the movements is included, e.g., the movement of axis from one position to the other takes 500 ms. So that the UML SC model is adequately realistic as the real plant model. This specification can be used for I/O tests of mechanical, electrical, and pneumatic equipment. Therefore basic control functionality is testable.

The *Control Behavior Model* describes the logical behavior and defines sequences of a component and/or control system which can be tested. This model definition can be used to test the implementation of the application.

In the following the *Test Case Generation Process* will be explained. Figure 1a illustrates the overall process flow for the test case generation, starting with the UML specification of the System under Test (SuT) and completing with an executable FB network. The FB network can be based on IEC 61131-3 or IEC 61499 standard depending on individual SuT implementations. At the beginning of the test process, the specification has to be described by UML diagrams. We used UML SC diagram models for test description.

The test generation aims at covering all branches of the state chart diagram. In order to reduce complexity, the UML model is traversed in a way that each branch is visited once by evaluating each transition. The resulting test sequences are stringed together and form individual test cases for the System under Test.

A test case consists of preconditions, actions, post conditions, expected results, and actual results, which are presented in the meta-model definition, see Fig. 1b. The preconditions are parameters which are used for the initial test case state. Then the parameters which are defined as action parameters will be applied on the SuT to execute the test case. The post conditions defines parameters so that test cases end in a defined state. The expected results are defined for the reporting process, so that an actual theoretical comparison is possible. In most cases, post conditions and

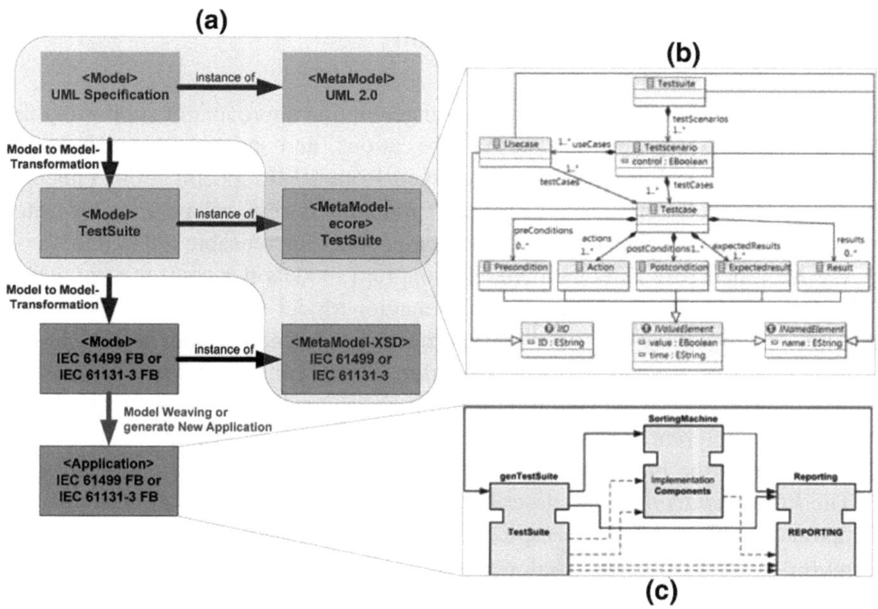

Fig. 1 Overview of the transformation process to generate an automation control applications: **a** Workflow for automated test case generation; **b** Test suite meta-model definition; **c** Function block network based on IEC 61499 standard

expected results have the same parameters. However, for certain applications the expected result is not a safe state of the system. For these post conditions, which are different from the expected results, ensures a transition from the test result state into a safe system state where the test case is finished. This safe system state may be the starting point for other/next test cases. Therefore post conditions may have additional parameters compared to the expected results. Finally they will be used for the reporting process e.g., online feedback.

The schematic overview of the used Test Suite meta-model is shown in Fig. 1b. This Test Suite meta-model, the specified UML diagrams, and the UML 2.0 meta-model are necessary for the Test Suite Model generation and represent the basic structure of the Test Suite. The Test Suite Model is automatically generated through a model-to-model transformation and includes all test cases. For the model transformation process we defined several rules. The rules are defined by Xtend, a transformation language provided by the Eclipse Modeling Project.

Based on the generated Test Suite Model, the automation model (IEC 61499 FB or IEC 1131-3 FB model) is also generated automatically through the model-to-model transformation process. Therefore the Test Suite model, the Test Suite meta-model, and the IEC meta-model are needed. The aim of this transformation is to generate a control FB which includes the Test Suite. Therefore, the same transformation language (Xtend) as the Test Suite generation is used. To get the definition of the IEC standard, the IEC 61131/IEC 61499 Meta Models are used. IEC 61131-3 does not provide a general meta-model definition. Each development tool uses its own meta-model. Our meta-model definition for the IEC 61131-3 implementation is based on the PLCopen specification [23]. The IEC 61499 [21] standard organization provides a general meta-model definition. These meta-models include the structure definition of the automation model which is necessary for the transformation process.

The final step of the automated test case generation for control applications is the creation of a test application. The generated FBs, the SuT, and the reporting component are connected by its data- and event connections. The FB network is shown at the schematic overview in Fig. 1c. As a special case, if the test framework should include an existing application, the generated test framework will also be connected with the SuT, which is called “model weaving”, otherwise a new application will be generated.

5 Prototype Implementation of the Proposed Test Case Generation

This section presents the prototype implementation of the proposed test case generation approach using a sorting machine. The basic functions of the sorting machine include the transportation of palettes via a conveyor belt and the sorting of parts from the palette into boxes by handling units which consists of axis and vacuum gripper components.

Figure 2 shows a selection of the implemented and used UML SC diagrams for the test case generation process. Therefore the UML SC description for the axis component and the vacuum gripper component are considered in more detail. This UML SC diagrams include the required information for the test case and the use case generation. UML SC diagrams consist of states and transitions. Individual test cases can be derived directly from transition parameters of the state charts for test case generation. These state charts don't contain necessary information for complete automated control code generation. Because, if the same state charts specification is used for the control code generation and the test case generation only the model transformation procedure can be verified.

A testing process which is only based on single test cases (TC1.1), see Fig. 3, is not always sufficient for testing industrial automation systems, because system states are highly interwoven and the complexity of industrial automation systems tends to be tremendously high. There is the need for SuT testing with sequences of test cases (e.g., TS1.1). In order to form such test sequences the combination of single test cases has to fulfill the following rule: The post condition parameters of a test case n has to be equal to the precondition parameters of the next test case n + 1 in the test sequence.

This paper shows the implementation based on the IEC 61499 standard. The test suite is included in one IEC 61499 Basic Function Block (BFB). Therefore the interface of the BFB, the used algorithms, and the internal Execution Control Chart (ECC) are generated automatically. An example implementation of a single test case and a test sequence is shown in Fig. 3. The ECC implements finite state machines based on Moore machines [24].

There are two necessary strategies for the implementation of the ECC, one for test cases and one for use case tests. Figure 3a shows the ECC of a single test case implementation with the test cases $ID = 1.1$ and $ID = 1.2$. The selected test case ID can be parameterized through the Test Suite BFB which is generated automatically (Fig. 1c). At the beginning, the test case mode must be selected (e.g., $Mode\_TC = true$), and the test case ID must be defined (e.g., $TCselect = 1.1$). This test case represents the transition at the axis control behavior state chart in Fig. 2c from the state *IN* to the state *OUT* with the action parameter $moveIn = 0$. After initializing the FB through an event trigger, the generated algorithm of the BFB

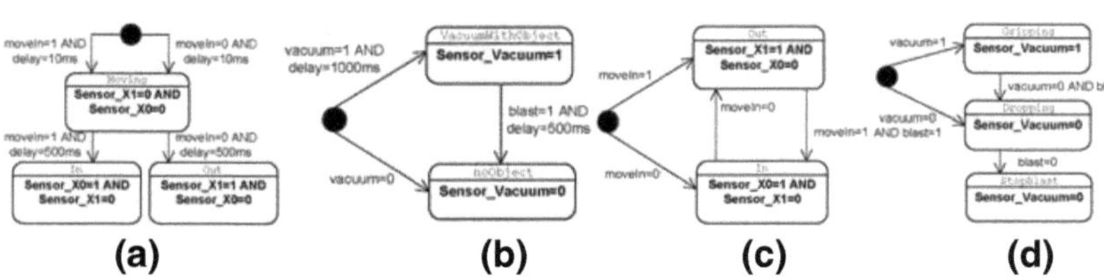

Fig. 2 State chart definition/specification by UML state charts of the axis component and the vacuum. **a** Axis plant behavior. **b** Vacuum gripper behavior. **c** Axis control behavior. **d** Vacuum gripper control behavior

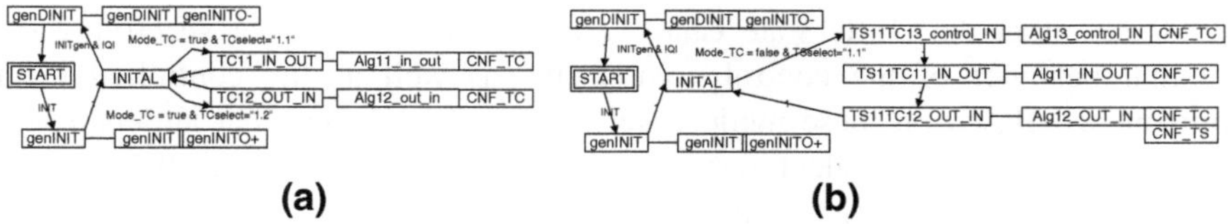

Fig. 3 Results of the automated generated execution control chart (*ECC*), which is included in one IEC 61499 basic function block (*BFB*). **a** Single text case, ID = 1.1 and ID = 1.2. **b** Test sequence 1.1 includes test case 1.3, test case 1.1 and test case 1.2

(e.g., *Alg11*) is executed and an output event *CNF_TC* is sent. The results of the test case can be positive or negative, which are reported to the "*REPORTING*" FB.

The second strategy of the ECC implementation is used for the sequence/use case tests. Figure 3b shows an example of a test sequence. A test sequence is defined by a specific set of test cases which are executed in a predefined consecutive order. In this case the test case mode must be selected (e.g., *Mode_TC* = *false*) and the test sequence ID must be parameterized, (e.g., *TSselect* = *1.1*). After initializing the FB through an event trigger the generated algorithms are executed ordered like "*Alg13-Alg11-Alg12*". If each algorithm is executed an output event *CNF_TC* is sent. After that, if all test cases are executed, an output event *CNF_TS* is sent. Therefore the test sequence process is finished and the next test case or test sequence is ready for the execution.

There are some limitations at this time. The specification of the plant behavior state chart diagrams include timing information, e.g., how much time needs the axis from one end position to the other end position. This information is currently not used. We extracted only positive test cases for the test case generation from the UML specification. No negative test cases and special test cases are generated. Hametner et al. [3] explains the terms "positive, negative, and special test case" in more detail. The test results of our implementation have shown that a detailed specification between the test suite interface and the SuT is of great importance. This test approach is applicable for testing industrial automation component libraries, because it is easy to specify control and the plant behavior by UML SC diagrams. Furthermore, this approach supports testing of existing and new libraries.

6 Conclusion and Further Work

Ensuring software quality of industrial control systems while at the same time reducing the engineering effort is of crucial importance for industrial automation systems. In previous investigations we developed first approaches for a test-driven development process for industrial control systems. However these approaches are elaborate and can only be partly performed in an automatic way.

In order to improve the testing process of industrial control systems we investigated the application of a model-based testing process in this article. We could show that the UML State Charts are a suitable model for specifying the nominal behavior of the control components as well as the controlled plant components. Based on these models we developed a method that automatically derives platform independent test cases, e.g., IEC 61131-3 and IEC 61499. Depending on the used target control platform a test model platform as well as specific test code is generated, that allows an automatic execution and report on the test cases. By applying this approach to an example control application we could show the feasibility and advantages of this approach. The automatic test case generation enables quick feedback loops at low cost during development. Our current implementation uses Boolean data types. Support for continuous data is already given, for which test cases can be generated automatically. In the ladder case the number of resulting test cases increases by the value range of the test data values.

Future work includes (a) a larger application context at our industry partners, (b) extending the state chart diagram with timing information and (c) the automated test case generation with negative test cases.

References

1. I. Sommerville, *Software Engineering*, 9th edn. (Addison-Wesley, Reading, 2010)
2. T. Linz, H. Schäfer, A. Spillner, *Software Testing Foundations: A Study Guide for the Certified Tester Exam*, 3rd edn. (Rocky Nook, 2011), 284p, ISBN-10: 1933952784, ISBN-13: 978-1933952789
3. R. Hametner, D. Winkler, T. Östreicher, S. Biffl, A. Zoitl, The Adaptation of Test-Driven Software Processes to Industrial Automation Engineering. in *8th IEEE International Conference on Industrial Informatics* (2010)
4. G. Booch, J. Rumbaugh, I. Jacobson, *The Unified Modeling Language Reference Manual* (Addison-Wesley, Reading, 1999)
5. M. Broy, B. Jonsson, J.-P. Katoen, M. Leucker, A. Pretschner, *Model-Based Testing of Reactive Systems* (Springer, Berlin, 2005)
6. K. Thramboulidis, G. Frey, Towards a model-driven IEC 61131 based development process in industrial automation. J. Softw. Eng. Appl. **84**, 217–226 (2011)
7. M. Utting, B. Legeard, *Practical Model-based Testing* (Morgan Kaufmann, Los Altos, 2007)
8. P. Chevalley, P. Thevenod-Fosse, Automated Generation of Statistical Test Cases from UML State Diagrams. in *Computer Software and Applications Conference. COMPSAC. 25th Annual International* (2001), (pp. 205–214)
9. R. Hametner, D. Winkler, T. Östreicher, N. Surnic, S. Biffl, Selecting UML Models for Test-Driven Development along the Automation Systems Engineering Process. in *IEEE International Conference on Emerging Technologies and Factory Automation* (2010)
10. C. Seidner, O. Roux, Formal methods for systems engineering behavior models. Ind. Inf. IEEE Trans. **4**(4), 280–291 (2008)
11. S. Panjaitan, G. Frey, Combination of UML Modeling and the IEC 61499 Function Block Concept for the Development of Distributed Automation Systems. in *Emerging Technologies and Factory Automation. IEEE Conference on* (2006), pp. 766–773
12. T. Hussain, G. Frey, UML-based Development Process for IEC 61499 with Automatic Test-case Generation. in *Emerging Technologies and Factory Automation. ETFA'06. IEEE Conference on* (2006), pp. 1277–1284

13. W. Zhang, W. Halang, C. Diedrich, Specification of function block applications with uml. in *Robotics and Automation. ICRA. Proceedings of the IEEE International Conference on* (2005), pp. 4002–4007
14. A. Petrenko, N. Yevtushenko, Testing from partial deterministic fsm specifications. IEEE Trans. Comput. **54**, 1154–1165 (2005)
15. T. Chow, Testing software design modeled by finite-state machines. Trans. Softw. Eng. **SE-4**(3), 178–187 (1978)
16. R. Binder, *Testing Object-Oriented Systems: Models, Patterns, and Tools* (Addison-Wesley, Reading, 2000)
17. C. Meinel, C. Stangier, Modular Partitioning and Dynamic Conjunction Scheduling in Image Computation. in *Proceedings of the 2002 IEEE/ACM International Workshop on Logic and Synthesis (IWLS02)*, New Orleans, USA. Citeseer, (2002)
18. D. Winkler, R. Hametner, S. Biffl, *Automation Component Aspects of Efficient Unit Testing*. IEEE, ETFA, Mallorca, Spain, September (2009)
19. T. Moser, G. Dürr, and S. Biffl, Ontology-Based Test Case Generation for Simulating Complex Production Automation Systems. in *Proceedings of the 22nd International Conference on Software Engineering* and *Knowledge Engineering* (SEKE'2010), Redwood City, San Francisco Bay
20. IEC 61131-3, IEC 61131-3 Standard—Programmable controllers—Part 3: Programming Languages, 2nd edn. International Electrical Commission (2003)
21. IEC 61499-1, Function blocks—Part 1: Architecture. Geneva: International Electrical Commission (2005)
22. I. Hegny, M. Wenger, A. Zoitl, IEC 61499 based Simulation Frame-work for Model-Driven Production Systems Development. in *IEEE Conference on Emerging Technologies and Factory Automation (ETFA)*, (2010), pp. 1–8
23. PLCopen Technical Committee 6, XML Formats for IEC 61131-3, Version 2.01—Official Release, PLCopen, Technical Report, May 2009
24. A. Zoitl, T. Strasser, A. Valentini, Open source initiatives as basis for the establishment of new technologies in industrial automation: 4DIAC a case study. in *IEEE International*
25. R. Lewis, *Modeling Control Systems Using IEC 61499—Applying Function Blocks to Distributed Systems* (The Institution of Electrical Engineers, London, 2001), 208p, ISBN-10: 0852967969, ISBN-13: 978-0852967966

A Realistic Simulator for Humanoid Soccer Robot Using Particle Filter

Yao Fu, Hamid Moballegh, Raúl Rojas, Longxu Jin and Miao Wang

Abstract This work presents a realistic simulator called Reality Sim for humanoid soccer robots especially in simulation of computer vision. As virtual training, testing and evaluating environment, simulation platforms have become one significant component in Soccer Robot projects. Nevertheless, the simulated environment in a simulation platform usually has a big gap with the realistic world. In order to solve this issue, we demonstrate a more realistic simulation system which is called Reality Sim with numerous real images. With this system, the computer vision code could be easily tested on the simulation platform. For this purpose, an image database with a large quantity of images recorded in various

---

Based on <Reality Sim: a realistic environment for robot simulation platform of humanoid robot>, by <Yao Fu, Hamid Moballegh, Raúl Rojas, Longxu Jin, Miao Wang> which appeared in the proceedings of the 5th International Conference on Automation, Robotics and Applications (ICARA2011). © 2011 IEEE.

Y. Fu (✉) · L. Jin
ChangChun Institute of Optics, Fine Mechanics and Physics,
Chinese Academe of Sciences, ChangChun, China
e-mail: yao.fu.felicity@gmail.com

L. Jin
e-mail: jinlx@ciomp.ac.cn

Y. Fu
Graduate University of Chinese Academe of Sciences, Beijing, China

Y. Fu · H. Moballegh · R. Rojas · M. Wang
Department of Mathematics and Computer Science, Freie Universität Berlin,
14195 Berlin, Germany

H. Moballegh
e-mail: moballegh@gmail.com

R. Rojas
e-mail: Rual.Rojas@fu-berlin.de

M. Wang
e-mail: miao.wang@fu-berlin.de

G. Sen Gupta et al. (eds.), *Recent Advances in Robotics and Automation*,
Studies in Computational Intelligence 480, DOI: 10.1007/978-3-642-37387-9_21,

camera poses is built. Furthermore, if the camera pose of an image is not included in the database, an interpolation algorithm is used to reconstruct a brand-new realistic image of that pose such that a realistic image could be provided on every robot camera pose. Systematic empirical results illustrate the efficiency of the approach while it effectively simulates a more realistic environment for simulation so that it satisfies the requirement of humanoid soccer robot projects.

1 Introduction

The research in this paper is based on the humanoid league of soccer competition in RoboCup [1]. As an international robotics competition, RoboCup aims to foster AI and intelligent robotics research by providing a standard problem where a wide range of technologies can be integrated and examined.

A soccer robot project is comprised of many sub-systems, e.g., vision, planning and motion. Simulation platforms [2, 3] are contributive in developing, testing and improving Control, Computer Vision and other relevant aspects in robot soccer competition. Nevertheless, the simulation environment and the real world have large difference which attracts many researchers to put effort to lessen it. Jakobi, N [4] proposed minimal simulations for evolutionary robotics within which controllers that meet these conditions will successfully transfer into reality. Tom Ziemke [5] discussed significance of a robot simulator in experimentation with active adaptation of non-trivial environments. Juan Cristóbal Zagal et al. [6–8] developed a simulator which is called UCHILSIM for RoboCup four-legged league, aiming at solving the so-called reality gap. Josh C. Bongard et al. [9]. presented a new co-evolutionary approach, by which algorithm, they automatically adapt the robot simulator using behavior of the target robot and adapt the behavior of the robot using the robot simulator. In literature [10], some more methods are proposed to assess discrepancies between the simulation and the reality.

Nevertheless, the gap between reality and simulation platform still exists. The papers mentioned above mostly focus on the controlling function between them. Actually, the gap in computer vision aspect is obviously not realistic enough, i.e. the testing environment is commonly a simulated image, but not the real image. In this chapter, we develop a novel subsystem, which is called Reality Sim for reducing the gap between the reality and simulation especially in vision aspects.

First, an image database was built with huge mount of real images. The images are captured by the robot from FUmanoids Team [11]. The team attended the Humanoid league of RoboCup champion each year from 2006 and had excellent records of two times 2nd place and one time 3rd place. Secondly, in order to record the images by three-dimension camera poses, the Particle Filter method is applied for recording camera poses of the images. The camera pose recording problem is similar to the well-known robot self-localization problem. Thirdly, although thousands of images recorded by camera pose are in the database, it is hardly possible to incorporate all camera poses. Every camera pose includes three dimension variables-x, y coordinates and yaw angle hence it is too hard to incorporate so many three dimension points on a $400 \times 600\ \text{cm}^2$ size soccer field just by building a database. Once a camera pose is not included in the database, an interpolation method is brought to reconstruct a brand-new image. By these approaches, the realistic image at every robot pose could be found from the database or retrieved by an image interpolation algorithm. Consequently, Reality Sim is of great use especially in developing and testing vision algorithms. Likewise, the simulator is more approximate to the realistic world.

The remaining of the paper is organized as follows, the camera pose recording method of images is explained in the second section, among which, the Particle Filter algorithm is adopted. The third section introduces image interpolation process. After that, experiments are shown in the fourth section, followed by conclusions drawn finally.

2 Camera Pose Calculation

In the Reality Sim subsystem, with the purpose of returning an image at designated pose to simulator, the camera pose was recorded previously. Camera pose calculation is similar with conventional mobile robot self-localization [12] in soccer robot scenario. For Self-localization problem, the robot's pose which generally includes location and orientation relating the environment is estimated from sensor data. In a humanoid robot system, Self-localization is a crucial problem since many other modules e.g., world model, behavior control depends on it. The camera pose computation could be regarded as a robot self-localization problem.

2.1 Particle Filter Algorithm

Many strategies have been applied to robot self-localization issue, e.g., Kalman Filter [13], grid-based method [14], Multiple Model Kalman Filters [15] and Particle Filter and so on. Among these, Particle Filter is the most popular approach due to its excellent performance in solving robot global localization problem. As the so-called Sequential Monte Carlo (SMC) method [16–18], Particle Filter represents the estimation the state of a dynamic system by a set of "particles", which are differently-weighted samples of the state space.

The Sequential Monte Carlo (SMC) method is mainly composed of motion, observation, resampling and state estimation. The steps of camera pose calculation, or rather, self-localization method in this paper is depicted as follows,

- *Initialization with N samples*: In the initialization, the weight of every sample is assigned a same initial value. One sample includes three variables, x, y coordinates and yaw angle.
- *Prediction*: As the so-called motion model, in prediction step, the "predictive" particles are got to predict the current position. Suppose x_t is state (location) at time t, y_t is sensor readings at time t, u_{t-1} is control command, like action, velocity and so on at time $t-1$. u_{t-1} is used to predict the position at time t. In this chapter, the samples generated from previous frame camera pose are introduced to predict the current position at t.
- *Observation model* (*Update*): In the robot soccer scenario, the observation model calculates what the robot observes when it is at specific location by the sensor. The observation model $P_t(y_t|x_t)$ is used to update the posterior belief $p_t(x_t)$ in each time step. The belief $p_t(x_t)$ is represented by some weighted samples. The evaluation of the weight ω_t^i (i = 1–N) will be explained in section B.
- *Resampling*: Resampling aims at eliminating samples that have small weights and concentrating on the ones with large weights thereby reducing effect of degeneracy problem. The posterior belief $p_t(x_t)$ is resampled for N_s times, afterwards, a new sample set $\{\hat{x}^i\}_{i=1}^N$ is got and $\omega_i^k = 1/N_s$.
- *State Estimating*: Many measures e.g., overall averaging, best particle, etc., could be used in estimating state [19]. In other words, computing a robot pose via given sample set. In this chapter we select winner-take-all strategy, i.e., best particle method.

Figure 1 illustrates the camera pose calculation process via particle filter method for one image frame. T is the maximum iteration time. The iteration begins with N initialized particles. In each iteration, each sample is given a weight according to the evaluation method which will be mentioned below and the resampling is afterwards adopted. With these steps, the particles gradually focus on the most probable states.

2.2 Observation Model

In the whole process, the observation model is essential to estimate the weight of samples. In this chapter, we separately put forward two models, i.e. distance-based model and Gauss-based model for seeking a more efficient model.

2.2.1 Distance-Based Model

In this model, weight of every sample is updated by an evaluation function. Further more, the evaluating process mainly includes three steps. Firstly, Projecting the field lines onto the image by camera model [20] from vision subsystem of

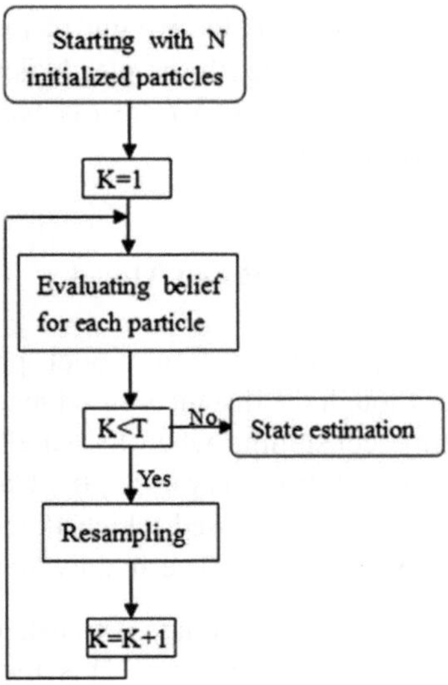

Fig. 1 The chart of camera pose calculating process

FUmanoids project according to its absolute coordinate. The projection green lines which are projected onto the real field white lines are shown in Fig. 2. This step comprises a translation from absolute coordinate in real world to image coordinate on the image via the robot pose. Secondly, calculating the distance between projection of field line and the real field line on the image by color feature in two opposite directions. The distance comparing process is also shown in Fig. 2. The short green and blue lines direct from the projection to the real field line are in two opposite directions. As a soccer playing field is a relatively simple environment, not too much feature could be utilized; as a result, the white line on the field is one of the most obvious features in this environment. In the meanwhile, contour of the field is found to eliminate effect of irrelevant data outside the field. Finally, as is shown in Fig. 2, for a projection line, there are many line-pairs from the projection to real field lines, selecting a shorter one of the line pair and saving its distance value. The sum of distance values are used to calculate the weight of each sample. The weighting method is shown in Eq. (1) as follows,

$$\omega_i = e^{\sum_{j=1}^{M} \frac{Dis_{ij}}{\alpha}}, i \in (1, N) \tag{1}$$

Fig. 2 The distance between the projection and real field line

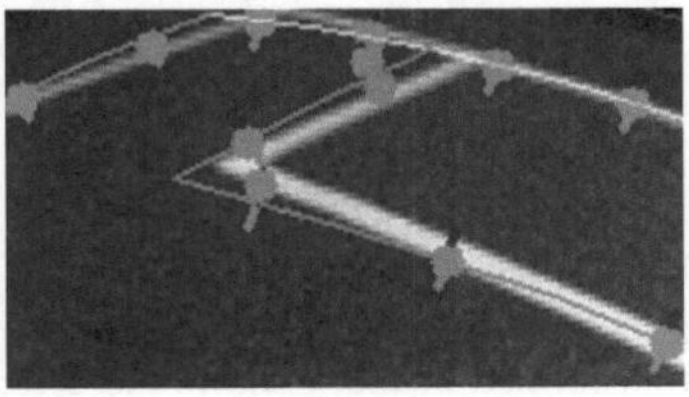

where ω_i is the weight of the ith sample, N is sample number, Dis_{ij} is the jth distance of the ith sample, M is the distance number and α is a factor for adjusting range of weight. Afterwards, weights of samples are normalized so that the sum of weights equals 1.

2.2.2 Gauss-Based Model

Another observation model proposed in this chapter is Gauss-based model. Although the distance-based model is effective in calculating the weight value, it is time-consuming when the smaller distance of the two orthorhombic directions is detected. For this reason, the Gauss-based observation model for the particle filter method is proposed which is more universal for every frame and faster than the previous model. The mainly steps of this model is shown as follows,

1. The original image is translated into a gray image. Then Gauss blurring method is used afterwards. This method could effectively reduce noise and camera artifacts. The image after Gauss blurring is shown in Fig. 3e.
2. Detecting border of the field is used to further decrease the irrelevant data. The red line around the border of the field in Fig. 3a is the detected border line. The border detection function is included in the computer vision subsystem of the FUmanoids [11] project.
3. Projecting the camera model line onto the image for every particle. The image after projection is shown in Fig. 3a, b–d demonstrate the detail information of red frames from the image in Fig. 3a in which the green projection lines are shown.
4. Calculating accumulated pixel values on the projected lines. The accumulated value is signed to every particle as the weighting value. The weighting formula is shown in Eq. (2),

$$\omega_i = e^{\sum_{j=1}^{M} \frac{Pix_{ij}}{\alpha}}, i \in (1, N) \tag{2}$$

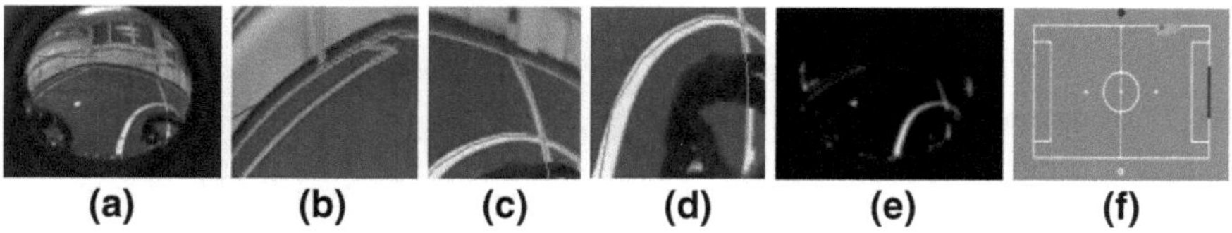

Fig. 3 The Gauss-based observation model of the particle filter algorithm. **a** is the image after projection; **b**, **c**, **d** shows the detail information in red frame of the image (**a**); **e** is its Gauss-transformed image; **f** shows the detected robot pose on the field by particle filter algorithm

in which ω_i is still the weight of the *i*th sample, *N* is sample number, Pix_{ij} is the *j*th pixel value on the image of the *i*th sample, *M* is the quantity of pixel points at which the pixel value is calculated. α is the factor for adjusting range of weight as Eq. (1). Afterwards, sum of sample weights is normalized to 1.

2.3 Resampling

The resampling step is essential to avoid the problem of degeneracy of the algorithm after several iterations and its basic idea to eliminate particles with small weights and concentrating on particles with large weights. The typical resampling algorithm namely Residual resampling [21] is adopted in this chapter. Considering a representation of current sample set $\{x_i^k, \omega_i^k\}$, k $= 1\ldots N_s$, which represents the probability density function of current robot pose, a new set $\{x_i^{\prime k}, \omega_i^{\prime k}\}, k = 1\ldots N_s$ is resampled so that $x_i^k = x_i^{\prime l}, k, l \in [1, N_s]$ and weight $\omega_i^{\prime k} = 1/N_s$.

3 Image Reconstruction

Although a great amount of images are stored in image database, it is yet impossible to comprise all camera poses. Since for the whole 400 $\times$ 600 cm^2 size soccer field and three dimensions pose-*x*, *y* coordinates and yaw angle, 720,000 of poses are needed. It is hard to traverse every camera pose in such a huge state space just by building an image database. Once the simulation platform calls for an image which is not included in the database, an image interpolation is invoked to interpolate a brand-new image and return it back. In mathematical field, Interpolation means a construction of new data points within the range of a discrete set of known data points. The data are obtained by sampling or experimentation. In this chapter, regression analysis [22] is adopted to interpolate a new image by its nearby images.

The reconstructed image will be interpolated by its nearest neighbor images. In order to find out the correlation of camera poses between the two images, a regression function is defined, i.e. a mathematic model is built. After that, unknown parameters in the model are estimated from sample data.

A regression model relating *Z* to a function of *X*, *Y* and β is shown as follows,

$$Z = f(X, Y, \beta) \tag{3}$$

Z is a dependent variable, *X* and *Y* are independent variables, β is unknown parameter matrix, *f* is model function. In order to estimate unknown parameters β, a solution to minimize the distance between the measured and predicted values of the variable *Y*, Least-Squares [23, 24] is used. Once β is estimated, *Y* could be calculated by the model function.

According to this model, the function *f* for interpolating among images is given as follows,

$$\begin{bmatrix} X_2 \\ Y_2 \end{bmatrix} = \begin{bmatrix} \alpha_0 & \beta_0 \\ \alpha_1 & \beta_1 \\ \alpha_2 & \beta_2 \\ \alpha_3 & \beta_3 \\ \alpha_4 & \beta_4 \\ \alpha_5 & \beta_5 \end{bmatrix}^{\mathrm{T}} \begin{bmatrix} 1 \\ X_1 \\ Y_1 \\ X_1 Y_1 \\ X_1^2 \\ Y_1^2 \end{bmatrix} \tag{4}$$

As is shown in Eq. (4), (X_1, Y_1) are pixel points sampled from camera pose of nearby image. Whereas, (X_2, Y_2) are corresponding points on current robot pose. The α_i and β_i are respectively unknown parameters. Firstly, the corresponding sample point pair (X_1, Y_1) and (X_2, Y_2) are separately selected from the nearby image and image of current camera pose. After finding the relation between the image coordinate and absolute coordinate, the pixel values on image whose absolute coordinate value are the same are used as sample data. Secondly, Least squares method [22, 23] is adopted for estimating model parameters α_i and β_i subsequently. After the parameters are estimated, the model is built. Thirdly, (X_2, Y_2) points are calculated from the model and they are furthermore interpolated onto current robot image. Afterwards, the correlation between the two camera poses is found. Furthermore, if some pixels on image which belong to current robot pose do not have correlation with nearby image, it could not be assigned into any value. For this kind of pixel points, a reverse model, namely a model from current camera pose to its nearby image is built and the corresponding RGB values of pixels from nearby image is assigned to these pixels subsequently. The reconstructed images is shown in Fig. 4b.

Up to now, the main frame of the system is depicted. Previously, building an image database in which images are recorded by camera pose. Then when the simulation platform calls an image at some camera pose from Reality Sim, an image in the database or a reconstructed new image is returned.

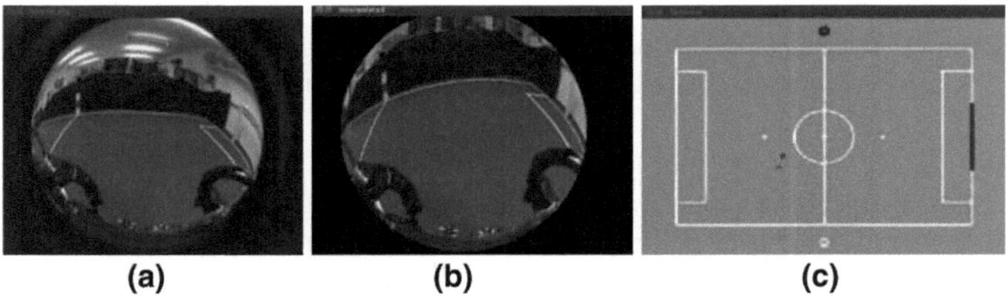

Fig. 4 **a** Is the interpolated image; **b** is its nearby image; **c** shows the robot pose on the field

Fig. 5 System chart

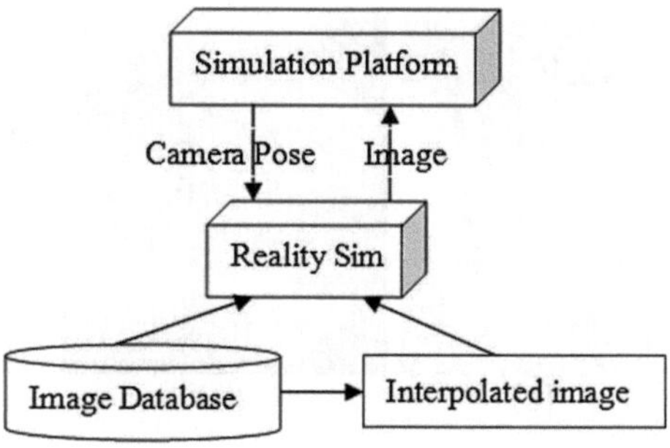

The system chart is shown in Fig. 5. This system could be efficiently utilized in testing the vision code on simulator.

4 Experimental Results

In the experiment, both of the two robot platforms-2009 and 2011 platform of FUmanoids soccer robot project which separately participated the RoboCup 2009 [11], 2010 [11] and RoboCup 2011 [25] are adopted. One robot in the FUmanoids team 2009 is shown in Fig. 6a and the image captured by this robot is shown in Fig. 6b. Figure 6c and d shows the robot-Anna in FUmanoids team 2011 and the captured image. Both of the captured images are 480×640 pixel size. The experiment environment is shown in Table 1. Figure 7 is the simulation platform of FUmanoids project.

An image database is built at first. The robot soccer field [26] which is 400×600 cm^2 size is shown in Fig. 8. The images are captured by a fish-eye camera equipped on the robot which covers a full range of $180 \times 90°$. For building database, a robot from FUmanoids soccer robot team moved successively and randomly on the soccer field. In the meanwhile, are captured images. To measure the validity of our approach, a ground truth database, the images in which are captured every 30 cm interval and manually recorded by camera pose is utilized as ground truth.

In the database about 10,000 images are indexed by camera pose. They are tagged by pose information, x, y coordinates, yaw, pitch and roll angle. As is depicted in part II, Particle filter algorithm is putted to use for tagging the images by robot pose.

Figure 9 illustrates the process of particle number selection for particle filter algorithm. The accuracy of the method increases with the number of samples. However, increased particle quantity causes more processing time. While the particle number is up to 1,000, the varying of the average error is not obvious. Considering time consuming and precision, sample number is set to 800. For this experiment, the Gauss-based observation model is adopted.

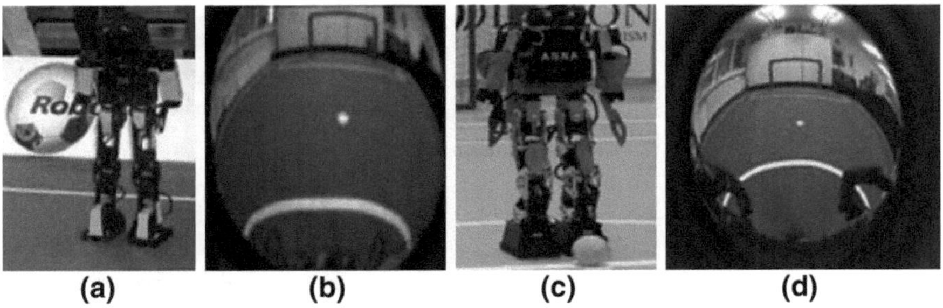

Fig. 6 **a** Robot-Eve of FUmanoids 2009 platform; **b** The image captured by Eve; **c** robot-Anna of FUmanoids 2011 platform; **d** The image captured by Anna

Table 1 Experimental environments

Developing environment	Parameters
PC configuration	Intel i5 2.5 GHz, RAM 4 GB
Operation system	Ubuntu 10.10
Programming language	C++ & OpenCV 1.0
Development environment	Eclipse Helios 3.6

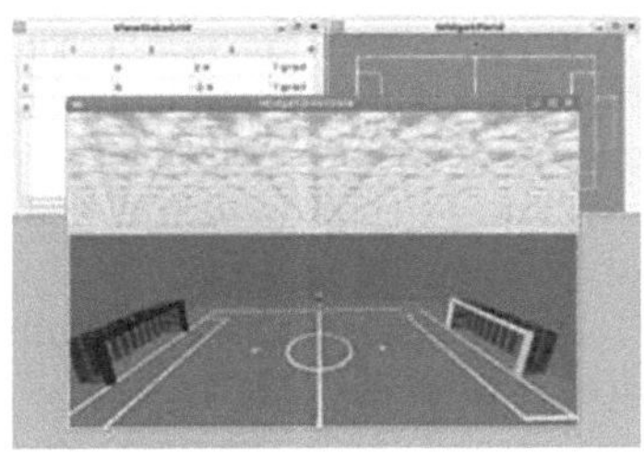

Fig. 7 The simulation platform of FUmanoids project

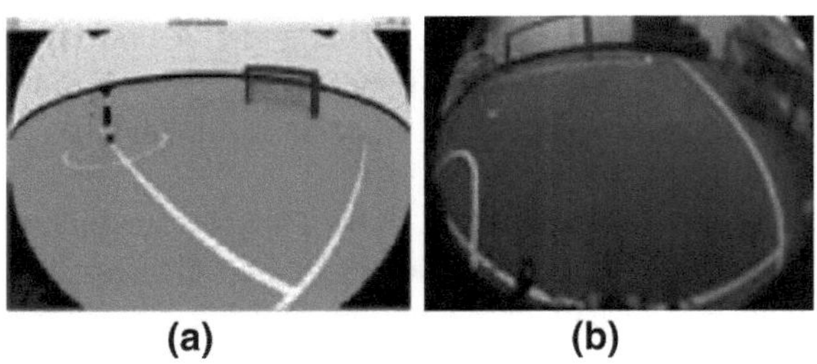

Fig. 8 A result of camera pose recording

To evaluate the effects of two observation models, the corresponding experiment results are shown in Table 2. The Gauss-based observation model shows a smaller average error and obviously shorter computation time. For this experiment, the particle number equals 800. In the mean while, $-1{,}000$ is assigned to factor α in formula (1) and (2). The initial values of the samples are set according to pose of the first frame of an image sequence.

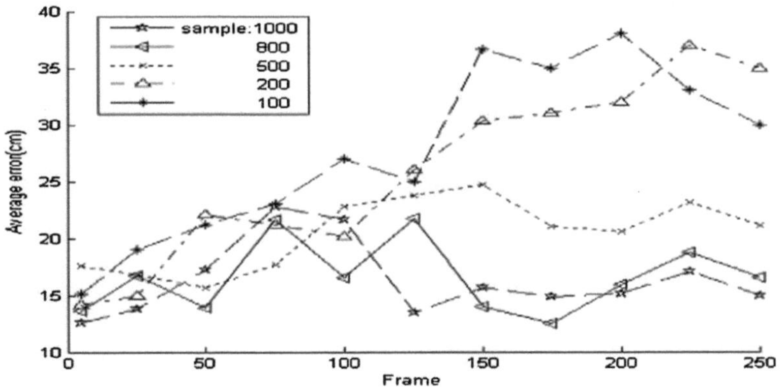

Fig. 9 Selection of particle number

Table 2 Comparison of two likelihood measures

Method	Average error (cm)	Average time (s)
Distance-based	18.5	1.76
Gauss-based	18.1	0.85

Fig. 10 **a** A sample of the simulated image; **b** A sample of the interpolated image

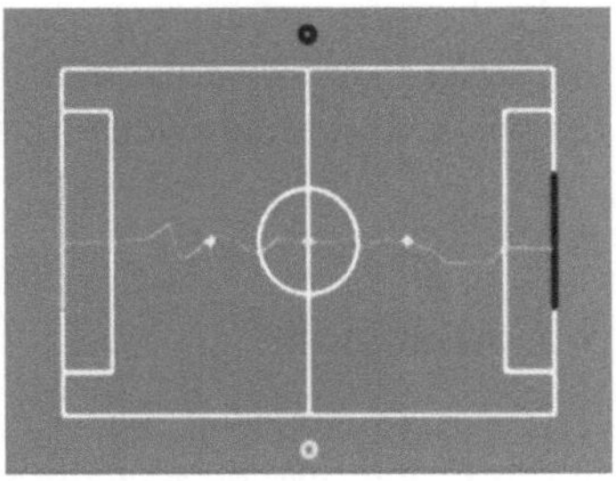

Another experiment for evaluating recording method by particle filter algorithm, a robot moving route is drawn in Fig. 10. The robot moves from the middle of one side of the field to the other side of the field straightly just like the route of ground truth data. Contrasting with ground truth data, the average distance error is 18.51 cm. An essential aspect which causes the error is that if lines are very far away from the robot, it is hard to recognize the color of the line exactly.

The following experiment is relevant to image interpolation. Figure 8 shows the comparison of the simulated image and the interpolated image. Compared with the real image, the interpolated image is obviously more realistic.

The point-pair (X, Y) for interpolation is selected by scanning the whole soccer field at 25 cm × 25 cm intervals. In this experiment, a route of robot pose is selected as is shown in Fig. 11. In the experiment, the robot moves from (0, 3, 0) to (8, 6, 4) continuously. The variables of 3-D pose respectively stands for x, y coordinates and yaw angle. The images at every robot pose are shown in following Fig. 12.

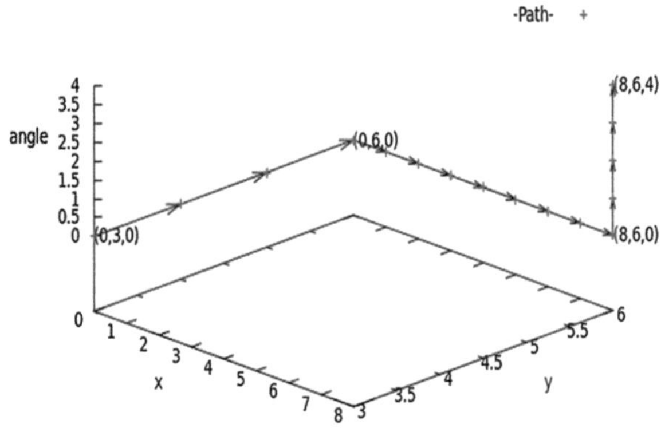

Fig. 11 The path of robot pose

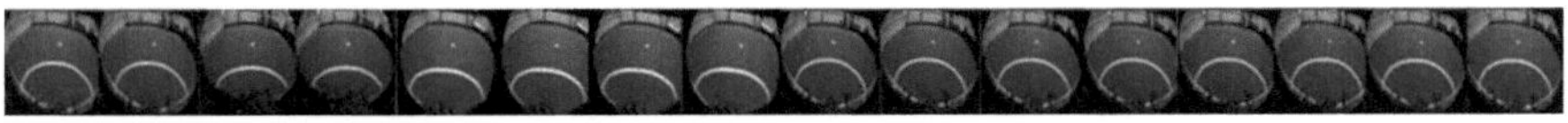

Fig. 12 An image sequence when the robot poses changes successively from (0, 3, 0) to (8, 6, 4). (the sequence is from left to right, images are from the FUmanoids platform 2009.)

As is shown in Fig. 12, when robot pose moving continuous from (0, 3, 0) to (8, 6, 4), the image at current pose is returned to the system. By images interpolation, the images vary continuously and smoothly. The above experiments demonstrate the whole system provides a more realistic world for the simulation platform.

5 Conclusion

In this chapter we presented a simulation subsystem-Reality Sim which is utilized mainly in testing vision code in simulation platform of a Humanoid soccer robot project. In this system, great quantities of real images are captured by the robot camera and stored in an image database. Besides recording the camera pose by camera model and Particle Filter method, an image interpolation method is adopted to get the images which are not included in database. With this system when a robot moves smoothly on the simulated soccer field, an image at current pose will be returned immediately. This work efficiently satisfies the need of vision code testing and extends the application area of simulation platform. In the future, we will enrich this system to make it even more realistic.

Acknowledgments The authors gratefully acknowledge Daniel Seifert for his knowledge of the project and other members of FUmanoid Team for providing the software base for this work. A video which is relevant to the chapter is linked: http://www.youtube.com/watch?v=TjjBYVMxZak.

References

1. H. Kitano et al. RoboCup: a challenge problem for AI and robotics. RoboCup-97: Robot Soccer World Cup I, (Springer, Heidelberg 1998), pp. 1–19
2. K. Asanuma, K. Umeda, R. Ueda, T. Arai, in *Development of a Simulator of Environment and Measurement for Autonomous Mobile Robots Considering Camera Characteristics*. Proceedings of robot soccer world cup VII (Springer, Heidelberg, 2003)
3. T. Ishimura, T. Kato, K. Oda, T. Ohashi, in *An Open Robot Simulator Environment*. Proceedings of robot soccer world cup VII (Springer, Heidelberg, 2003)
4. N. Jakobi, Minimal simulations for evolutionary robotics. PhD thesis, University of Sussex, 1998
5. Ziemke, On the role of robot simulations in embodied cognitive science. AISB J. **1**(4), 389–399 (2003)
6. M. Young, The Technical Writer's Handbook. Mill Valley, CA: Juan Cristobal Zagal and Javier Ruiz-del-Solar. Combining simulation and reality in evolutionary robotics. J. Intell. Robot Syst. **50**(1), 19–39 (2007)
7. J.C. Zagal, J. Ruiz-del-Solar, in *UCHILSIM: A Dynamically and Visually Realistic Simulator for the RoboCup Four Legged League*, vol. 3276 . RoboCup 2004: Robot soccer world cup VII, lecture notes in computer science (Springer, Berlin, 2004), pp. 34–45
8. J.C. Zagal, J. Ruiz-del-Solar, P. Vallejos, in *Back-to-Reality: Crossing the Reality Gap in Evolutionary Robotics*. IAV 2004: Proceedings 5th IFAC symposium on intelligent autonomous Vehicles, Elsevier Science Publishers B.V. AISB J. 1(4), 389–399 (2004)
9. J.C. Bongard, H. Lipson, in *Once More Unto the Breach: Co–Evolving a Robot and its Simulator*. Proceedings of the ninth international conference on the simulation and synthesis of living systems (ALIFE9), pp. 57–62
10. L. Iocchi, F. Dalla Libera, E. Menegatti, in *Learning Humanoid Soccer Actions Interleaving Simulated and Real Data*. Proceedings of the second workshop on humanoid soccer robots IEEE-RAS 7th international conference on humanoid robots, Pittsburgh, 2007
11. B Fischer et al. FUmanoid team description paper 2010. (Workshop Robocup Singapore 2010)
12. S. Thrun, D. Fox, W. Burgard, F. Dellaert. Robust Monte Carlo localization for mobile robots. Artif. Intell. **128**(1–2), 99-141 (2001)
13. R.E. Kalman, A new approach to linear filtering and predictionproblems. J. Basic Eng. **82**(1), 35–45 (1960)
14. M.J. Quinlan, R.H. Middleton, in *Comparison of Estimation Techniques Using Kalman Filter and Grid-Based Filter for Linear and Non-Linear System*. Proceedings of the international conference on computing: Theory and applications technique for RoboCup soccer (*ICCTA2007*) (1960)
15. M.J. Quinlan, R.H. Middleton. Multiple model kalman filters: a localization technique for RoboCup soccer. Lect. Notes Comput. Sci. **5949**, 276–287 (2010)
16. S. Thrun, W. Burgard, D. Fox, *Probabilistic Robotics* (MIT Press, Cambridge, 2005)
17. A.De Doucet, N. Freitas, N.J. Gordon, *Sequential Monte Carlo Methods in Practice* (Springer, Heidelberg, 2001)
18. M.S. Arulampalam, S. Maskell, N. Gordon, and T. Clapp, A tutorial on particle filters for on line nonlinear/non-gaussian bayesian tracking. IEEE Trans. Sig. Process. **50**(2) (2002)
19. T. Laue, T. Röfer, in *Pose Extraction from Sample Sets in Robot Self-localization-a Comparison and a Novel Approach*. Proceedings of the 4th European conference on mobile robots—ECMR'09, (Mlini/Dubrovnik, Croatia, 2009), pp. 283–288
20. T. Langner, Selbstlokalisierung für humanoide Fußballroboter mittels Mono-und Stereovision. Master thesis. FU Berlin, FB Mathematik und Informatik, Berlin. September 2009 (in German)

21. R. Douc, O. Cappe, E. Moulines, in *Comparison of Resampling Schemes for Particle Filtering*. ISPA 2005. Proceedings of the 4th international symposium on image and signal processing and analysis (2005), pp. 64–69
22. A. Desrosières, *The Politics of Large Numbers: a History of Statistical Reasoning*, Trans. Camille Naish (Harvard University Press, United State, 2004)
23. A. Björck, *Numerical Methods for Least Squares Problems* (SIAM, Philadelphia, 1996)
24. J. Nocedal, J. Stephen, *Wright Numerical Optimization* (Springer, Heidelberg, 1999)
25. D. Serfert et al. FUmanoid team description paper 2011. Workshop RoboCup Istanbul (2011)
26. RoboCup soccer humanid league rules and setup, http://www.tzi.de/humanoid/bin/view/Website/Downloads

Simulation of Humanoid Robot Vertebra

Mouna Souissi, Vincent Hugel and Pierre Blazevic

Abstract This chapter deals with 2D simulations of a humanoid robot equipped with backbone pitch joints to study the advantages of having such a mechanism for daily human-like movements. The movements under investigation here involve knee flexion for sitting down on a chair or picking up objects on the floor. The trunk is decomposed into a thorax and a lumbar part. As the lumbar region is the most mobile part in the human vertebral column, vertebrae are only placed in the robot's lumbar part. Simulations are carried out in the sagittal plane to investigate the influence of the number of vertebra pitch joint on the movements. Results show that a number of two pitch joints is a good tradeoff in matter of work at hip and thorax inclination. A parallel mechanism is proposed for the design of humanoid vertebra. This mechanism is inspired by a flight simulator system, and has been adapted and optimized to enable pitch and roll motion of a humanoid trunk at reduced energy cost.

Based on *Influence of the number of humanoid vertebral column pitch joints in flexion movements*, by Souissi et al., which appeared in the Proceedings of the 5th International Conference on Automation, Robotics and Applications (ICARA 2011).

M. Souissi (✉)
School of High Studies in Engineering, Site Balsan, 2 allée Jean Vaillé, 36000 Chateauroux, France
e-mail: mouna.souissi@hei.fr

V. Hugel · P. Blazevic
Versailles Engineering System Laboratory (LISV), University of Versailles, 10/12 avenue de l'Europe 78140 Vélizy, France

G. Sen Gupta et al. (eds.), *Recent Advances in Robotics and Automation*, Studies in Computational Intelligence 480, DOI: 10.1007/978-3-642-37387-9_22,

1 Introduction

Humanoid robots are expected to live with humans to assist them in their daily tasks. Safety and versatility will be a key feature for such applications. Current humanoid robots are not fully suitable for working in our daily environment. They can do a very few limited tasks compared with what a human does in the daily life. Humans unconsciously use their spine to achieve many movements like trunk flexion/extension, side bending, and yaw rotations.

Today humanoid robots equipped with degrees of freedom (DOF) in the trunk can be classified into three categories. The first category regroups robots designed to walk. Serial rotary joints were added in the trunk to enable them to achieve other movements like sitting down, lying down and standing up. These robots have one DOF [1–4], two DOF [5–7] or four DOF [8] in the trunk. All of them feature a yaw rotary joint, then one pitch joint and roll joints. The second category of robots have no legs but a wheeled base. They are equipped with an articulated trunk whose kinematics is different from the kinematics of walking robots and was developed to increase manipulation skills. The DOF in the trunk are therefore task-dependent. In addition to the yaw joint they generally feature more than two pitch joints [9, 10]. The third category regroups robots in development equipped with a more sophisticated spine, that tends to mimic human spine flexibility [11–14]. They feature vertebrae with a shape similar to human vertebrae. Tendons are used to drive the body like muscles. Unfortunately, it is difficult to make these robots walk.

This chapter investigates the influence of the number of pitch joints in flexion movements and tries to determine how many vertebrae are needed for a humanoid robot to execute the task of sitting down. In all walking humanoids equipped with an articulated spine the mechanism in the trunk is composed of serial links. In this study we propose to use a parallel mechanism that can be placed in the vertebral column of humanoid robots taking into account bending constraints. This mechanism was imagined for the design of a flight simulator [15] and is adapted for a humanoid vertebra to reduce energy costs.

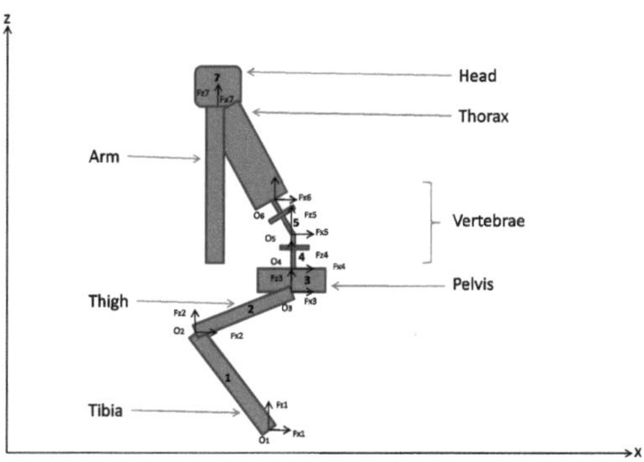

Fig. 1 The robot' model is constituted by rectangular segments, namely head, trunk, pelvis, femurs, tibias and arms. There is no foot. Foot mass was taken into account in the tibias. Arms are connected to the shoulders at the top of the thorax. The trunk is composed of a thorax and a lumbar part

Table 1 Masses and heights of body parts ($\delta_j = 1$ if $N_j \neq 0$).

Part	Masses (Kg)	Height (m)
Head	$0.08\ M_T$	$0.07\ H_T$
Arms(2)	$0.1\ M_T$	$0.47\ H_T$
Tibias(2)	$0.12\ M_T$	$0.27\ H_T$
Thighs(2)	$0.18\ M_T$	$0.22\ H_T$
Pelvis	$0.02\ M_T$	$0.04\ H_T$
Trunk	$(0.5 + 0.01\delta_j)\ M_T$	$0.4\ H_T$
Vertebra pitch joint	$0.01\ M_T$	$\frac{1}{12}\ H_{Trunk}$
Lumbar part	$N_v\ 0.01\ M_T$	$H_{Trunk} - H_{Thorax}$
Thorax	M_{Trunk}	$(1 - N_v \frac{1}{12})H_{Trunk}$

N_v is the number of vertebrae. Total height H_T is 1.40 m. Total mass M_T is 40 Kg without pitch joints in the vertebral column. This mass increases by 0.01 times M_T for each pitch joint. The mass of the thorax is equal to the mass of the trunk when the robot has no articulated spine. In the case where there is no joint in the trunk, the pelvis is fixed to the thorax

The chapter is organized as follows. Section 2 deals with the robot's model. Section 3 is dedicated to trajectory planning and inverse kinematics. Section 4 presents simulation and results and Sect. 5 describes the parallel mechanism proposed for the vertebra.

2 Robot's Model

Figure 1 depicts the robot model used for simulation in the sagittal plane (x, z). Because human vertebrae in the lumbar part are responsible for most of trunk movements [16] rotary joints are only placed inside the lumbar part of the humanoid trunk. Masses and heights of the different body parts are derived from a human kid-size model (Table 1).

In order to investigate the influence of the number of vertebrae on sitting movement and knee flexion, we conduct a dynamics study of the model in the sagittal plane using Newton-Euler equations. Table 2 explicits the notations used.

For segment i the forces applied are listed below:

$$-\boldsymbol{F}^i, \boldsymbol{F}^{i+1}, \boldsymbol{F}^{i+1,+}, M^i\boldsymbol{g}$$

where $\boldsymbol{F}^i$ is the force exerted by segment i on segment $i-1$, and $\boldsymbol{F}^{i+1,+}$ is the additional force exerted by arms on segment i:

$$\boldsymbol{F}^{i+1+} = \delta_{i,thorax}.M^a.(\boldsymbol{g} - \ddot{\boldsymbol{x}}_{G^a} - \ddot{\boldsymbol{z}}_{G^a}) \tag{1}$$

where *thorax* is the index of the thorax part. It is assumed that arms remain always vertical. As as consequence, arms' acceleration is the same as acceleration of attachment point at shoulder which is point O^{head}.The dynamics that involves forces and acceleration for segment i is written as:

Table 2 Notations

Symbol	Signification
M^i, M^a	Mass of segment i, resp. both arms
L^i	Length of segment i
$I^i_{G^i}$	Moment of inertia of segment i at G^i
$I^i_{O^i}$	Moment of inertia of segment i at O^i
α^i	Angle of segment i with vertical
β^i	Joint angle, between segment $i-1$ and i
$\ddot{\alpha}^i$	Absolute angular acceleration of segment i at O^i
G^i, G^a	Center of mass of segment i, resp. both arms
$x_{G^i}, z_{G^i}, \ddot{x}_{G^i}, \ddot{z}_{G^i}$	x and z position and acceleration of G^i
$x_{G^a}, z_{G^a}, \ddot{x}_{G^a}, \ddot{z}_{G^a}$	x and z position and acceleration of G^a
g	Gravitation constant
F^i_x, F^i_z	x, z components of force applied by seg. i to $i-1$
F^{i+}_x, F^{i+}_z	x, z components of force exerted by arms on seg. $i-1$
W^i	Torque at point O^i applied by segment i to $i-1$
$E\upsilon$	Energy of vertebrae joints

i refers to body segment numbered i

$$-\boldsymbol{F^i} + \boldsymbol{F^{i+1}} + \boldsymbol{F^{i+1,+}} + M^i\mathbf{g} = M^i(\ddot{\boldsymbol{x}}_{G^i} + \ddot{z}_{G^i}) \tag{2}$$

The projections onto x-axis and z-axis give:

$$-F^i_x + F^{i+1}_x - \delta_{i,thorax}.M^a.\ddot{x}_{G^a} = M^i\ddot{x}_{G^i}$$
$$-F^i_z + F^{i+1}_z - \delta_{i,thorax}.M^a.(g + \ddot{z}_{G^a}) - M^i g = M^i\ddot{z}_{G^i}$$

For segment i the applied torques and moments of external forces are listed below:

$$-\boldsymbol{W^i}, \boldsymbol{W^{i+1}}, \boldsymbol{M_t}(M^i.g), \boldsymbol{M_t}(F^{i+1}), \boldsymbol{M_t}(F^{i+1,+})$$

where

- W^i is torque applied by segment i on segment $i-1$,
- $\boldsymbol{M_t}(M^i\mathbf{g}) = \boldsymbol{O^iG^i} \times M^i\mathbf{g}$,
- $\boldsymbol{M_t}(F^{i+1}) = \boldsymbol{O^iO^{i+1}} \times \boldsymbol{F^{i+1}}$,
- $\boldsymbol{M_t}(F^{i+1,+}) = \boldsymbol{O^iO^{i+1}} \times \boldsymbol{F^{i+1,+}}$,

The dynamics that involves torques, moments of external forces and angular acceleration for segment i with reference point O_i is expressed as:

$$\begin{aligned} I^i_{O^i}\ddot{\alpha}_i = & -\boldsymbol{W^i} + \boldsymbol{W^{i+1}} + \boldsymbol{M_t}(M^i\mathbf{g}) \\ & + \boldsymbol{M_t}(F^{i+1}) + \boldsymbol{M_t}(F^{i+1,+}) \end{aligned} \tag{3}$$

where $I^i_{O^i}$ is the moment of inertia of segment i relative to O_i: $I^i_{O^i} = I^i_{G^i} + M^i(\frac{L^i}{2})^2$. The projection onto lateral axis gives:

$$W^{i+1} = W^i - M^i g(x_{G^i} - x_{O^i}) + I^i_{O^i}\ddot{\alpha}_i + (x_{O^{i+1}} - x_{O^i})(F_z^{i+1} + F_z^{i+1,+}) - (z_{O^{i+1}} - z_{O^i})(F_x^{i+1} + F_x^{i+1,+}) \tag{4}$$

Static torques can be obtained by removing acceleration in Eq. 4.

3 Trajectory Planning and Inverse Kinematics

In the initial position the robot is standing up, with knees slightly flexed.

The first motion consists of sitting down on the chair. It is composed of two phases, one phase of sitting with the torso bent forward, and a second phase to put the torso straight again once seated. In the first phase the trajectory of the pelvis is given as input so that the pelvis goes to the seat. The second motion consists of bending down while flexing knees to be able to pick up some object on the floor. For all movements the following constraints are used:

- the pelvis trajectory is defined in the (x, z) plane. Pelvis acceleration is generated using a 3rd order polynomial to have continuity of jerk. Velocity and position are obtained by integration using initial and final zero velocities, and distance between initial and final positions.
- the pelvis is kept horizontal along motion. This is always possible except when there is no joint in the vertebral column. In this case the robot must use its hip joints to bend torso forward.
- robot's mass center longitudinal coordinate x_G is kept constant.

The height of the chair is 80 % of tibia length, and the length of the chair is 80 % of thigh length. The chair is placed close to the ankle at a distance of 5 % of the tibia length.

The motion algorithm uses the pseudo-inverse technique. The input trajectory to be tracked is defined as

$$F = \left[x_G, x^{pelvis}, z^{pelvis}, \theta^{pelvis}\right] \tag{5}$$

Given the input above, the algorithm must compute all joint angles (β^i):

- At t = 0, calculate initial $\beta^i_{t=0}$, $F_{t=0}$, and Jacobian $J_{t=0}$.
- For each next position F_{next} of input trajectory,

  1. $\Delta F \leftarrow F_{next} - F(t)$.
  2. Calculate pseudo-inverse $J^+$from Jacobian J.
  3. $\Delta\beta \leftarrow J^+ \, \Delta F$.
  4. $\beta(t + \Delta t) \leftarrow \beta + \Delta\beta$.
  5. Calculate $F(t + \Delta t)$ using $\beta(t + \Delta t)$ from forward geometric model.
  6. Calculate Jacobian J using $\beta(t + \Delta t)$ for next step.

The algorithm tends to minimize both quantities $||\Delta\beta||^2$ and $||\Delta F - J\Delta\beta||^2$.

4 Simulation and Results

Simulations aim to determine the influence of the number of vertebra pitch joints in matter of hip work and thorax inclination. The same experiment of sitting down was carried out with 0, 1, 2, 3, 4, 5, 6, 7 and 8 joints in the spine. Figure 2 is related to the execution of the motion of sitting down on a chair—1st phase. In the first phase of sitting, the upper part of the robot is tilted forward to keep stability. The movement is continued in the second phase to put the upper body straight while being seated. In this case the final coordinate of center of mass x_G is set to $x_G = 0.08\ \text{m}$ to reach a final upright position of the trunk.

Hip work. This work is more important for the robot with rigid trunk except at the end of the movement compared with configurations where there are more than 2 pitch joints. A number of two pitch joints seems to be a good tradeoff (Fig. 3, left). However, these results may depend on the mass distribution in the trunk part. Here, it is assumed that half of the total mass is located in the thorax part in presence of vertebra joints.

Spine work. This work is less important for a column composed of a single joint. The more vertebrae, the more the work increases (Fig. 3, right). However the spine work reaches a peak when there are more than 4 joints.

Inclination of thorax. Figure 4 shows the trunk inclination increase as a function of bending angle.

According to the experiments we propose to keep two joints for the design of a robotic vertebral column and to place them in the lumbar part.

The simulations of knee flexion were carried out with only three vertebrae, i.e. four joints. Compared with previous experiments, we use the same algorithm but change the final position of the pelvis so that the robot can bend on its knees to reach the ground with its hands. This can be useful to grasp some objects from the floor. We obtain the same results as in the previous simulation of sitting down motion. In addition, the mass distribution between thorax and lumbar part is an

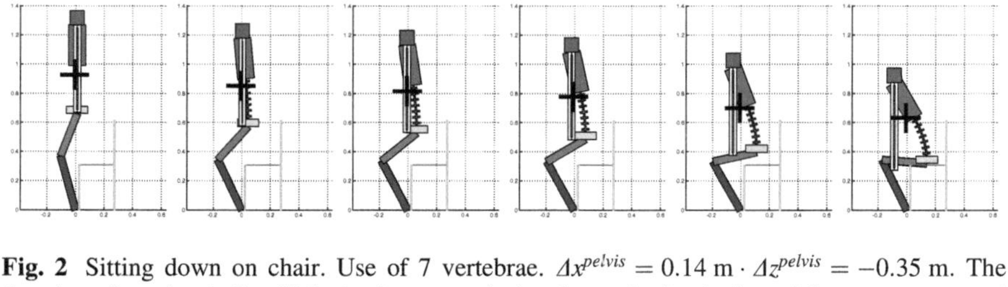

Fig. 2 Sitting down on chair. Use of 7 vertebrae. $\Delta x^{pelvis} = 0.14\ \text{m} \cdot \Delta z^{pelvis} = -0.35\ \text{m}$. The duration of motion is 3 s. Velocity increases during 1 s at the beginning of the movement, and decreases for 1 s at the end

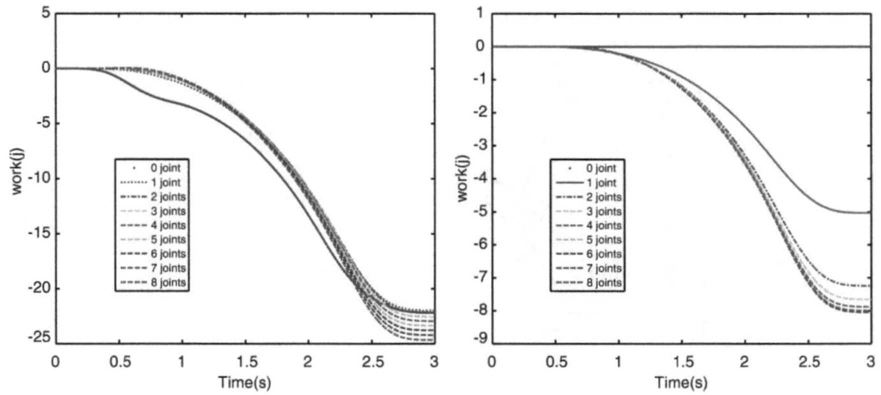

Fig. 3 Hip work (*left*) and spine work (*right*) for sitting down motion. The number of vertebrae joints varies from 0 to 8. Spine work is calculated as $E_\upsilon = \sum_{k=1}^{N\upsilon j} W_k \Delta\beta_k$

Fig. 4 Inclination (degrees) of thorax in sitting down motion for 0 to 8 joints. Bending angle increases significantly when the number of vertebra joints increases. From two joints, there is an increase of nearly 60 % of thorax inclination compared with the configuration without articulated spine

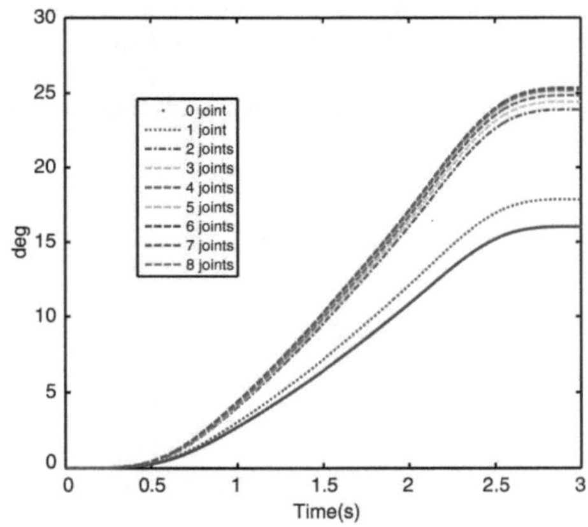

important factor that influences the inclination of the thorax. If the thorax weight is too heavy the robot will not be able to pick up objects on the floor because inclining the thorax accordingly would be impossible.

5 Model of Parallel Mechanism

In order to reduce the maximal torque required at each pitch and roll joints, we propose to use a parallel mechanism instead of a serial mechanism that equip most humanoid robots.

The parallel mechanism [15] is described in Fig. 5. The top platform can be pitched about fixed axis $\boldsymbol{j}$ by angle θ_{10}, and rolled about axis $\boldsymbol{i}'$ by angle θ_{21}. The

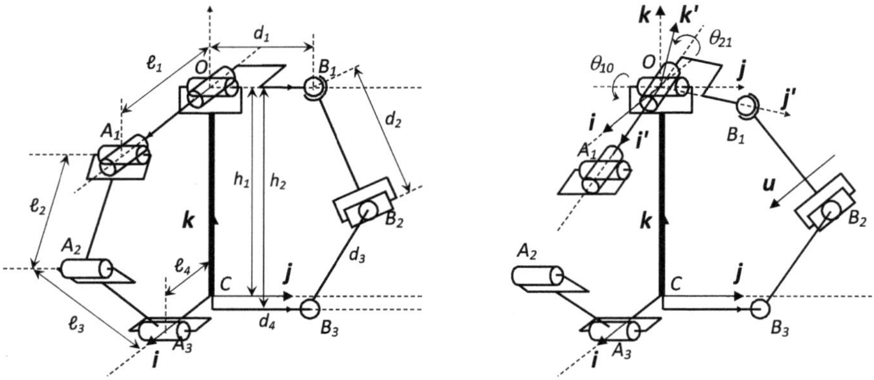

Fig. 5 Perspective view of parallel mechanism in initial position and after pitch and roll rotations of respectively θ_{10} and θ_{21} angles. The *central rod* is fixed and rigid. It is attached to the *top platform* by a Universal joint at O. The mechanism is actuated by two revolute joints, each of them is located at the *bottom* of each arm. The two arms are arranged at 90°. The arm $A_1A_2A_3$ is planar and remains in the (xz) plane. It is composed of two revolute joints at A_2 and A_3 and one Universal joint at A_1. The other arm $B_1B_2B_3$ is initially in the (yz) plane, and does not remain in this plane if the *top platform* rolls. The arm $B_1B_2B_3$ also includes two segments, one revolute joint at B_3, one Universal joint at B_2 and one ball joint at the attachment locus B_1 with the *top platform*

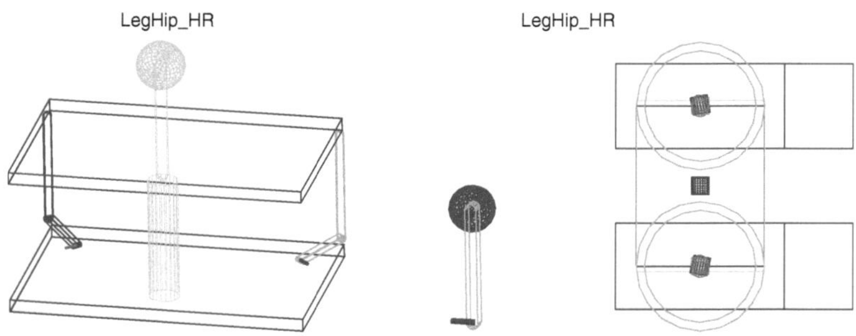

Fig. 6 ADAMS parallel model (*left*) and and serial inverted pendulum (*middle*). *Top view* of robot's trunk and footprints (*right*). For the parallel mechanism, the following parameters were adopted: $d_1 = l_1 = 0.17$ m, $d_2 = l_2 = 0.09$ m, $d_3 = l_3 = 0.04$ m, $d_4 = l_4 = 0.16$ m

two active joints are the revolute joints at A_3 and B_3. The associated rotation angles are denoted by α and β.

This mechanism is placed inside the lumbar part of the humanoid trunk in such a way that both arms are located at the back (Fig. 6). This enables to move the central rod forward to place it more in the center of the support polygon delimited by the footprints (see Fig. 6, right). This also enables to have longer lever arms. A sphere with a mass M is placed above the mechanism to simulate the upper body of the humanoid robot. The height between the top platform and the bottom platform in the parallel mechanism is set to 0.1 m. Lengths d_1 and ℓ_1 were set according to the space volume available in the lumbar part of the trunk. The parameters $d_2 = l_2$, $d_3 = l_3$ and $d_4 = l_4$ of the parallel mechanism were optimized manually by testing several configurations [17] to check which one was best in matter of torque magnitude and space occupancy. Each arm was tested separately by fixing the other one. We concluded that motor axes should be located more externally and that the upper part of each arm must be longer than its lower part. This enables to optimize the lever arm.

We aim to compare the parallel mechanism with the equivalent serial mechanism in terms of kinematics, and show that the parallel mechanism is better than the serial mechanism in matter of torque consumption. The serial mechanism is an inverted pendulum (Fig. 6, middle) where the pitch-roll joint is placed at the bottom. The distance between the sphere center and the pitch-roll joint center is the same as the distance joining the top platform and the sphere center in the parallel mechanism. The reference motion used for comparison is a circular motion, where the mass is bent 10° initially in the sagittal plane, then makes a complete turn about the center while keeping constant bending, and goes back to the initial position.

Angles trajectories are planned with Matlab-SIMULINK (Fig. 7), which is used to control the ADAMS model. To control the inclination of the spherical mass in Adams PID regulation was used for the motors.

As a result the maximal torque magnitude is five times greater in the serial mechanism than in the parallel mechanism (Fig. 8). The use of a parallel mechanism can lead to a significant reduction in the maximal torque required to drive the system. This study was carried out with one vertebra only. The piling up of vertebrae will require more power from the actuators of the spine if the trunk is to be inclined more than 10°. The next steps of this research will consist of the

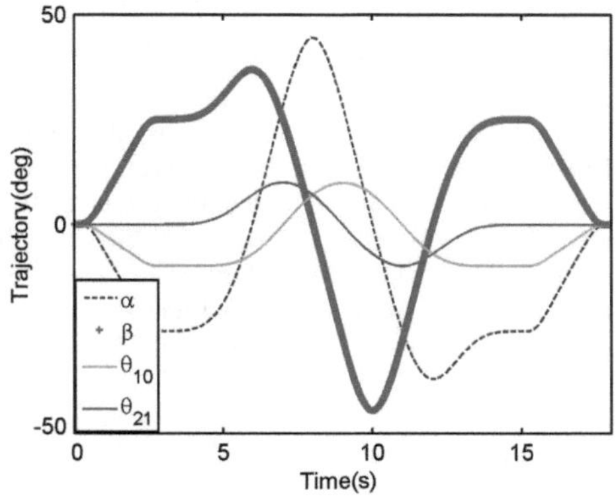

Fig. 7 θ_{10} and θ_{21} angle trajectories of platform for the reference motion and motor angles α and β for the parallel mechanism. α and β are calculated thanks to the inverse geometric model. In the case of the serial mechanism θ_{10} and θ_{21} are directly sent to the pitch and roll joint motors located at the *bottom point* of the pendulum

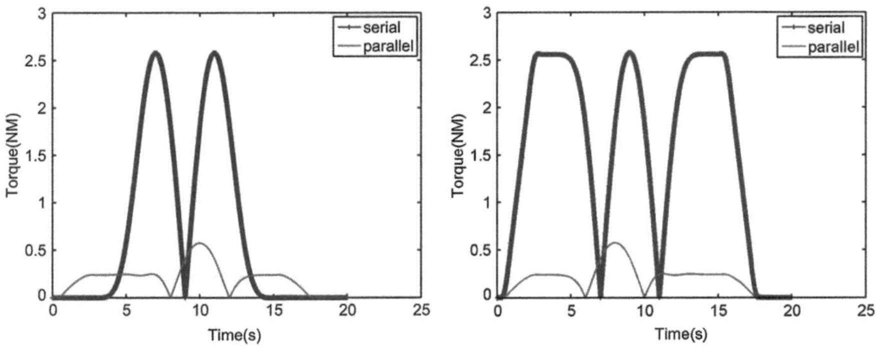

Fig. 8 Active torques in serial and parallel mechanism for the circular motion. *Left* pitch joint of serial mechanism and 2D arm motor joint of parallel mechanism. *Right* roll joint of serial mechanism and 3D arm motor joint of parallel mechanism

precise optimization of the parallel mechanism. The study will also be extended to take into account several piled parallel mechanisms to build a vertebral column for humanoid robots.

6 Conclusion

Simulation experiments of sitting down and knee flexion carried out in the sagittal plane have shown that two pitch joints in the vertebral column is a good tradeoff. The experiment of knee flexion for picking up objects on the floor has even emphasized the importance of trunk mass distribution between the thorax part and the lumbar part. Additional simulation experiments with a parallel mechanism compared with a serial mechanism show that the parallel mechanism allows to reduce active torques significantly after optimizing the layout of the parallel arms.

Acknowledgments This research is supported by the French Single Inter-Ministry Fund ROMEO. The ROMEO Project is a R&D project led by the French company Aldebaran Robotics. It involves twelve partners, including four SMEs, seven academic partners, and one user representative. Half of the 10€M global budget is funded by the French Government, the Ile de France region and the City of Paris. The aim of Romeo is to develop a 1.4 m high humanoid robot that is designed to be an assistant and a companion for elderly people at home.

References

1. Y. Sakagami, R. Watanabe, C. Aoyama, S. Matsunaga, N. Higaki, K. Fujimura, The intelligent ASIMO: system overview and integration, in *IEEE/RSJ International Conference on Intelligent Robot and System*, vol. 3 (2002), pp. 2478–2483

2. J. Kim, I. Park, J. Lee, M. Kim, B. Cho, J. Oh, System design and dynamic walking of humanoid robot KHR-2, in *IEEE/RSJ International Conference on Robotics and Automation* (2005), pp. 1431–1436
3. K. Loffler, M. Gienger, F. Pfeiffer, Sensor and control design of a dynamically stable biped robot, in *Proceedings of IEEE International Conference on Robotics and Automation* (2003), pp. 484–490
4. S. Lohmeier, T. Buschmann, H. Ulbrich, F. Pfeiffer, Modular joint design for performance enhanced humanoid robot lola, in *Proceedings of the IEEE International Conference on Robotics and Automation* (2006), pp. 88–93
5. R. Tellez, F. Ferro, S. Garcia, E. Gomez, E. Jorge, D. Mora, D. Pinyol, J. Poyatos, O. Torres, J. Velazquez, D. Faconti, Reem-B: an autonomous lightweight human-size humanoid robot, in *8th IEEE-RAS International Conference on Humanoids* (2008), pp. 462–468
6. K. Kaneko, F. Kanehiro, S. Kajita, H. Hirukawa, T. Kawasaki, M. Hirata, K. Akachi, T. Isozumi, Humanoid robot HRP-2, in *Proceedings of the IEEE International Conference on Robotics and Automation* (2004), pp. 1083–1090
7. E. Yoshida, J.P. Laumond, Motion planning for humanoid robots: highlights with HRP-2, in *French National Conference on Robotics Research, JNRR* (2007), pp. 105–118
8. Y. Ogura, H. Aikawa, K. Shimomura, H. Kondo, A. Morishima, Development of a new humanoid robot WABIAN-2, in *Proceedings of the 2006 IEEE International Conference on Robotics and Automation* (2006), pp. 76–81
9. T. Wimbock, D. Nenchev, A. Albu-Schaffer, G. Hirzinger, Experimental study on dynamic reactionless motions with DLR's humanoid robot justin, in *Proceedings of IROS* (2009), pp. 5481–5486
10. C. Torrey, How robots can help communication strategies that improve social outcomes. Dissertation. CMU-HCII-09-102 (May 2009)
11. I. Mizuuchi, S. Yoshida, M. Inaba, H. Inoue, The development and control of the flexible-spine of a human-form robot. Adv. Robot. **17**(2), 179–196 (2003)
12. T. Yoshikai, I. Mizuuchi, D. Sato, S. Yoshida, M. Inaba, H. Inoue, Behavior system design and implementation in spined musle-tendon humanoid 'Kenta'. J. Robot. Mechatron. **15**(2), 143–152 (2003)
13. I. Mizuuchi, Y. Nakanishi, Y. Sodeyama, Y. Namiki, T. Nishino, N. Muramatsu, J. Urata, K. Hongo, T. Yoshikai, M. Inaba, An advanced musculoskeletal humanoid Kojiro, in *Proceedings of the 2007 IEEE-RAS International Conference on Humanoid Robots (Humanoids)* (2007), pp. 294–299
14. I. Mizuuchi, A musculoskeletal flexible-spine humanoid Kotaro aiming at the future in 15 years' time, in *Mobile Robots Towards New Applications*, ed. by A. Lazinica (ARS/plV, Germany, 2006)
15. E. Sabrié, Analyse d'un mécanisme de simulation de vol sphérique et son contrôle en temps reel. Faculté des sciences et de génie universitaire, Laval, Québec (2004) (In French)
16. I.A. Kapandji, Physiology of the joints: the trunk and the vertebral column, vol. 3 (Feb 2007)
17. M. Souissi, V. Hugel, P. Blazevic, Design optimization of parallel joint mechanism for humanoid spine, in *The 16th IEEE Mediterranenan Electrotechnical Conference* (March 2012) (In Tunisia), pp. 997–1000. http://ieeexplore.ieee.org/xpl/login.jsp?tp=&arnumber=6196595&url=http%3A%2F%2Fieeexplore.ieee.org%2Fiel5%2F6191396%2F6196362%2F06196595.pdf%3Farnumber%3D6196595

Flexible Automation in Facility Logistics

Andreas Kamagaew, Jonas Stenzel and Michael ten Hompel

Abstract The concept of "Flexible Automation in Facility Logistics" with a Cellular Transport System shows the possibilities of automation and enhancement of flexibility and changeability in facility logistics systems. Additionally, the ease of use of complex decentralized control systems is shown by the utilization of Multi-Agent-Systems. This chapter shows how to improve these issues compared to conventional facility logistics systems, e.g. static conveyors or manual transportation, by using an autonomous vehicle swarm with a decentralized control architecture. Cellular Transport Systems are based on dedicated transportation entities (cells). Generally, these cells consist of Autonomous Transport Vehicles (ATVs) or autonomous conveying modules. Various functions such as advanced sensor/actuator interoperation, highly reliable communication, localization and energy management are implemented in each of these cells, facilitating different forms of adaptive, anticipatory and collective behavior.

Based on "Concept of Cellular Transport Systems in Facility Logistics", by Andreas Kamagaew, Jonas Stenzel, Andreas Nettsträter, and Michael ten Hompel which appeared in the Proceedings of the 5th International Conference on Automation, Robotics and Applications (ICARA 2011).

A. Kamagaew (✉) · J. Stenzel
Fraunhofer Institute for Material Flow and Logistics,
Dortmund, Germany
e-mail: andreas.kamagaew@iml.fraunhofer.de

M. t. Hompel
Chair for Materials Handling and Warehousing,
TU Dortmund—University of Technology, Dortmund, Germany

G. Sen Gupta et al. (eds.), *Recent Advances in Robotics and Automation*, 301
Studies in Computational Intelligence 480, DOI: 10.1007/978-3-642-37387-9_23,

1 Introduction

Productivity and flexibility are worldwide key challenges for industries and supply chains in terms of competitiveness and dynamics in global markets. Practical experience shows that highly structured industrial environments are very costly and time consuming in terms of changing or adapting such systems. Today, no system meets the tremendous challenge of continuously varying requirements for facility logistics and is simultaneously able to increase the degree of automation, flexibility and changeability. During the last years, research activities in decentralized material flow control connected with the Internet-of-Things architecture in logistics have been growing (cf. [1–3]). The basic idea of these concepts is the distribution of the prevailing central controls to a multiplicity of smaller self-organizing decentralized control units. Decisions are made autonomously by the decentralized units based on local and probabilistic information. Today, hierarchical structures are dissolved towards a mesh-like structure with self-containing entities. In material flow systems, these entities represent logistical objects: conveyors, unit loads or AutoID devices. Every entity acts autonomously and communicates with other entities. The communication between them is service-based. The holistic system function is then achieved through composition of different services from different entities. Encapsulation of construction and design together with the use of services as the systems interface ensures much higher flexibility. The entities are easily exchangeable and even replaceable by entities from different vendors. All previous Internet-of-Things projects were based on static conveyor systems in which unit loads are forced to use given ways and routes through the system. But the flexibility and autonomy is limited by the physical build-up of the conveyor system. The next step is the replacement of the conveyor system by a group of transport vehicles which are able to pick up unit loads and to have open path navigation. This so called Cellular Transport System is intended to overcome this trade-off by adding value to and by enhancing autonomy of sub-elements like Automated Guided Vehicles by the use of a Multi-Agent-System [4] and Swarm Intelligence [5].

A Cellular Transport System is an application of Swarm Algorithms to control the vehicles' interaction behavior. Moreover, optimization can be attained by using Swarm Algorithms for localization, navigation, collision avoidance, task allocation and transportation tasks. With this capability of cooperation in a swarm, different transport units can be carried in a collective of vehicles. So the flexibility and performance of transportation tasks is increased by using a swarm of vehicles for example by shared sensor information for predictive collision avoidance or interaction between the vehicles. As a main result, an entirely new degree of flexibility for designing logistical systems is achieved in terms of cost-efficiency and flexibility enabling a sustainable competitive advantage. An overview of this new system design will be given in the course of this chapter.

2 Technology

Facility Logistics is a multi-disciplinary domain and benefits from several fields of expertise like Mathematics, Computer Science, Automation, Robotics, different Engineering fields etc. A selective overview of our methods to build-up a Cellular Transport System is described in the following sections.

2.1 Cellular Transport Systems

Cellular Transport Systems are based on material handling or transportation entities. These are Autonomous Transport Vehicles or autonomous conveying modules. The control and the communication between these entities are fulfilled by a Multi-Agent-System (MAS). Cellular Transport Systems are flexible in terms of their topology and are able to adapt to environmental changes. Finally, this ensures the overall transport systems' performance due the interaction between the material handling entities [6].

2.2 Swarm Intelligence

Swarm Intelligence in artificial systems composes many individuals that operate using decentralized control and self-organization. It focuses on the behavior of the collective, which results from the interactions of the individuals with each other and with their environment. Like in natural swarms (e.g. fish swarm) no central entity is needed for the coordination of the individuals, so task organization and task allocation can be managed in a decentralized way [5].

2.3 Autonomous Behavior

The ability of making decisions based on given rules and observations defines the degree of autonomy in Robotics. The Autonomous Behavior of a system is based on autonomous and independent control of an entity in order to enable this entity to solve a task, e.g. transportation of goods or exploration of an area. With its limited level of intelligence a transport entity is able to accomplish tasks and react to unexpected changes of the environment without any interaction to other entities.

2.4 Sensor Fusion

The main purpose of Sensor Fusion is to achieve more precise measurements based on probabilistic approaches with several sensors than by particular evaluation of the sensors. So Sensor Fusion enables the reduction of sensors or the improvement of the measurement accuracy. It can be accomplished by Fuzzy Logic, Kalman, Extended Kalman, Particle Filter algorithms. Usually a sensor model is used for this purpose—especially for predictive algorithms [7]. Eventually, Sensor Fusion Algorithms can be used for localization, collision avoidance and object classification in the Cellular Transport Systems.

2.5 Multi-Agent-Systems

The main objective of Multi-Agent-Systems is to control complex systems without a central entity. These systems employ a set of agents that are autonomous, proactive, adaptive and reactive. Software agents own special competences to solve problems, acquire information, prevent conflicts etc. and are able to interact with other agents to solve the global problem [4].

3 System Architecture

Our approach to an adaptable, flexible and scalable facility logistics automation system is the MultiShuttle Move (MSM) (see Fig. 3). The MSM is a novel fusion of conventional shuttle and automated guided vehicle systems developed by Fraunhofer Institute for Material Flow and Logistics. The new vehicle is rail-guided while it is located in a racking system or a lift. Furthermore, the MSM is able to leave the rail-system and to operate as an AGV with open path navigation (the so called ATV). This navigation is based on Sensor Fusion of Round-trip Time-of-Flight ranging within a Wireless Sensor Network (WSN), measurements of two laser rangefinders and dead-reckoning. The MSM is provided with a differential drive which has a maximum speed of 1.1 m/s on the ground and 2.1 m/s in the rail-guided racking system—the higher velocity in the rail is realized by a gear. The maximum acceleration is limited to 2 m/s^2. A single vehicle weighs no more than 134 kg, but can transport payloads as high as 40 kg. Through the mutual concept different load handling devices (LHD) can be assembled on the ATV—at this stage a telescope LHD, a belt conveyor for double cycle LHD (simultaneous loading and unloading) and a passive non-powered LHD. The system topology and vehicle control architecture of the MSM are described in the following subsections.

3.1 System Topology

The system topology of the MSM environment is modeled as a set of different software agents (entities) where each agent has a unique, configurable behavior. These entities are

- *ATV*—each MSM represented by its control architecture that interacts with the following environmental subsystems
- *area*—mainly the whole floor area in the logistics facility where the MSM operates with Open-Path navigation
- *rack*—the high bay racking where the vehicles act like a conventional shuttle system
- *lift*—lifts in the racking system which are acting as vertical conveyors to serve all levels of the high bay racking
- *picking station*—interface between the MSMs and other facility logistic systems where the order picking takes place
- *job*—entity that represents a transport order for a respective unit load

The user of the Cellular Transport System is able to change the topology and behavior of the system entities, including the arrangement of waypoints, picking stations and the layout of the high bay racking. Waypoints may also include further topology attributes, e.g. a necessary reservation for this point. This information is editable through a graphical topology editor even during operation; it is stored as an XML "map" of the environment. For every transport order that is generated by a higher level system or the user itself, a job agent is initiated which passes in-formation like a sequencing priority to the transport entities whereas each entity is an ATV. Afterwards, each ATV moves a unit load from a source point to its destination while it communicates with other entities like lifts, picking stations or other ATVs. The ATV-to-ATV communication allows for predictive local collision avoidance, communication to the lift enables the ATV to move to different levels of the high bay racking (Fig. 2).

3.2 Vehicle Control Architecture

The layout of the software control of the ATV (cf. [6], [8, 9]) is split in three different layers (see Fig. 3), all implemented on an energy-saving industrial PC mounted on the ATV:

The sensor and actuator layer (SA layer) at the lowest level is responsible for all short-term tasks. This layer contains all tasks which are strongly coupled to the hardware and time critical closed-loop controllers, like fieldbus communication, sensor data acquisition and the closed-loop drive control for vehicle positioning. All safety relevant tasks are also located in this layer and operate decoupled from

Fig. 1 Multishuttle move system

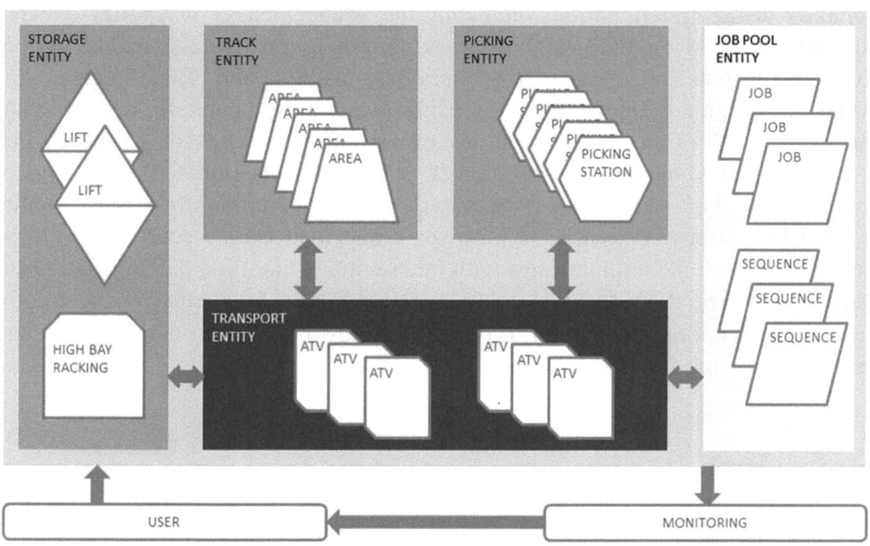

Fig. 2 Multi-agent-system based system overview

the rest of the control system to meet safety requirements. The tasks are executed in a software programmable logic controller.

The operational layer is responsible for mid-term tasks, like local collision avoidance, localization and local path planning. The collision avoidance identifies autonomously possible obstacles with the help of the sensor data from the SA layer and initiates a re-planning of the actual path or a reduction of the velocity

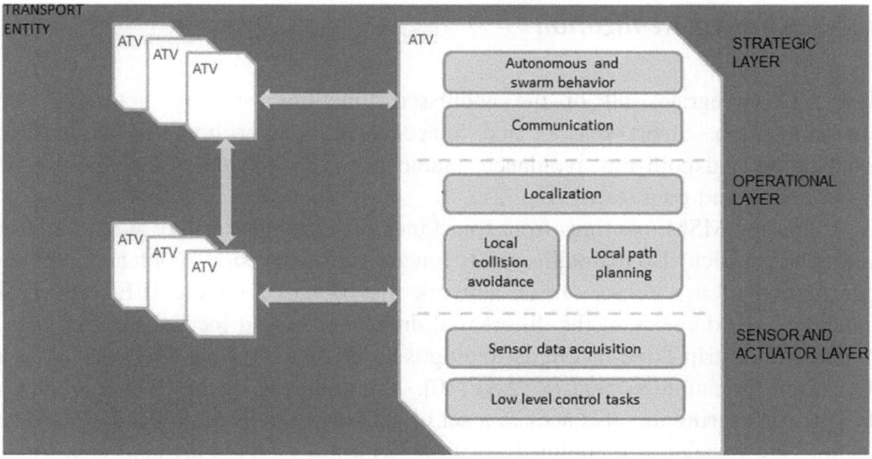

Fig. 3 Transport entity

depending on the actual situation. The actual state is available for the strategic layer or even external systems. Information of interest is the current state of charge, the actual position and velocity of the vehicle or the load condition. All tasks of this layer run as threads in the vehicle Operating System.

The strategic layer at the top is responsible for long-term operation and planning. The objective of this layer is the creation of an autonomous behavior, which is realized via a software agent. Therefore every vehicle is represented by a software agent, which is able to communicate and interact with other agents and external systems, like job management. The agent has a communication interface to all tasks of the operational layer. The connection to the SA layer is done via the ATV state task. The task autonomous behavior runs, like the operational layer tasks, in the vehicle Operating System.

An example for interaction of all three internal layers and the external layer is outlined in the following scenario: through the job management, the autonomous and swarm behavior task of one ATV receives information about a unit load, which needs to be transported. The task starts the routing based on the current position of the ATV and source-sink relations. The planning results define one or more destination and navigation-support points. These points are transferred to the path planning task (operational layer), which plans the trajectory and pose of the vehicle. The results of path planning are transferred into commands for position control (SA layer). The position control then generates control values for the drives of the ATV. This scenario shows how the MSM system represents the autonomous and decentralized concept of the Internet-of-Things architecture and the Cellular Transport System.

3.3 Technical Realization

The ATV integrates all of the addressed functions: sensing, actuating and communication. Smart sensors and Sensor Fusion algorithms for navigation, localization and collision avoidance enable the ATV to detect and analyze its environment and own status (Fig. 2).

Within the MSM, readings from four kinds of sensors are fused: A safety laser rangefinder which also maintains safety functions like collision avoidance, ranging information of a wireless sensor network that is based on the IEEE 802.15.4 standard, the odometry of the differential drive motors and inertial sensors.

The Round-trip Time-of-Flight ranging is based on a WSN which can work as a Real-Time Location System (RTLS) [10]. A mobile tag in this RTLS localizes itself by measuring the distances to a set of anchors used as reference points. The anchors are located at predefined positions within a global Cartesian coordinate system. The tag position can be calculated afterwards by trilateration. At least three anchors are needed for an unambiguous localization. The accuracy of the Round-trip Time-of-Flight ranging was improved by applying an Extended Kalman Filter (EKF) on the range measurement data within the MSM system. The implementation is described in [11].

3.4 Localization

The On-Rail localization of the vehicle is done by using dead-reckoning and light barriers for the slot detection in the multifunctional rail. When driving off the rail, localization is more complex. In order to reach a localization accuracy of up to several millimeters in this case, information of different sensor types has to be merged. Therefore, the localization algorithm is implemented as a Monte-Carlo Particle Filter (MPF) based on Bayesian models using dead-reckoning for the prediction phase, laser rangefinder readings and RF ranging for the update phase (cf. Fig. 4), following the approach described in [7]. This algorithm is applicable to both local and global localization problems. Furthermore, the modular architecture of the localization algorithm makes it easy to replace the existent sensors (e.g. replacing the laser rangefinder by a 3D-Camera) for future applications.

4 Application

The concept of Cellular Transport Systems is the answer to the unpredictable, highly volatile markets of contemporary facility logistics. Actual facility logistics systems consist of strongly coupled inflexible or non-automated systems and neither provide the necessary flexibility nor the high degree of automation to overcome actual industrial demands.

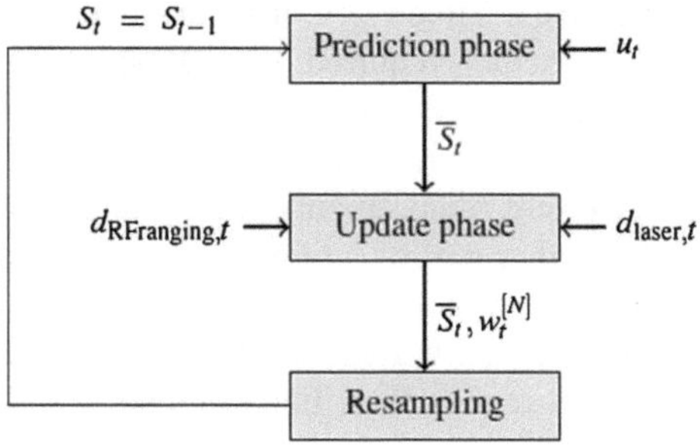

Fig. 4 Monte-Carlo particle filter flow chart

Therefore, the CTS shows a high degree of flexibility as it is able to adapt to changing topologies and demands. For example, if it is necessary to increase the number of storage locations, a new high bay racking has to be built up and additional ATVs have to be added in order to meet the new transportation requirements. In conventional systems, the complete control software has to be adapted or even changed. By the usage of a MAS, only changes in topology entities and new transport agents have to be (re-)initialized by the user (by adding a new storage and transport entity). Thus, the decentralized control structure is the enabler for a fast startup procedure, the ease of use of complex systems and scalability. It prevents the whole facility logistics system from breakdown times due to obsolete global software or hardware changes.

5 Conclusion and Outlook

At present the required invest for an enlargement of a material handling system based on the concept of a Cellular Transport System is proportional to the throughput without any additional costs for adapting (cf. Fig. 5) by meeting the

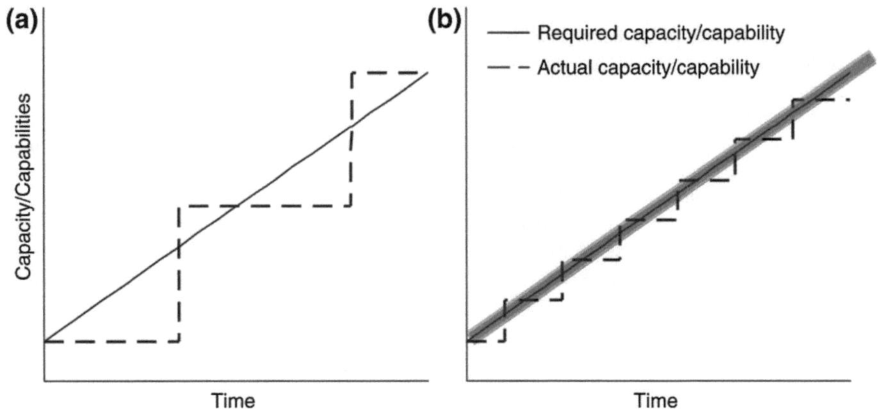

Fig. 5 Optimal adaption of a material handling system. **a** Inflexible system. **b** Flexible system(*gray* in **b**) cellular transport system

same costs. Effectively, it means that the maximum actual throughput is oriented on the actual required throughput. Through its flexibility, Cellular Transport Systems have economic benefits compared to non-automated systems or conventional conveyor systems, which have to be designed in an oversized way to meet future requirements or to retain singular transportation peaks.

A further part of future research is the comparison between the autonomy of an individual in the swarm and the Swarm Intelligence of the compound. The natural aspect, that an individual is not able to solve complex tasks cannot be applied to technical realizations. For example, simple transportation tasks have to be solved by an individual vehicle without any support from other vehicles—this is a contradiction to the Swarm Intelligence. So economical (how much costs autonomy?) and functional (how much autonomy is needed?) aspects set up the limits for the swarm and the autonomy level. These aspects are the major focus of the future research activities.

References

1. W. Günthner, M. ten Hompel (eds.), *Internet der Dinge in der Intralogistik* (Springer, Berlin, 2010)
2. H. S. Mayer, Development of a completely decentralized control system for modular continuous conveyors, Ph.D. dissertation, Universität Karlsruhe (TH), 2009
3. B. Scholz-Reiter, S. Schulz, Selbststeuerung in der Logistik, in Jahrbuch Logistik 2007, ed. by H. Wolf-Kluthausen. Korschenbroich: free Beratung GmbH, 2007, pp. 236–240
4. Z. Guessoum, Adaptive Agents and Multiagent Systems, IEEE Distrib. Syst. Online. **5**, 7 (2004)
5. M. Dorigo, E. Bonabeau, G. *Theraulaz, Swarm Intelligence—From-Natural to Artificial Systems* (Oxford University Press, Oxford, 1999)
6. M. ten Hompel, A. Kamagaew, *"Zellulare Intralogistik—in Zukunft nur noch Multishuttles?" in 20* (Deutscher Materialflusskongress, München, 2011)
7. S. Thrun, W. Burgard, D. Fox, *Probabilistic Robotics* (The MIT Press, Cambridge, 2005)
8. A. Kamagaew, E. Große, Zellulare Transportsysteme—Multishuttle Move, Multimodales Intralogistikkonzept. Hebezeuge Fördermittel **51**(4), 170–172 (2011)
9. A. Kamagaew, J. Stenzel, M. ten Hompel, Concept of Cellular Transport Systems in Facility Logistics in *Proceedings of the 5th International Conference on Automation, Robotics and Applications (ICARA)*, 2011, Wellington, New Zealand, Dec. 2011, pp. 40–45
10. C. Kirsch, C. Röhrig, Position Tracking Using Sensor Fusion of a Wireless Network and a Laser Range Finder, in *Proceedings of the 7th IEEE Workshop on Positioning Navigation and Communication (WPNC 2010)*, Dresden, Germany, March 2010
11. C. Röhrig, M. Müller, Indoor Location Tracking in Non-line-of-Sight Environments Using a IEEE 802.15.4a Wireless Network, in *Proceedings of the 2009 IEEE/RSJ International Conference on Intelligent Robots and Systems (IROS 2009)*, St. Lous, USA, Oct 2009, pp. 552–557

Application of Knowledge Driven Mobile Robots for Disassembly Tasks

Gottfried Koppensteiner, Christoph Krofitsch, Reinhard Hametner, David P. Miller and Munir Merdan

Abstract Considering the disassembly as a vital and prospective industry domain, we use the mobile robots to automate the disassembly process. In our system, each mobile robot has particular skills and is supervised by an agent with related objectives and knowledge. An agent has an ontology-based world model, which is responsible to maintain the knowledge about the robot's activities in relation to its environment as well as to its underlying software parts. The ontology is used to represent a specification of an agent's domain knowledge. The system functionality is tested with three mobile robots having a task to disassemble a particular Lego construct. Different rule-engines were benchmarked in order to enhance the systems performance.

Based on "Knowledge Driven Mobile Robots Applied in the Disassembly Domain", by Gottfried Koppensteiner, Reinhard Hametner, Rene Paris, Alejandro Moser Passani, and Munir Merdan which appeared in the Proceedings of the 5th International Conference on Automation, Robotics and Applications (ICARA 2011). © 2011 IEEE.

G. Koppensteiner (✉) · R. Hametner · M. Merdan
Institute of Automation and Control, Vienna University of Technology, Vienna, Austria
e-mail: koppensteiner@acin.tuwien.ac.at

R. Hametner
e-mail: hametner@acin.tuwien.ac.at

M. Merdan
e-mail: merdan@acin.tuwien.ac.at

C. Krofitsch
Practical Robotics Institute Austria, Vienna, Austria
e-mail: krofitsch@pria.at

D. P. Miller
Schools of AME and Computer Science, University of Oklahoma, Norman, OK, USA
e-mail: dpmiller@ou.edu

G. Sen Gupta et al. (eds.), *Recent Advances in Robotics and Automation*,
Studies in Computational Intelligence 480, DOI: 10.1007/978-3-642-37387-9_24,

1 Introduction

Disassembly has become a vital industry process due to the increasing necessity of optimizing resource usage. Currently, disassembly processes are partially automated only in a small set of cases such as: single use cameras [1], PCs [2], printed circuit boards as well as LCD monitors [3]. The main limiting factors are: non-uniformity of returned product models creating great uncertainty in the system control and structural configuration, etc. [4]. The rigid character and weak adaption capabilities of current implementations, due in part to their use of centralized hierarchical control structures, limits their ability to respond efficiently and effectively to dynamic changes.

Mobile robots offer new possibilities for flexible disassembly [5]. The state-of-the-art of mobile robot technology and predictions of future development give a clear view that mobile robots are going to be an essential part of every manufacturing process in future [6]. Mobile robots can support the system's flexibility and capability to handle dynamic changes. However, one of the obstacles to wider adoption of the mobile robots in the industry is the engineering cost to integrate mobile robots into systems. The field of robotics should develop in such a way to reduce the programming requirements and to increase the flexibility of mobile robots for different tasks [7]. In addition, mobile robots for disassembly should be (a) intelligent in the sense of path planning and able to communicate with other robots, (b) cooperative with other (stationary or mobile) robots, and (c) able to form a disassembly multi agent system, which is one of the future possibilities for reducing disassembly costs. Moreover, in order to avoid difficulties in communication in a heterogeneous robot environment, where each robot has its own kinematic structure and programming language, etc., it is necessary to develop standardized communication protocols and methods [5].

To cope with these requirements, we propose a knowledge-intensive multi-agent robot system, which enables ontology-based communication and cooperation among a set of autonomous and heterogeneous units (agents). In our system each agent supervises one particular mobile robot and, related to the robot's skills, has its own objectives and knowledge. In this context, ontologies allow the explicit specification of an agent's domain of application, increasing the level of specification of knowledge by incorporating semantics into the data and promoting knowledge exchange between agents in an explicitly understandable form [8]. An ontology based product model is used [9] to link product designs, disassembly planning and scheduling processes, as well as required disassembly equipment, possessed by a particular mobile robot, in a way that enables automatic reasoning as well as wide data integration. Consequently, on the one side, a vision system can use this model to reason about the content of a captured image. On the other side, an agent controlling a mobile robot can extract required disassembly information from this model to select and perform the necessary actions [10]. The architecture is based on agents that have a rule-based behavior. Rules are considered as if–then statements applied to the knowledge base. The application of this kind of decision-making mechanism supports a knowledge capture in a more modular and explicit way.

2 Agent Architecture

In previous work [11–13] a generic agent architecture was developed to facilitate the design of multi-agent control systems. This architecture clearly separates the control software into two layers: the high level control (HLC) and the low level control (LLC). The LLC layer is in charge of controlling the hardware directly. It is responsible for performing all necessary operations in real-time and is based on the IEC 61499 Standard [14]. The HLC layer is responsible for more complex tasks such as coordination, monitoring or diagnostic, which might require a longer computation time. This allows using one HLC-Architecture with different LLCs caused by heterogeneous mobile robots. In the rest of the chapter, we will present the structure and functionality of the HLC adapted for the control of and negotiation between mobile robots. To enable agents to understand the transmitted messages between the LLC and HLC, a representation data type is as well as the *MessageContent* concept and its sub-concepts is included in the agent's ontology [15]. Figure 1 shows the architecture of the system which provides the HLC for the mobile robots, with focus on the interplay of the used technologies.

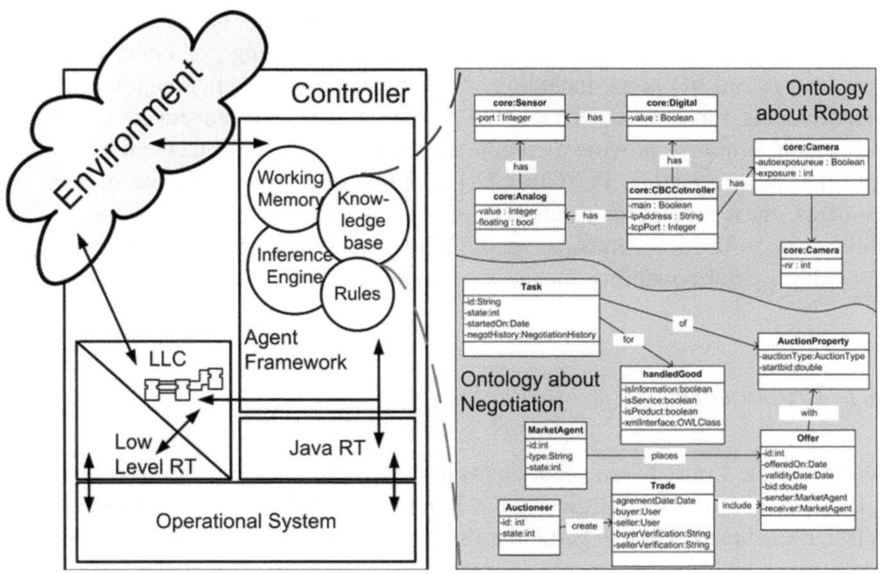

Fig. 1 Agent architecture includes the HLC and LLC overview [24] applied on mobile robots in the disassembly (negotiation) domain

The whole system environment is based on an ontology which is copied to each of the controllers. The framework uses its ontology for multiple purposes. First of all, it serves as the basis for the rule based control system, but additionally, it serves as a place for configuring the robot and the intercommunication. The ontology describes typical components from robots (e.g. sensors, motors) as well as the robots communication including coordination, cooperation and negotiation) with all necessary protocols. The ontology can be easily extended to support application specific needs, e.g. for mapping disassembly structures. This supports the agents with the necessary domain knowledge about the robots and the negotiation mechanisms between them in order to provide the robots with a framework for collaboration. The knowledge of the HLC is defined in rules which are different for each robot type. The knowledge base is periodically updated. This is realized through an updating process which is performed between the specific system-libraries of the controller and the ontology.

3 System Integration

The High Level Robot Control is implemented on the CBCv2 [16] which includes an ARM 7 based DAQ/Motor control system, an ARM 9-based CPU/Vision processor running Linux, an integrated color display as well as a touch screen. It is being used by thousands of middle and high school students in the educational robotics program Botball [17]. Due to the aim of the sparkling science initiative [18], it was one major reason to take a controller system which is well-suited for a collegiate robotics lab. Therefore, the Atmel ARM7 32 bit MCU running at 48 MHz, because of its speed, availability, and wide range of embedded communication and I/O ports, including 16 10-bit ADCs, is good enough to meet the requirements for a robotics Lab. On the other side, the embedded Linux and reloadable firmware provide a complete package for the enhancement with agent and rule based systems. The CBCv2 is a USB host (allowing the use of standard cameras, mass storage and network interfaces) and can also be used as a USB device for software downloads. At the USB port a Wi-Fi Stick can be used; this provides a good possibility for the communication between mobile robots.

3.1 Framework

On top of the Linux-based robot controller, explained in the previous section, the CBCJVM [19] is used as the system-library. CBCJVM uses a lightweight Java Virtual Machine (JVM), in particular the JamVM, which is designed for embedded platforms which makes it perfect for the use on the CBC. It includes libraries for working with the low level components of the CBC, including reading sensors and controlling motors, which are used by the updating process to synchronize the ontology objects representing low-level components. In this process, the controller's own components mapped in the ontology are synchronized with the values from the system-libraries (e.g. sensor values) at fixed update intervals. On top of this, the JADE-Leap platform takes place. JADE [20] simplifies the implementation of multi-agent systems through a middle-ware that complies with the FIPA specifications [21]. The LEAP libraries allow obtaining a FIPA-compliant agent platform with reduced footprint and compatibility with mobile Java environments. Altogether, this is the host for the Agents, which finally control the robots. On the one side, they are responsible for listening for ontology changes and activating the rule engine appropriately, on the other side they handle the intercommunication between other robots, which is further described in the next section. This architecture enables writing rule-based behaviors using the full knowledge from the ontology, one of the key aspects of this framework. The rule-based behaviors are written in different rule-based languages and are fit to the mechanical possibilities of the robot and the tasks the robot is intended for. Currently, the applicable rule engines are Jess [22] and OPSJ.

3.2 *Agent Communication*

The agent communication ensures the possibility to share ontology values between different controllers. That enables defining rules for foreign ontology values or using knowledge from other robots in order to provide reliable data for robot collaboration or about a deal within the negotiation process.

Figure 2 shows two different negotiation procedures in the disassembly process. The task allocation with auctions starts with a "need for some service". This order could be transmitted to only one robot which is able to fulfill the requested operation, or as broadcast within an auction procedure to evaluate the best situated robot. In order to ensure that every agent in the system talks about the same part of interest, parts of the ontologies could be transmitted within these messages (see message "Object: yellow part") and in order to be able to reason about it, mapped to the agent's ontology. If more than one service is necessary for this operation, the requesting agent starts one "auction" (request) for each service. Every robot which is able to perform some particular service sends a bid or the offered service back. After all subjects of the deal are negotiated, a deal-commitment is sent and the operation is started. Within such operations, spread between different robots, coordination might be necessary in order to schedule each step of the operation. This is done by negotiation about the necessary operations. In this procedure two or more robots, working together to solve one requested task, exchange messages about necessary steps. Therefore a fast reaction time is necessary: in the case of the LLC to be able to react quickly to sensor values (for instance to avoid crashes); in the HLC to be able to reason as fast as possible about incoming messages.

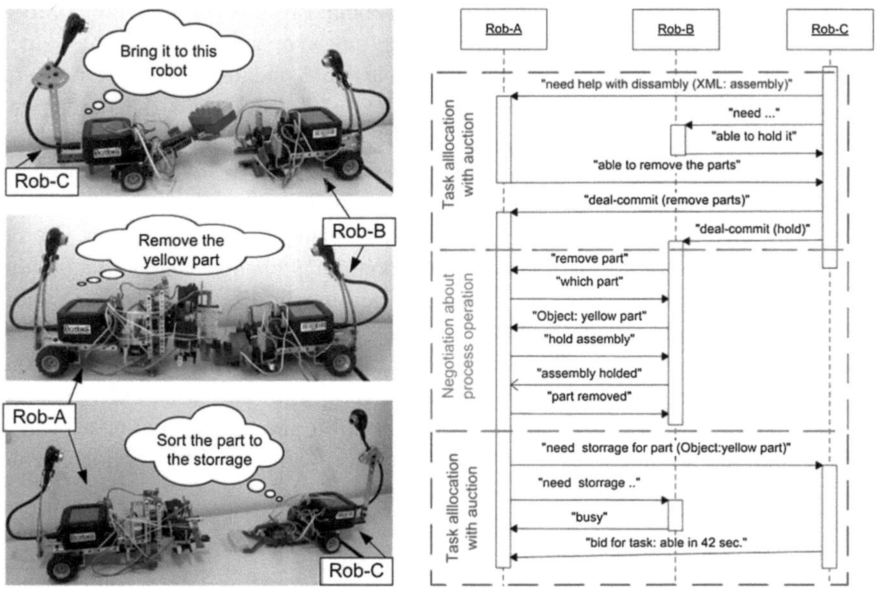

Fig. 2 Robot negotiation

4 System Implementation

Due to memory constraints of the CBCv2, it was not possible to use the Protégé ontology API [23] at runtime. Instead, the ontology is translated into Java classes and objects which get stored to a file at build time. The access to the serialized file is then possible on the CBCv2. The mapping is done in two stages: firstly, the classes are designed in the ontology and are converted into Java classes. Secondly, after the generated classes are compiled, the objects contained in the ontology are read, instantiated, and written to a file via Java serialization. At runtime, the objects are read into the framework via the ontology interface, and added to the rule engine's fact base.

4.1 Customizability and Flexibility

The HLC framework implementation was focused on two important characteristics: (1) great amount of customizability for being applicable in various fields and tasks, and (2) flexibility for different software- and hardware-environments through exchangeability. Beside definition of application rules, the customizability was realized by (1) providing almost unlimited ontology extension with new classes and instances, (2) implementing and executing new software agents, and (3) adding user-commands for executing specific pieces of code from rules or command line. This is possible through holding the framework configuration and interfaces in one single object, which is globally accessible. In terms of flexibility, the following framework components can be fully exchanged:

- Multi-Agent System: Instead of JADE, another FIPA-compliant Agent Platform might be used.
- Rule Engine: The whole rule-based system including inference engine and rule language can be exchanged. Currently, Jess and OPSJ might be used.
- Hardware interface: Instead of using the CBCv2, another robot controller might be used by implementing the low-level library connection.

In this system, plugins representing single components can be implemented in Java projects containing classes implementing specific Java interfaces defined in the core project of the framework. The usage of plugins can be statically configured in a specially implemented configuration format (e.g. when changing the right plugin class in the configuration file, OPSJ rule engine can be used instead of Jess without any other efforts). Example rules for both rule engines can be found in the next chapter in the context of performance measurements.

5 Performance Benchmarking

When implementing this framework in its first version, Jess was the only applicable rule engine. In most cases, the rule firing time was insufficient due to the limited resources of the CBC, and some measurements showed that the engine used in Jess consumes most of the controller's resources. Considering the importance of fast reaction on particular events it was necessary to investigate possible system improvements. Trying out another rule engine might solve this problem and this was actually the reason to implement the plugin into the system which was introduced the section before. For comparing Jess and OPSJ rule engines, two different performance benchmarks are used and their results presented later in this section. Both tests are executed in the framework environment, which means that JADE and the ontology might have an effect on the results.

5.1 Methodology

The first benchmark is separated into two parts. It is intended for measuring the time the different rule engines consume to load Java objects into their working memory. After loading, the allocated heap memory is measured. Therefore, this benchmark shows the data loading time and the heap memory consumption dependent on the number of objects to load. Every object's state consists of ten integer variables and one string variable; the maximum heap memory on the CBC is limited to 32 megabytes. The execution is planned as follows: Each run loads a number of objects, beginning with 10 for the first run up to 5,000 for the last run. After loading has finished in each run, the allocated heap memory will be measured. Every run will be iterated ten times and averaged for the result.

The second benchmark illustrates the optimal way of using a rule engine in the framework. It simulates the usage of one analog sensor in the framework, which means that the sensor will be read periodically. When the sensor's value is in a specific range, the time it takes to fire a rule, whose left-hand side matches this range, will be measured (reaction time). In particular, the reaction time consists of two parts: (1) the time for updating the appropriate rule engine fact with the new sensor value and (2) the time the rule engine takes to fire the rule. Because the sensor's value changes continually, every update leads to an invocation of the rule engine. This could overload the CBC and causes longer rule engine reaction times. This benchmark measures the overall reaction time for different update intervals and reveals the most efficient update interval configurations for the framework. The execution is planned as follows: Each run, using a specific update period beginning at 120 ms for the first run down to 0 ms for the last run, measures the reaction time 250 times, i.e. fires the rule 250 times. Figure 3 shows the rules for this performance benchmark in Jess- and OPSJ language. In the right-hand side, a static method will be called, which measures the time and prepares for the next iteration.

5.2 Results

The results of the first benchmark are presented as follows (Fig. 4):

The object loading time is roughly linear for both rule engines; for better visualization the loading time is scaled logarithmically. OPSJ is significantly faster than Jess, on average by a factor of 38. The heap memory plot is smoothed slightly in order to compensate side effects caused by JADE or other disturbance. Here is OPSJ performing better as well, the average memory savings compared to Jess are 2.6 Megabytes.

```
(defrule testrule
  (AnalogSensor {value > 500})
  =>
  (call main.Main ruleFired)
)
```

```
rule testrule
if {
    d: AnalogSensor (
      d.getValue() > 500
    );
} do {
    Main.ruleFired();
}
```

Fig. 3 Rule used in reaction time performance benchmark: Jess (*left*) and OPSJ (*right*)

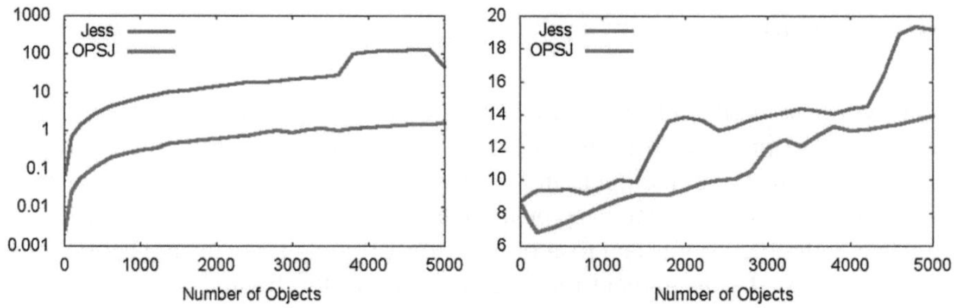

Fig. 4 Results of the data performance benchmark: object loading time on number of objects (*left*) and allocated heap memory on number of objects (*right*)

Figure 5 shows the results of second performance benchmark. The 250 measurements, which were taken in each run, are visualized as a boxplot. They illustrate the range of reaction times which are possible for a specific update interval. As can be seen, every range goes from nearly zero (min) up to at least the update interval (max). This is explained by the randomness of the sensor's value: When the value passes the threshold, the next update could be immediate or long time off.

As expected, a low update interval (10 and 0 ms) when using Jess takes the controller to overload. In the graph, this can be seen in the raising maximum at 10 ms which indicates unstable reaction time. At 0 ms, even the average reaction time raises. In case of OPSJ, reaction time does not increase significantly with low update interval. Note, that this benchmark simulates just one analog sensor which is insufficient in most robotic applications. When using an appropriate number of sensors and actuators, both rule engines might cause serious performance issues with low update intervals; Jess more than OPSJ.

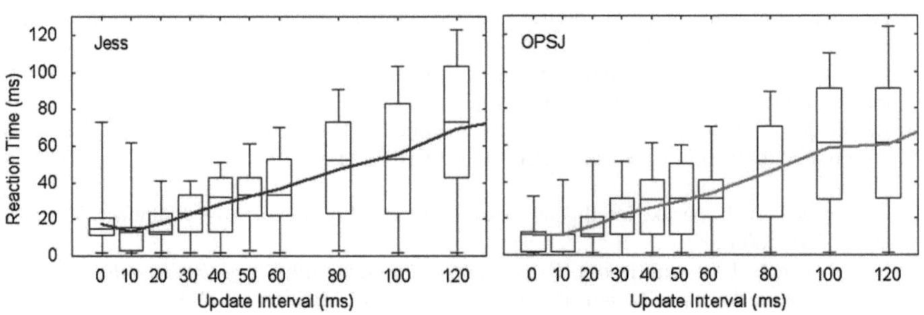

Fig. 5 Results of the reaction time performance benchmark: reaction time on update interval; visualization of raw data as boxplots and averaged as line

6 Conclusion

In this work, an intelligent MAS based on ontologies for a distributed mobile robot system applied in the disassembly domain was presented. The architecture provides robots, which are intelligent in the sense of path and task planning as well as cooperative with other (stationary or mobile) robots. They are also able to communicate with other robots and to coordinate disassembly tasks using negotiations. In order to face the insufficient reaction time due to the limited resources of the CBC a customizable architecture was developed, which allows the replacement of agent platforms, rule engines or hardware devices. Currently, Jess and OPSJ are the applicable rule engines, which differ significantly in their resource requirements. Jess takes a lot more time for data loading and allocates more heap memory than OPSJ. Furthermore, the lightweight OPSJ engine fires rules faster than the Jess engine, which mostly comes out when updating sensor and actuator values in low interval. This clearly points out the benefits of OPSJ in allocated memory and processing time, which both are the main limiting factors on mobile controllers. In this work the HLC was benchmarked. In order to provide the agents with faster reactive behaviors for sensors and motors a similar set of tests and evaluations of the LLC should be performed.

Acknowledgments The authors would like to acknowledge the support by the Sparkling Science program, an initiative of the Austrian Federal Ministry of Science and Research. We also want to thank all partners involved in the DISBOTICS Project, especially the students at Vienna Institute of Technology (TGM), department for Information-Technology, and the KISS Institute of Practical Robotics.

References

1. M. Matsumoto, Business frameworks for sustainable society: a case study on reuse industries in Japan. J. Cleaner Prod. **17**(17), 1547–1555 (2009)
2. F. Torres et al., Automatic PC disassembly for component recovery. Int. J. Adv. Manufact. Technol. **23**(1), 39–46 (2004)
3. H. Kim, S. Kernbaum, G. Seliger, Emulation-based control of a disassembly system for LCD monitors. Int. J. Adv. Manuf. Technol. **40**(3), 383–392 (2009)
4. J.R. Duflou, G. Seliger, S. Kara, Y. Umeda, A. Ometto, B. Willems, Efficiency and feasibility of product disassembly: a case-based study. CIRP, Manufact. Technol. **57**(2), 583–600 (2008)
5. P. Kopacek, B. Kopacek, Intelligent, flexible disassembly. Int. J. Adv. Manufact. Technol. **30**(5), 554–560 (2006)
6. F.M. Asl, A.G. Ulsoy, Y. Koren, *Dynamic Modeling and Stability of the Reconfiguration of Manufacturing Systems*. Technical report (University of Michigan, 2001)
7. A. Lazinica, B. Katalinic, Self-organizing multi-robot assembly system. Int. Symp. Robot. **36**, 42 (2005)
8. M. Merdan, *Knowledge-based Multi-Agent Architecture Applied in the Assembly Domain.* PhD Thesis, Vienna University of Technology, (2009), http://www.ub.tuwien.ac.at/diss/AC05040230.pdf

9. K. Kyoung-Yun, D.G. Manley, H. Yang, Ontology-based assembly design and information sharing for collaborative product development. Comput. Aided Des. **38**(12), 1233–1250 (2006)
10. M. Merdan, W. Lepuschitz, T. Meurer, M. Vincze, in *Towards Ontology-Based Automated Disassembly Systems*. IEEE international conference on industrial electronics control and instrumentation, (2010), pp. 1386–1391.G
11. G. Koppensteiner, M. Merdan, W. Lepuschitz, I. Hegny, in *Hybrid Based Approach for Fault Tolerance in a Multi-Agent System*. IEEE/ASME international conference on advanced intelligent mechatronics, (Singapore, 2009)
12. M. Merdan, M. Vallee, W. Lepuschitz, A. Zoitl, Monitoring and diagnostics of industrial systems using automation agents. Int. J. Prod. Res. **49**(5), 1497 (2011)
13. M. Vallee, H. Kaindl, M. Merdan, W. Lepuschitz, E. Arnautovic, P. Vrba, in *An Automation Agent Architecture With a Reflective World Model in Manufacturing Systems*. IEEE international conference on systems, man, and cybernetics (SMC09), (San Antonio, Texas, USA, 2009)
14. A. Zoitl, *Real-Time Execution for IEC 61499*. ISA-o3neidaA, USA, ISBN: 978193439-4274 (2009)
15. M. Merdan, W. Lepuschitz, I. Hegny, G. Koppensteiner, in *Application of a Communication Interface Between Agents and the Low Level Control*. Proceedings of the 4th international conference on autonomous robots and agents, Wellington, New Zealand, 2009
16. D.P. Miller, M. Oelke, M.J. Roman, J. Villatoro, C.N. Winton, in *The CBC: A LINUX-based low-cost mobile robot controller*. IEEE international conference on robotics and automation (ICRA), pp. 4633–4638, 3–7 May 2010
17. KISS Institute of Practical Robotics, *The Botball Season*. http://www.botball.org, Accessed May 2011
18. Sparkling Science, BMWF. http://www.sparklingscience.at/en. last viewed June 2011
19. B. McDorman, B. Woodruff, A. Joshi, J. Frias, in *CBCJVM: Applications of the Java Virtual Machine with Robotics*. Global conference on educational robotics, Edwardsville (2010)
20. Telecom Italia Labs, *Java Agent Development Framework*. http://jade.tilab.com/. Accessed March 2011
21. The Foundation for Intelligent Physical Agents, *FIPA* Specifications. http://www.fipa.org/specifications/index.html. last viewed July 2011
22. Sandia National Laboratories, *Jess: The Rule Engine for the JavaTM Platform*. Available at: http://herzberg.ca.sandia.gov/. last visited May 2011
23. Stanford Medical Informatics, *Protégé Ontology Editor*. Stanford University. Protégé Website. http://protege.stanford.edu. Accessed May 2011
24. G. Koppensteiner, M. Merdan, I. Hegny, W. Lepuschitz, S. Auer, B. Groessing, in *Deployment of an Ontology-Based Agent Architecture on a Controller*. 8th IEEE international conference on industrial informatics, Japan, 2010

Power Distribution Control Using Multi-Agent Systems

Munir Merdan, Alexander Prostejovsky, Ingo Hegny, Wilfried Lepuschitz, Filip Andrén and Thomas Strasser

Abstract Availability and functionality of reliable and efficient electric energy systems are prerequisites for economic and social welfare. In the last decades, more and more stress is put on the electricity supply and infrastructure increasing also minimal electricity networks requirements. Multi-agent systems represent a promising approach able to realize and implement the above described functionalities and services in a power distribution network. In this chapter an automation agent approach and the related system architecture for handling such dynamic networks is presented and discussed. The agents provide monitoring and diagnostics abilities required for the robust functioning of components in the distributed environment. Furthermore, the system also provides means to coordinate

Based on "Multi-Agent system for self-optimizing power distribution grids", by Munir Merdan, Wilfried Lepuschitz, Thomas Strasser, and Filip Andren which appeared in the Proceedings of the 5th International Conference on Automation, Robotics and Applications (ICARA 2011).

M. Merdan (✉) · A. Prostejovsky · I. Hegny · W. Lepuschitz
Vienna University of Technology, Automation and Control Institute, Vienna, Austria
e-mail: merdan@acin.tuwien.ac.at

A. Prostejovsky
e-mail: prostejovsky@acin.tuwien.ac.at

I. Hegny
e-mail: hegny@acin.tuwien.ac.at

W. Lepuschitz
e-mail: lepuschitz@acin.tuwien.ac.at

F. Andrén · T. Strasser
Energy Department, AIT Austrian Institute of Technology, Vienna, Austria
e-mail: filip.andren@ait.ac.at

T. Strasser
e-mail: thomas.strasser@ait.ac.at

G. Sen Gupta et al. (eds.), *Recent Advances in Robotics and Automation*,
Studies in Computational Intelligence 480, DOI: 10.1007/978-3-642-37387-9_25,

energy generation and consumption, keep nodal voltages within predefined bounds and avoid overstressing of equipment. The presented approach is currently in the development phase and will be tested in the AIT power distribution laboratory.

1 Introduction

The continuous growth of electric energy consumption as well as the upcoming large-scale integration of distributed energy production units, based on renewable energy resources (photovoltaic systems, wind turbines, etc.) result in an increasingly complex electric network [1]. However, today's grids form a vertically integrated scheme with centralized generation, distributed consumption, limited interconnection capabilities between the control areas, and commercial and regulatory frameworks that are not harmonized for mutual advantage [2]. The current control approach of using a centralized Supervisory Control and Data Acquisition (SCADA) system—also known as Energy Management System (EMS) for the transmission domain and Distribution Management System (DMS) for distribution grids—has deficiencies in flexibility and adaptability and is no longer sufficient for certain control operations [3]. Conventionally, energy management is ensured by a central facilitator whose program is based on long series of conditional branches. Even if this solution achieves a constant supply of the load, it cannot fulfill other objectives such as fault tolerance of an element [4]. Consequently, the search for a control strategy gets more and more difficult with increasing system size. Besides, in present power system networks, a single fault can lead to massive cascading effects affecting power supply and power quality [5]. Moreover, applied monitoring, analysis, and control technologies are sometimes too slow, limited to protecting only specific components. The process of system restoration is mainly based upon the operators' experience and results from offline studies [6].

Results of actual research and demonstration projects show that the realization of an active distribution network, also referred to as Smart Grid, is a promising way to overcome the limitations of state-of-the-art approaches. Realizing smart grid demands requires the establishment of an innovative system architecture in order to provide coordinated monitoring and control capabilities in a heterogonous grid environment. The system architecture defines how well the system components are utilized and in what manner the application's functionalities are integrated and the services are delivered. The changed demands on distribution grids require the application of intelligent methods to keep the system state within allowed bounds at any time. Multi-Agent Systems (MASs) are considered as an approach able to realize/implement the above described functionalities and services for intelligent devices in a Smart Grid [7]. MAS offers a convenient way to cope with the dynamics in large complex systems, making the control of the system decentralized, thereby reducing the complexity, and increasing flexibility, as well as enhancing fault tolerance [8]. Agents have the ability to coordinate several different devices like sources, loads or switches in a decentralized manner and are capable of finding a technically and economically optimal operation point under consideration of all kinds of constraints [9]. In this context, a distributed system architecture enables local data processing and minimizes the need for massive data exchanges (e.g. capable of providing local fail proof, feeder level forecasts aggregated at substation level). A distributed system can enable the high performance needed for preventing or containing rapidly evolving adverse events [8]. In electric power systems agents are used for diverse complex services, e.g. for controlling, communication and protection of the grid [10], as well as monitoring, visualizing and coordinating energy use within the home [11]. The issue of fault diagnosis and reconfiguration is addressed in [12]. In this context, Liu et al. [13] have described the need for fault detection in naval shipboard power systems; Huang et al. [14] have presented a multi agent approach for fault detection using nonlinear parameter identification techniques. Solanki et al. [7] propose different approaches for restoration of the power system by distributed reconfiguration. Chouhan et al. have presented a multi-agent based system for distribution network reconfiguration [15]. However, despite the growing awareness and the on-going research activities of MAS technology in the power distribution area, there are still some requirements that remain to be considered, such as

- the interaction between distinct conceptual entities, the ability to locally undertake an analysis/decision without a central facilitator,
- the adaptability of the system to easily integrate new elements, and
- the ability to provide self-corrective reconfiguration and restoration, and to handle randomness of loads and market participants in real-time [4, 16].

Preliminarily, an automation agent architecture was developed and recently successfully implemented at the ACIN's laboratory palette-transfer system showing valuable advantages in self-diagnostics [17], robustness [18] and self-reconfiguration [19]. This chapter presents the concept of the on-going Multi-Agent-based Smart Grid (MASGrid) project, which aims to finalize the development of the automation agent and demonstrate its proven benefits for active power distribution in order to realize a Smart Grid. The MASGrid approach aims on providing abilities to automatically reconfigure the power system topology based on changed conditions of the system and coordinate energy consumption, allowing important savings of energy compared to the state-of-the-art, as well as sophisticated monitoring and diagnostics suitable for distributed grid environments.

This chapter is structured as follows: In the next section a motivation example is introduced. The third section presents the architecture of the MAS. In section four, we describe in more detail the architecture of agents controlling physical components of the grid and their abilities. Section five presents the reconfiguration ability of the system. A short description of the agent framework and the simulation environment is given in section six. Finally, section seven concludes this chapter and provides an outlook.

2 Motivation Example

Promising application examples for the proposed MAS architecture are so-called Micro-Grids [20]. In Fig. 1 an overview of a small Low-Voltage (LV) distribution network with three customers is provided. It is assumed that each customer can operate in a controlled islanding mode. Hence, each customer can be connected to or disconnected from the grid. Each customer has a different number of distributed generators, storages and loads that are connected to the MAS using agents. On the left hand side in Fig. 1 a typical future customer from a residential area (i.e. a private home consuming and producing energy) is depicted, including an Electric Vehicle (EV), a Photovoltaic System (PV) and the building. In the middle of Fig. 1, a commercial customer is depicted, also equipped with a PV system as well as a parking lot with charging stations for EVs. Larger commercial customers are also expected to include more controllable loads, which make the local energy management more challenging than for a private home. On the right hand feeder a commercial building representing a critical infrastructure with a battery backup system is shown. In order to keep the distribution system operating or parts of it in case of faults, each customer can be disconnected forming an island (i.e. a Micro-Grid). To manage and control such an island the proposed MAS provides a very good basis.

For example, "Island 1" in Fig. 1 represents such a Micro-Grid only for a residential customer. Using the Micro-Grid MAS consisting of the Line Agents (LNAG), the Switch Agent (SWAG), the Load and Storage Agent (L/STAG) as well as the Supply Agent (SAG) it would be possible to manage and control such a local grid.

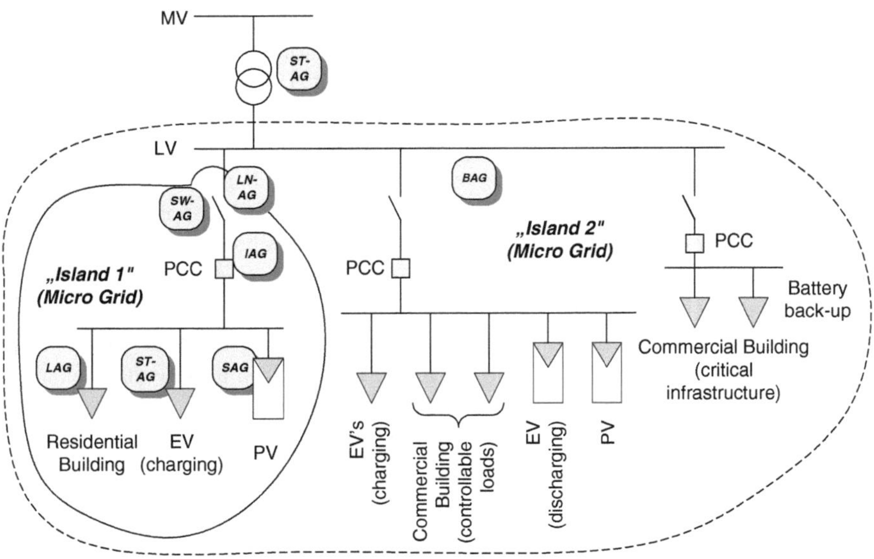

Fig. 1 Micro grid controlled by MAS

The big advantage of such a system topology is related in its flexibility and scalability. Using the above described agents provides a generalized way of describing energy systems, which means other types of customers, as well as other grid parts, can be described using the same method. Moreover, using the same methodology larger Micro-Grids can be created by a combination of smaller Micro-Grids, from the ICT point of view. In Fig. 1 the low-voltage part of the grid can be combined into a Micro-Grid using the same agent architecture, where the customers are seen as loads and distributed generation. In addition the provided MAS concept allows an improved condition monitoring of the local grid as well as the whole distribution grid. Based on a more comprehensive knowledge about the local islands advanced control strategies can be applied.

3 System Architecture

The traditional electrical power system consists of generation, high voltage transmission grids and distribution systems. Such large and complex, nonlinear systems are prone to cascading failures due to a single fault in transmission and/or distribution lines. In this frame, SCADA systems monitor and control transmission, substation, and distribution field devices using remote terminal units at geographically separated sites.

In order to reduce its complexity, but to stay consistent with current solutions, the control of a power system can be split in two "hierarchically" ordered layers which incorporate specific agent types. The Management Layer is responsible for the system's overall network stability and gives control over its functionality and acts as the interface to human operators via a SCADA. Two types of agents, the System Agent and the Data Agent, are employed at this layer which exchange information with the subjacent execution layer as well as with the SCADA system in order to perform global grid tasks. The agents of the Execution Layer directly control the power and energy system's components and perform local monitoring tasks by using information collected from sensors or exchanged with adjacent system components. Different requirements such as sophisticated communication means or real-time capabilities have to be met by those so-called Automation Agents. In the following, the different types of agents and their general activities are described.

3.1 System Agent

The system agent specifies agent roles and responsibilities, including the monitoring of system voltage and frequency to detect contingency situations or grid failures. It possesses extensive information regarding the network and has the capability of changing load flows. This agent is able to find the best stable configuration of the power system by rerouting the power in the case that one part of the system breaks. It also informs the human observer over a SCADA system about performed reconfiguration activities. Whenever changes happen in the power system, like a change of the grid topology or the addition of a new device into the power grid, this agent updates its world model and merges the model with the SCADA system in order to keep the consistency. The reconfiguration algorithm is embedded inside the agent with the objective function to always supply the critical load without exceeding the line capacity while maintaining the voltage limits.

3.2 Data Agent

A main responsibility of this agent is to conduct data mining procedures on the collected historical data in order to monitor and analyze the electricity consumption. It is able to forecast the possible peaks and advise the System Agent how to manage them. A data agent enables the human observer to control the status of loads based on priority predefined by a user. The information to be processed includes the main parameters of the various network components, fault records as well as the planned maintenance schedule.

3.3 Automation Agents

Automation agents are the key components of our architecture, as they represent the physical topology items and provide particular functions and services as well as a communication interface to cooperate with other agents. The agent is able to send and receive measured information, requests and responses to requests. Using all measured and transmitted information and its knowledge database the agent is able to make decisions on its own, which can be sent to other agents or may be directly executed by its effectors. The following types of Automation Agents are distinguished [21].

- Bus Agent (BAG,): Each physical bus is represented by at least one BAG which carries out facilitating functions, allowing the bus to operate as an independent "Micro-Grid". At commissioning time, the agents of the grid devices connected to the bus are assigned by the System Agent to the corresponding BAG, to which they report their status and state their demands. The BAG keeps track about the power flow situation in its neighborhood along with line and generation constraints. In case of an event in the grid that requires intervention (e.g. a line fault) the BAG takes measures to maintain as much functionality of the devices in its responsibility as possible by negotiating with other BAGs or consulting the System Agent.
- Line Agent (LNAG): By utilizing measuring equipment and data from the assigned BAGs, the LNAG observes the power flow and physical condition.

- Supply Agent (SAG): The SAG is the agent interface for distributed generators, for which it keeps track on the various parameters (i.e. current output, maximum output) and reacts to BAG requests accordingly. By comparing its nominal condition with the observed real one, the SAG incorporates diagnostic and monitoring functions.
- Load Agent (LAG): The LAG's connected device is a power sink for which it features likewise services as the SAG.
- Storage Agent (STAG): Storage units are key devices in Smart Grids as they are able to either supply or require electric power for a limited amount of time (e.g. an electric car as suggested by the Vehicle-to-Grid concept). Therefore, the STAG incorporates the features of both the SAG and the LAG.
- Grid/Substation Agent (GAG): The GAG represents the interconnection to other grid segments not covered by the MAS. As the power flow through the intersection also depends on the load and generation situation of the external grid and can only be influenced by devices like tap-changing transformers, the GAG mainly features diagnostic functions.
- Switch Agent (SWAG): Switchers and breakers are necessary for topology changes and fault handling. As they are similar in their behavior, both are represented by the same agent.

4 Automation Agent Architecture

In order to ease the design of this type of agents, we introduced a generic architecture for automation agents in [22] distinguishing hardware, Low Level Control (LLC) and High Level Control (HLC) within each agent (Fig. 2a). The LLC is responsible for the control of the hardware and has to carry out some of its tasks under real-time constraints. The real-time capability is especially important for monitoring/diagnostic tasks to avoid slow reactions in the case of detected anomalies. The HLC of an automation agent incorporates an ontology-based world model that provides an explicit representation of the agent's immediate environment and supports reasoning about its state, diagnostic activities and the coordination with other agents. Real-time data and services are considered as well as operational and business data and services (e.g. condition, configuration, schedules).

The generic three-layered architecture of automation agents allows independently designing and developing the control layers. However, clearly defined interfaces between the layers are necessary to foster the reuse of existing implementations and to allow the exchange of implementation technologies and sub-architectures. Inter-layer interfaces are mostly service and data interfaces. Nevertheless, interfaces within the layers must not be neglected. In the Hardware-Layer these interfaces are the electrical coupling of components in the grid, allowing energy exchange and being responsible for the stability of the grid. Therefore the physical structure of the grid is represented within the ontology.

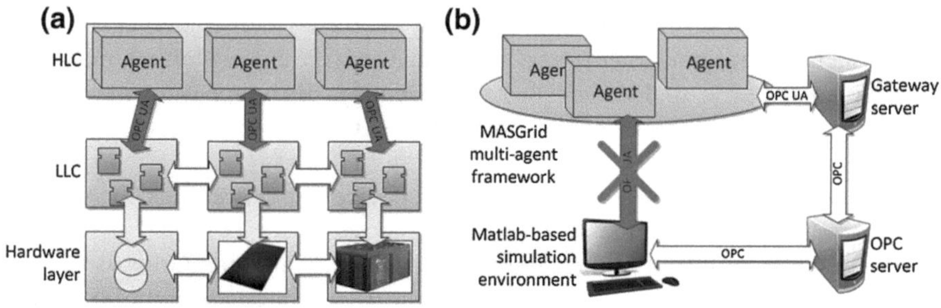

Fig. 2 Three-layered architecture of automation agents including all interfaces (**a**) and the simulation tool-chain (**b**)

For the independent implementation and validation of high-level functionalities the interface between HLC and LLC has a vital role. In order to gain flexibility in the choice of LLC technologies, industrially accepted protocols shall be used. Due to the consistent representation and integration of data and services the use of the emerging OPC Unified Architecture (OPC UA) seems promising. Its predecessor OPC DA is widely used in industrial control environments. All major PLC vendors provide OPC servers to access relevant process data. However, only OPC UA provides means for consistently modeling data. Furthermore OPC UA fosters arbitrary grouping of data and services in its model, e.g. hierarchical modeling. Another advantage is the event-based data exchange in the OPC UA paradigm.

Alarms can be instantaneously propagated to the decision-making HLC, and action sequences chosen by the HLC to stabilize or optimize the grid can be propagated and executed without delay by the LLC. OPC UA provides modeling capabilities to model all communication aspects, hence there is no need to circumvent the model and implement hidden interfaces.

The strict interface provides important possibilities: existing LLC-implementations with different system architectures (i.e. legacy systems) are covered by the MASGrid approach. Furthermore, since all components and services of the LLC and hardware-layer are encapsulated within the interface, simulation environments may be coupled to the HLC implementation to facilitate and a continuously validate the development. The interfaces between the LLC and the HLC are highly dependent on the control equipment and the controlled hardware. Model-driven design of control systems is a promising approach to abstractly describe the involved interfaces and gain flexibility in the choice of implementing technologies [23]. The implementation is facilitated by providing a simulation environment based on behavior models of the interface and the controlled plant [24].

For development and execution of the HLC, JADE,[1] a multi-agent framework and runtime environment, is utilized. In order to support the development of the software agents, the LLC and Hardware-Layer are represented by the Matlab-

[1] www.jade.tilab.com

based SimPowerSystems[2] toolbox which used for simulating the electrical networks according to the various defined use cases like the motivation example in Sect. 2. The current tool-chain is depicted in Fig. 2b.

5 Reconfiguration

The process of power network reconfiguration has to be differentiated into global and local tasks as they vary in their demands. Local reconfiguration is applied whenever immediate actions are necessary, such as in case of a line fault, in order to autonomously recover into a stable operation point and maintain the load supply to the best possible extent. In MASGrid, local reconfiguration is performed by the Automation Agents and their corresponding equipment without guidance from the System Agent (see Chaps. 2 and 3) to minimize the reaction time. On the contrary, the global reconfiguration problem aims towards shaping the load and generation profiles to ensure stable operation of the network [15, 24], as well as to optimize the grid for given optimization criteria such as power loss minimization [25]. As the Automation Agents only obtain local information, the system-wide reconfiguration processes have to be executed by the System Agent, which keeps track of complete grid status. The problem itself is formulated as an objective function satisfying the system constraints. A successful implementation of this functional principle on a pallet-transportation system is described in [26].

6 Conclusion and Outlook

The continuous growth of electric energy consumption as well as the upcoming large-scale integration of distributed energy production units, based on renewable energy resources (photovoltaic system, etc.) results in an increasingly complex electric network. Current monitoring, analysis, and control technologies for distribution networks will not be able to meet these increasingly diverse future challenges. In this chapter a MAS architecture for power distribution grids which supports self-reconfiguration and self-diagnostics is presented. Agents gather data from a variety of sensors combining them with information from different interpretation algorithms and multi-agent interactions to generate an overall diagnostic conclusion and inform other responsible agents. Respectively, the MAS automatically reconfigures the power system topology based on changed conditions of the system. Additionally, the conjunction of the Smart Grid with a SCADA system leads to less required efforts for human operators to supervise the grid.

2 http://www.mathworks.com/products/simpower/

The future work will be concentrated to demonstrate the advantages of the presented MAS framework in the AIT power distribution laboratory for testing advanced Smart Grid services. Controller-Hardware-in-the-Loop (CHIL) and Power-Hardware-in-the-Loop (PHIL) methodologies, provided at AIT, will be used to test the capabilities of the MASGrid system in a realistic environment.

References

1. A. Molderink, V. Bakker, M.G.C. Bosman, J.L. Hurink, G.J.M. Smit, Management and control of domestic smart grid technology. smart grid, IEEE Trans. **1**(2), 109–119. doi:10.1109/TSG.2010.2055904
2. European Smart Grids Technology Platform: Vision and Strategy for Europe's Electricity Networks of the Future, European Commission, ISBN 92-79-0414-5 (2006), **44**
3. S.D.J. McArthur, E.M. Davidson, V.M. Catterson, A.L. Dimeas, N.D. Hatziargyriou, F. Ponci, T. Funabashi, Multi-agent systems for power engineering applications—part I: concepts, approaches, and technical challenges. Power Syst., IEEE Trans. **22**(4), 1743–1752 (2007)
4. J. Lagorse, D. Paire, A. Miraoui, A multi-agent system for energy management of distributed power sources. Renew. Energy **35**(1), 174–182 (2010)
5. K. Nareshkumar, M.A. Choudhry, J. Lai, A. Feliachi, in *Application of Multi-agents for Fault Detection and Reconfiguration of Power Distribution Systems*. Power and energy society general meeting, 2009. PES '09. IEEE, 1–8. doi:10.1109/PES.2009.5276005
6. P. Zhang, F. Li, N. Bhatt, Next-generation monitoring, analysis, and control for the future smart control center. Smart Grid, IEEE Trans. **1**(2), 186–192 (2010). doi:10.1109/TSG.2010.2053855
7. J.M. Solanki, S. Khushalani, N.N. Schulz, A multi-agent solution to distribution systems restoration. IEEE Trans. Power Syst. **22**(3), 1026–1034 (2007)
8. N. Jennings, S. Bussmann, Agent-based control systems: Why are they suited to engineering complex systems? Control Syst. Magazine, IEEE **23**(3), 61–73 (2003)
9. M. Wolter, S. Brenner, T. Isermann, L. Hofmann, in *Application of Adaptive Agents in Decentralized Energy Management Systems for the Purpose of Voltage Stability in Distribution Grids*. North American power symposium (NAPS), 2009. North American power symposium (NAPS), pp. 1–5 (2009)
10. J. Jung, C.C. Liu, in *Multi-agent System Technologies and an Application for Power System Vulnerability*. IEEE Power Engineering Society General Meeting (2003)
11. L. Cristaldi, A. Monti, R. Ottoboni, F. Ponci, in *Multiagent Based Power Systems Monitoring Platform: A Prototype*. Power Tech Conference Proceedings, 2003 IEEE Bologna, vol. 2, June 2003, pp. 5
12. P. Vytelingum, T. Voice, S.D. Ramchurn, A. Rogers, N.R. Jennings, in *Intelligent Agents for the Smart Grid*. Proceedings of AAMAS, pp. 1649–1650, 2010
13. J. Hossack, S.D.J. McArthur, J.R. Mcdonald, J. Stokoe, T. Cumming, in *A Multi-agent Approach to Power System Disturbance Diagnosis*. Proceedings of international conference on power system management and control, April 2002, vol. 488, pp. 317–322
14. L. Liu, K.P. Logan, D.A. Cartes, S.K. Srivastava, Fault detection, diagnostics and prognostics: software agent solutions. IEEE Trans. Veh. Technol. **56**(4), 1613–1622 (2007)
15. S. Chouhan, W. Hui, H.J. Lai, A. Feliachi, M.A. Choudhry, in *Intelligent Reconfiguration of Smart Distribution Network using Multi-agent Technology*. Power and energy society general meeting, 2009. PES '09. IEEE, pp. 1–6, 26–30 July 2009
16. S. McArthur, E. Davidson, V. Catterson, A. Dimeas, N. Hatziargyriou, F. Ponci, T. Funabashi, Multi-agent systems for power engineering applications—part II: technologies,

standards, and tools for building multi-agent systems. Power Systems, IEEE Trans. **22**, 1753–1759 (2007)
17. M. Merdan, M. Vallee, W. Lepuschitz, A. Zoitl, Monitoring and diagnostics of industrial systems using automation agents. Int. J. Prod. Res. **49**(5), 1497 (2011)
18. M. Vallee, M. Merdan, W. Lepuschitz, G. Koppensteiner, Decentralized reconfiguration of a flexible transportation system. Ind. Inf. IEEE Trans. **7**(3), 505–516 (2011). doi:10.1109/TII.2011.2158839
19. W. Lepuschitz, A. Zoitl, M. Vallee, M. Merdan, Towards self-reconfiguration of manufacturing systems using automation agents. IEEE Trans. Syst. Man Cybern. Part C Appl. Rev. **41**(1), 52–69 (2011)
20. N. Hatziargyriou, Microgrids—the key to unlock distributed energy resources? IEEE Power Energy Mag. **6**(3) (2008)
21. A. Prostejovsky, W. Lepuschitz, T. Strasser, M. Merdan, in *Autonomous Service Restoration in Smart Distribution Grids using Multi-agent Systems*. electrical and computer engineering (CCECE), IEEE Canadian Conference on May 2012, pp. 1–5
22. M. Vallee, H. Kaindl, M. Merdan, W. Lepuschitz, E. Arnautovic, P. Vrba, An automation agent architecture with a reflective world model in manufacturing systems, IEEE Int. Conf. Syst., Man, Cybern. (SMC'09), (San Antonio, Texas, USA, 2009), pp. 305–310
23. T. Strasser, M. Rooker, I. Hegny, M. Wenger, A. Zoitl, L. Ferrarini, A. Dede, M. Colla, in *A Research Roadmap for Model-Driven Design of Embedded Systems for Automation Components*. Proceedings of the 7th IEEE international conference on industrial informatics (INDIN 2009), pp. 564–569, June 2009
24. I. Hegny, M. Wenger, A. Zoitl, in *IEC 61499 Based Simulation Framework for Model-Driven Production Systems Development*. Proceedings of the 2010 IEEE conference on emerging technologies and factory automation (ETFA), Sept 2010
25. H. Schmidt, N. Ida, N. Kagan, Fast reconfiguration of distribution systems considering loss minimization, IEEE Trans. Power Syst., **20**(3), Aug 2005
26. I. Zabet, M. Montazeri, in *Decentralized Control and Management Systems for Power Industry via Multiagent Systems Technology*. 4th International power engineering and optimization conference (PEOCO), 2010, pp. 549–556

Human Upper Body Pose Region Estimation

Sara Bilal, Rini Akmeliawati, Amir A. Shafie and Momoh J. E. Salami

Abstract The objective of this chapter is to estimate 2D human pose for action recognition and especially for sign language recognition systems which require not only the hand motion trajectory to be classified but also facial features, Human Upper Body (HUB) and hand position with respect to other HUB parts. We propose an approach that progressively reduces the search space for body parts and can greatly improve chance to estimate the HUB pose. This involves two contributions: (a) a fast and robust search algorithm for HUB parts based on head size has been introduced for real time implementations. (b) Scaling the extracted parts during body orientation was attained using partial estimation of face size. The outcome of the system makes it applicable for real-time applications such as sign languages recognition systems. The method is fully automatic and self-initializing using a Haar-like face region. The tracking the HUB pose is based on the face detection algorithm. Our evaluation was done mainly using 50 images from INRIA Person Dataset.

1 Introduction

Many Human to Computer Interaction (HCI) applications require the detection and tracking of HUB parts such as human activity analysis, in which human motion and gestures are detected, tracked and recognized from cameras. Many systems have been developed to detect HUB parts but the challenge is to find effective

S. Bilal (✉) · R. Akmeliawati · A. A. Shafie · M. J. E. Salami
Department of Mechatronics Engineering, International Islamic University Malaysia (IIUM), Jl Gombak 53100, Kuala Lumpur, Malaysia
e-mail: smosb@hotmail.com

R. Akmeliawati
e-mail: rakmelia@iium.edu.my

G. Sen Gupta et al. (eds.), *Recent Advances in Robotics and Automation*,
Studies in Computational Intelligence 480, DOI: 10.1007/978-3-642-37387-9_26,

methods for detecting, labeling and tracking the HUB parts for a particular application. The developed system for HUB detection should be robust to avoid occlusion and to recover easily from tracking errors when the object is not in the scene. Therefore, for applications such as Sign Language (SL), it is very important to find hand location with respect to head, shoulder and chest which convey a lot of meaning. The HUB parts can be used as reference for static and dynamic gestures see Fig. 1a and b from Malaysian SignLanguage (MSL) database.

In this chapter, a new approach has been introduced to detect HUB parts. Features have been extracted to segment HUB and ellipses were fitted to each segment.

The chapter is organized as follows; Sect. 2 presents the research background. The proposed method for modeling the HUB parts are described in Sect. 3. Tracking the HUB methodology is discussed in Sect. 4. The experimental results are presented in Sect. 5. The system performance is shown is Sect. 6. Section 7 states the conclusion.

2 Previous Work

Many approaches have been developed for human pose recognition using computer vision and pattern recognition approaches. Hyeon et al. [1] built a model of HUB by segmenting the upper body regions of humans in images and then compared the edges in the image with a predefined curvature model. Another HUB modeling approach using Ω model was developed by [2, 3]. Their system makes use of Ω model which has different scales to describe the HUB. Nicolas Burrus and Justus Piater [4] have developed top-down approaches based on pictorial models for Dutch Sign Language. The approach simultaneously models the geometry of human parts, the appearance of each part and the temporal continuity in a unified statistical framework. The modeling methods include the use of template-based distance measures, such as Chamfer Distance as in [1] and parameterized approaches, such as Mixture Density Models. Beymer et al. [3] continuously segment the image using stereo range. They build a model for human torso and Kalman filter was used for tracking. Build a model as presented in the previous approaches to represent the HUB can have high detection rate but slow the overall computational process time. Another work using stereo vision was presented by Darrell et al. [5]. Their system has been developed using skin colour and face

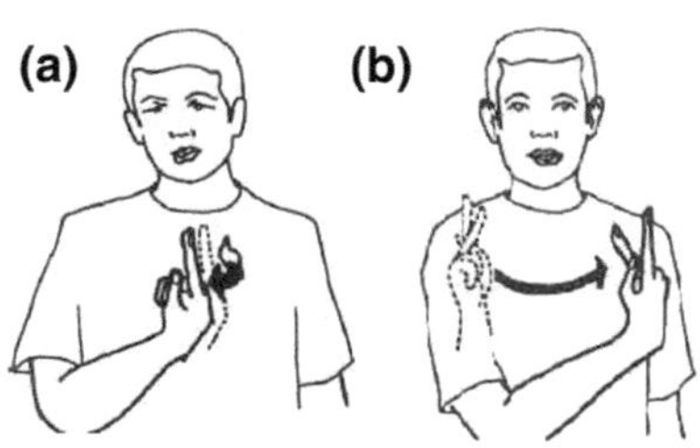

Fig. 1 **a** A sign 'Alone' from MSL database. **b** A sign 'Us' from MSL database

position as a starting point to identify human pose. Far from modeling the HUB, Govindaraju et al. [6–8] have detected head-and-shoulder based on the geometry properties of face profile. Prior information was required and their system has a high error rate. In Ref. [9] an Adaptive Combination of Classifiers (ACC) has been built to detect head-and-shoulder and partially occluded people in a static image. Yi Sun et al. [10] have proposed a head-and-shoulder detection algorithm based on wavelet decomposition technique and support vector machine (SVM). Using such techniques in low level feature extraction may slow the overall process [9, 10]. Also Ferrar et al. [11] proposed Calvin HUB detector to estimate a 2D human pose as a spatial configuration of body parts in TV and movie video shots. Their approach progressively reduces the search space for body parts by using a generic detector which using a weak model of pose to substantially reduce the full poses search space. Then a 'grabcut' initialized on detected regions proposed by the weak model, to further prune the search space. Moreover, they have integrated spatiotemporal model covering multiple frames to refine pose estimates from individual frames, with inference using belief propagation.

The goal of this chapter is to detect HUB parts in a sequence of video frames for real-time applications. The HUB detection system must be fast and robust to fulfil the requirements of real-time applications. Even though, the proposed algorithm is based on a figure of human body adjust for artists, but appears to give significantly good performance on pedestrian images.

3 Modeling Human Upper Body

Many existing methods which use geometric modeling, boosted classifiers and SVM have been introduced [6–10]. But real-time applications, such as SL recognition systems, require a fast, compatible and synchronized method for HUB parts detection and tracking. Therefore, the main contribution of this work is the introduction of a fast and robust search algorithm for HUB parts based on the study in [12]. It assumes that all body parts can be measured with head size. Face location has been used as an initialization for the system. Head size differs from face size with only few measurements as shown in Fig. 2. The head measurement is found first by dividing the head into half, then, defining that 1/3 below the nose is the lip. By using that ratio, one can estimate that forehead is 2/3 the upper pupil half. Then 1/3 above the forehead is the hair end. The algorithm for finding head measurements is as follows

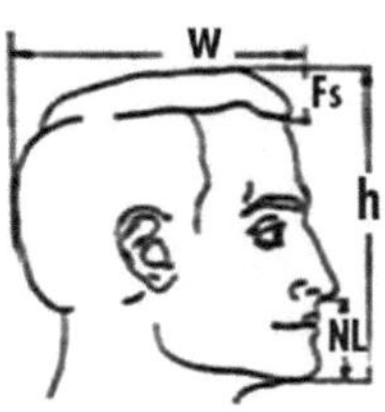

Fig. 2 Head and face dimensions

- The pupils are in the middle of the head, top to bottom equal to $1/2 \times h$.
- The bottom of the nose is between the pupils and bottom of the chin.
- Below the nose and between the lips $(NL) = 1/3 \times 1/2 \times h$.
- 2/3's below the nose is the chin crease.
- Forehead size $(Fs) = 1/3 \times 1/2 \times h$.

where h = head height and w = head width.

3.1 Head and Face Detection with Haar-like Features

Initialising the system with a face will provide an adequate search region for other HUB parts. Neural network methods, support vector machines and other kernel methods have been used for face posture detection [13–16]. However, most of these algorithms use raw pixel values as features which make the system sensitive to noise and changes in illumination. Instead, other approaches such as Haar-like features, which are similar to Haar basis functions has been proposed by [17]. The features encode differences in average intensities between two rectangular regions, and they can extract textures independent on absolute intensities. Our proposed method uses Haar-like features [17, 18] and AdaBoost algorithm [19, 20] to find face region. After the face region has been detected in the image, other HUB regions were found as described next in Sect. 3.2.

3.2 Shoulders, Chest, Abdomen and Hip Line Detection

HUB parts such as neck, shoulders chest and hands are very crucial to understand human gestures. The detected face using Haar-like and AdaBoost algorithm has been used as an initial position to find these HUB parts. A study in [12] shows that the human body can be measured using head dimension. A human figure adjusted for artists based on accurate 8 head size is shown in Fig. 3. It assumes that:

1. The neck space is 1/4 of one head-length, and it starts under the chin of that top first head.
2. The second head is the shoulders head. It is the top of three trunk heads and is drawn under this neck space.
3. One quarter of one head down in this second head is the shoulder line. This allows space for the neck-support muscles above the clavicle.
4. The shoulder line is two head-lengths (two widths on a female) wide and is the top line of the torso triangle that extends down to the space between the legs, or the chest triangle that only extends down to the hip line.
5. One head down the chest is the abdomen that also extends one head down to hip line.

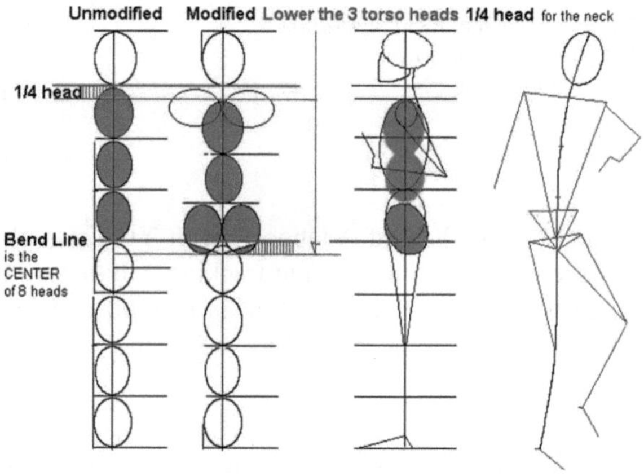

Fig. 3 Accurate 8-head-high adult male human figure, adjusted for artists [12]

3.3 HUB Scale Dimensions Based on Face Size

When the human stands in a side view, see Fig. 4, the dimensions of the oriented head and face are not following the other HUB part orientation. In such case, first, the system calculates the size of the head in a front view. Then, if the size is reduced to 1/3 of the head size, the other detected HUB parts will keep its preliminarily size. This is done because for SL recognition systems, the whole HUB parts must be completely covered. Signs could be started or end from/at head, shoulder or chest.

4 Tracking HUB Region

Pose detection (or initialisation) is typically performed on an initial video frame followed by pose tracking where the pose parameters obtained from the current frame is used as a starting value for the subsequent video frame. Many methods exist to track pose using motion histories [21] and optical flow [22] technique. CAMSHIFT [23] is designed for face and coloured object tracking using probability distribution. CAMSHIFT is very sensitive to the change of the face colour over time. Therefore the tracking in this work is based on the face centre point which is obtained from the skin binary image. The method used to find skin pixels has been adapted from Bilal et al. [24] which combines the appearance-based

Fig. 4 A sign 'Voice' from MSL database

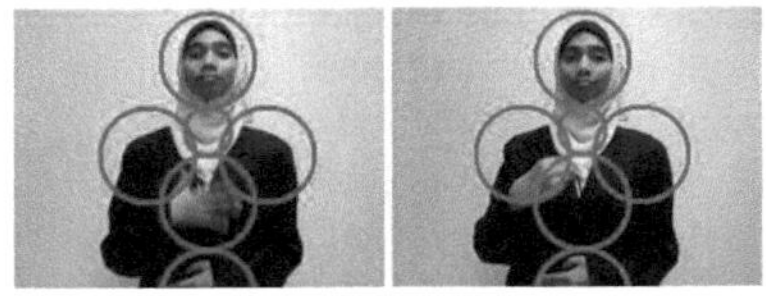

Fig. 5 Shows the tracking of HUB for two frames from sign 'White' using [24]

method (Haar-like and Adaboost) with YC_bC_r to obtain skin areas. After finding the face in the skin image, the tracking is based on the position of the face in the image frame as shown below. This method of tracking performs better than the CAMSHIFT tracking method as shown in Figs. 5 and 6.

5 Experimental Results for Detecting the Human Upper Body

Detecting HUB for SL was achieved using VC++, OpenCV and the assumptions stated in Sect. 3.

5.1 Head Detection

First, the system has used Haar-like feature for extracting information from the face using two line features, four edge features, one centroid feature and one diagonal feature, followed by AdaBoost algorithm for learning. After the face has been detected, see Fig. 7, we have obtained the head size from the face by finding Eq. (1):

$$HeadSize = \frac{1}{3} \times \left(\frac{facesize}{2}\right) + facesize \tag{1}$$

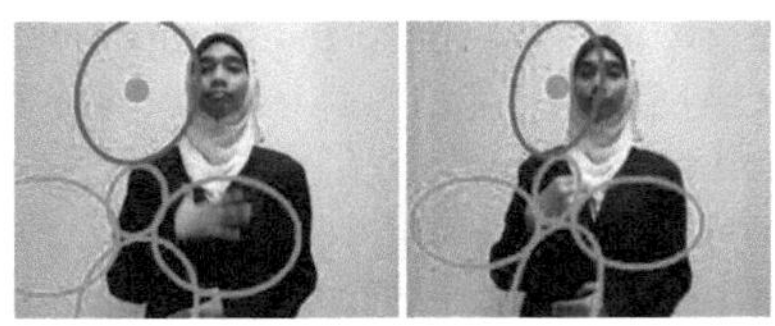

Fig. 6 Shows the tracking of HUB for two frames for sign 'White' Using CAMSHIFT [23]

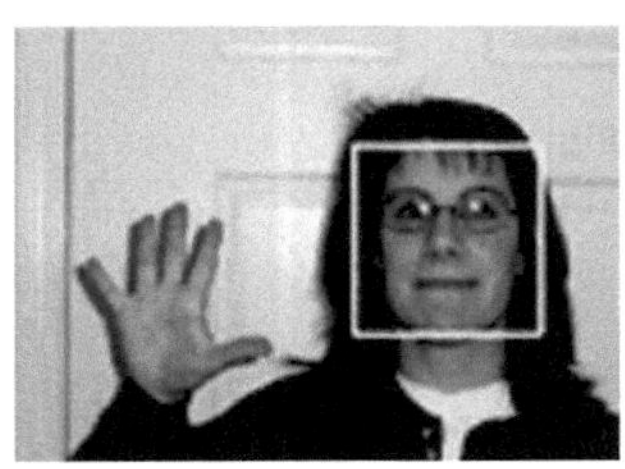

Fig. 7 Haar-like and AdaBoost algorithm used for detecting face

5.2 Neck, Shoulder and Chest Detection

The HUB parts were located using the head size algorithm as stated in Sect. 4 and shown in Fig. 7. We have applied our pose estimation technique over 50 images (see Fig. 8) extracted from INRIA person dataset [25]. Also some other images (see Fig. 9) extracted from Season 5 Episode 2 of "Buffy the Vampire Slayer", has been used for system accuracy evaluation. The system fails to detect the HUB as appears in Fig. 10 due to the failure of the face detector algorithm. This can happened because the images in Fig. 10 have some parts that could lower the face detector performance which is based on a white and black template.

6 System Performance

The system performance has been measured using 50 images and Table 1 shows the performance of the detector.

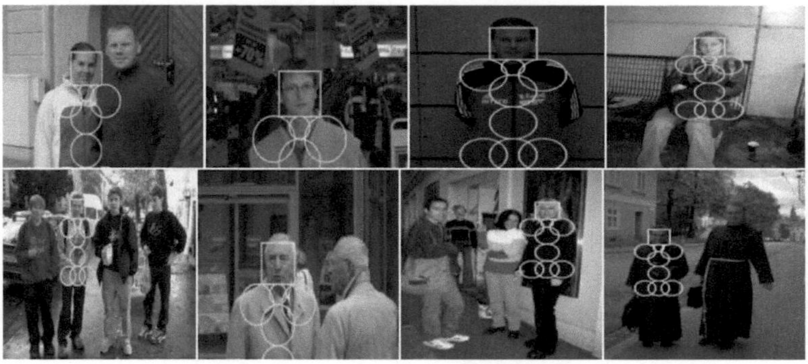

Fig. 8 HUB detection system performance using INRIA person dataset [25]

Fig. 9 HUB detection system performance using 3 extracted images from season 5 episode 2 of "buffy the vampire slayer"

Fig. 10 HUB detection system failure

Table 1 Comparison results between the proposed method and the Calvin detector

Method	System settings	False positives/50	True positives/50	Time in (ms)	System pitfalls	System advantages
Proposed method	A visible person in scene	6	44	90	Face detection failure cause HUB detection failure	High HUB detection performance
Calvin method [11]	The upper body must appear large in scene	30	20	98	Detection failure in small images.	Person in both scene: standing or sitting

6.1 Results from Calvin Group Method

The performance of the Calvin system using INRIA person dataset [25] in shown in Fig. 11. The results in Fig. 12 which have used the Buffy data set show better performance. These accuracy variations happened between the two dataset because the upper-bodies appear smaller in the INRIA person dataset.

6.2 Comparisons Result

Ferrari et al. [11] detection method was build to find the upper body parts in TV and movie video shots. Their approach progressively reduces the search space for body parts and has high accuracy when tested on the images from Season 5 Episode 2 of "Buffy the Vampire Slayer". Their system is applicable to estimate the HUB while the person is facing the camera and whether he is in standing or sitting position. They have tested their system on INIRA person dataset which showed less accuracy due to the size of the images. The system proposed in this

Fig. 11 HUB detection system performance using INRIA person dataset [25]

Fig. 12 HUB detection system performance using extracted images from season 5 episode 2 of "Buffy the Vampire Slayer" [11]

chapter is able to find the HUB parts region in any dataset as long as the human face is visible in the view. Table 1 shows the proposed system performance and the Calvin detector performance.

7 Conclusions and Future Work

The location of hand with respect to the head, and other HUB parts conveys a lot of meanings to understand SL. Therefore, tracking HUB parts as well as scaling the size while rotating is an important issue while classifying hand gestures. Many methods have been developed for HUB detection and tracking but real-time application requires fast computational time as well as the storage requirements is extremely important. In this chapter, a fast and robust HUB parts detection algorithm has been introduced to detect and track the HUB parts region. These performances make it applicable for applications such as SL recognition systems, interaction with robotics, etc.

Future work could include further improvements for body part segmentations. This is possible to find the shoulder, arms and other HUB part dimensions.

Acknowledgments This work is supported by Research Matching Grant Scheme RMGS 09–03, International Islamic University Malaysia (IIUM).

References

1. D.H. Hyeon, et al., in *Human Detection in Images using Curvature Model.* Presented at the international conference on circuits/systems computers and communications (ITC-CSCC), 2001
2. A. Broggi et al., in *Shape-based Pedestrian Detection.* The Proceedings of the IEEE Intelligent Vehicles Symposium, 2000
3. D. Beymer, K. Konolige, in *Real-time Tracking of Multiple People using Continuous Detection.* Presented at the international conference on computer vision (ICCV) frame-Rate workshop, 1999
4. N. Burrus, J. Piater, in *Monocular Human Upper Body Pose Estimation for Sign Language Analysis.* Talk at the 4th Multitel Spring Workshop, Mons, Belgium, 2009
5. T. Darrell, et al., in *Integrated Person Tracking using Stereo, Color, and Pattern Detection.* In the Proceedings of IEEE conference on computer vision and pattern recognition, 1998, pp. 601–608
6. V. Govindaraju, et al., in *A Computational Model for Face Location.* The Proceedings of the IEEE Third International Conference on Computer Vision, 1991, pp. 718–721

7. V. Govindaraju, et al., in *Locating Human Faces in Newspaper Photographs*. In the Proceedings of the IEEE computer society conference on computer vision and pattern recognition, 1989
8. V. Govindaraju, et al., in *A Computational Model for Face Location Based on Cognitive Principles*. The Proceedings of the American Association for Artificial Intelligence (AAAI), 1992
9. A. Mohan et al., Example-based object detection in images by Components. IEEE Trans. Pattern Anal. Mach. Intell. **23**, 349–361 (2001)
10. Y. Sun, et al., in *Head-and-shoulder Detection in Varying Pose*. Presented at the advances in natural computation, First international conference, ICNC, Part II, Changsha, China, Proceedings, 2005
11. Ferrar, et al., in *A progressive Search Space Reduction for Human Pose Estimation*. The Proceedings of the IEEE Conference on Computer Vision and Pattern Recognition (2008)
12. F.R.C.W. Course, in http://www.realcolorwheel.com/human.htm. 2011
13. K.K. Sung and T. Poggio, Example-based learning for view-based human face detection. IEEE Trans. Pattern Anal. Mach. Intell. **20**, 39–51 (1998)
14. H.A. Rowley et al., Neural network based face detection. IEEE Trans. Pattern Anal. Mach. Intell. **20**, 23–28 (1998)
15. E. Osuna, et al., in *Training Support Vectormachines: an Application to Face Detection*. The Proceedings of CVPR, 1997, pp. 130–136
16. B. Heisele, et al., in *Face Detection in Stillgray Images*. A.I. Memo 1687, Center for Biological and Computational Learning, Ed 2000
17. P. Viola and M. Jones, in *Robust Real-time Object Detection*, Cambridge Research Laboratory Technical Report 2001
18. Q. Chen, et al., in *Real-time Vision-based Hand Gesture Recognition using Haar-like Features*. Instrumentation and measurement technology conference—IMTC, 2007
19. R. Lienhart and J. Maydt, in *An Extended Set of Haar-like Features for Rapid Object Detection*. In the proceedings of IEEE international conference on image processing ICIP, 2002
20. Y. Freund and R.E. Schapire, A Short Introduction to Boosting. J. Japan. Soc. Artif. Intell. **14**, 771–780 (1999)
21. G.R. Bradski and J.W. Davis, Motion segmentation and pose recognition with motion history gradients. Mach. Vis. Appl. **13**, 174–184 (2002)
22. V.v. Megen etal., in 3D pose tracking using optical flow, 2010
23. G.R. Bradski, Computer vision face tracking for use in a perceptual user interface. Intel Technol. J. **2**, 2–21 (1998)
24. S. Bilal et al., in *A Hybrid Method using Haar-like and Skin-color Algorithm for Hand Posture Detection, Recognition and Tracking*, Presented at the international conference on mechatronics and automation (ICMA), Xi'an, China, 2010
25. Images for evaluating UB frontal detector, extracted from INRIA person dataset, http://pascal.inrialpes.fr/data/human/

MIX
Papier aus verantwortungsvollen Quellen
Paper from responsible sources
FSC® C105338

If you have any concerns about our products,
you can contact us on
ProductSafety@springernature.com

In case Publisher is established outside the EU,
the EU authorized representative is:
Springer Nature Customer Service Center GmbH
Europaplatz 3, 69115 Heidelberg, Germany

Printed by Libri Plureos GmbH
in Hamburg, Germany